# Perspectives in
# STRUCTURAL CHEMISTRY
## VOLUME IV

# Perspectives in
# STRUCTURAL CHEMISTRY

## Editors:

Professor J. D. DUNITZ
Laboratorium für Organische Chemie,
Eidg. Technische Hochschule,
Zurich, Universitätstrasse 6,
Switzerland

Professor J. A. IBERS
Department of Chemistry,
Northwestern University,
Evanston, Illinois, U.S.A.

### Contributors to Volume I:

Dr. DOYLE BRITTON
Minneapolis, U.S.A.
Dr. A. McL. MATHIESON
Clayton, Victoria, Australia
Dr. P. J. WHEATLEY
Zurich, Switzerland

### Contributors to Volume II:

Professor J. D. DUNITZ
Zurich, Switzerland
Dr. B. R. PENFOLD
Christchurch, New Zealand

### Contributors to Volume III:

(the late) Dr. A. D. WADSLEY
Melbourne, Australia
Professor STEN ANDERSSON
Stockholm, Sweden
Professor R. MASON
Sheffield, England
Professor M. R. CHURCHILL
Cambridge, Mass., U.S.A.
Dr. J. J. DALY
Zurich, Switzerland

### Contributors to Volume IV:

Dr. HOWARD T. EVANS, Jr.
U.S. Geological Survey,
Washington, D.C. 20242
U.S.A.

Professor O. BASTIENSEN
Department of Chemistry,
The University of Oslo,
Oslo 3, Norway

Dr. H. M. SEIP
Department of Chemistry,
The University of Oslo,
Oslo 3, Norway

Dr. JAMES E. BOGGS
Department of Chemistry,
The University of Texas,
Austin, Texas 78712, U.S.A.

Professor F. H. HERBSTEIN
Department of Chemistry,
Technion-Israel Institute of
Technology,
Haifa, Israel

# PERSPECTIVES IN

# STRUCTURAL CHEMISTRY

## VOLUME IV

### Edited by J. D. DUNITZ & J. A. IBERS

**1971**

**JOHN WILEY & SONS,** New York - London - Sydney - Toronto

# Preface to Volume IV

Nearly all of chemistry is concerned with the interrelationships between structure and energy, between structure and reactivity. Thus nearly all of chemistry is in a sense structural chemistry. For the purpose of this series, we have taken a more restrictive point of view and rather arbitrarily defined structural chemistry as that part of chemistry which deals with the metrical aspects of structure.

When we began this series, in 1967, we felt that an apology was in order for introducing yet another series of reviews into the scientific literature. However, the response to the first two volumes, on the part of readers, reviewers, and contributors, has shown that we were not alone in believing that a series of reviews covering the field of structural chemistry, as defined above, would help to fill a genuine need.

We are conscious that it is impossible to satisfy all tastes, we are aware that there are serious deficiencies in the selection of topics that have appeared, but we had hoped that we could continue to provide readers with reviews that would be useful to them both in research and teaching. It is with regret, therefore, that we have to announce that this will probably be the last volume of the series.

J. D. Dunitz

J. A. Ibers

July, 1971

# Contents

# Perspectives in
# STRUCTURAL CHEMISTRY
## *VOLUME IV*

# Heteropoly and Isopoly Complexes of the Transition Elements of Groups 5 and 6

HOWARD T. EVANS, JR., U.S. Geological Survey, Washington, D.C. 20242, U.S.A.

## I. HISTORICAL INTRODUCTION

The existence of polynuclear oxometallate complexes in aqueous solution has been known since early in the nineteenth century. As long ago as 1826 Berzelius[1] described the compound ammonium phosphomolybdate. Five years later Berzelius, writing on the chemistry of vanadium, described orange-red crystalline salts which he formulated, for example, as $(NH_4)_2V_4O_{11} \cdot 2H_2O$, but which were undoubtedly decavanadates. Only gradually, in subsequent decades, were chemists able to accumulate enough knowledge and experimental information to begin to understand the molecular nature of these and a host of similar compounds. In the so-called heteropoly complexes, the proportion of the hetero element in the complex is so low that chemists for a long time believed that it constituted an impurity in their crystalline preparations.

The heteropoly complexes were established firmly as true compounds by the extensive work of Marignac[2] on the silicotungstates in 1864. He relied heavily on crystallographic studies, since one of the outstanding properties of these complexes is their ability to form good crystals readily from aqueous solutions. His analytical work was evidently also of the highest quality, inasmuch as he was able to formulate the complexes correctly without expressing any qualifications concerning the ratios of the metal atoms in the molecules.

After Marignac's work had been published chemists could pursue the study of this type of complex with greater confidence, and discover what types of discrete formulations and molecular sizes could be established and what types of hetero atom could enter into such complexes. In this work all attention was concentrated on molybdate and tungstate systems. Conjectures concerning the structures of these complexes were often made, but necessarily on a purely hypothetical basis. Even when Werner accomplished his historic discoveries concerning the theory of structures of inorganic complexes, no light was cast on the nature of the heteropoly molybdates and tungstates. Valiant attempts in this direction were made by Miolatti and Rosenheim in a long series of investigations and publications, but to no avail. Nevertheless, Rosenheim's group added enormously to our knowledge of the chemistry of this class of compound.[3] He first used the terms heteropoly and isopoly complexes to apply to the polynuclear oxo complexes of molybdenum and tungsten, respectively with and without a second cation playing a central role in the core of the molecule. Rosenheim's main interest centered on the former, the heteropoly complexes.

After Rosenheim's work, the heteropoly complexes were generally considered to consist of a central cation or mononuclear oxo anion enclosed in a cage of oxomolybdate or oxotungstate polyhedral groups. Pauling,[4] recognizing the significance of the common cubic symmetry of phosphotungstic acid and its analogs, suggested a cage structure of $MoO_6$ octahedra linked by corners into a shell enclosing the $PO_4^{3-}$ ion, the whole having $T_d$ symmetry. The idea was in the right direction, although later it was found that the cage, being based on edge-sharing instead of corner-sharing, is much more compact, and requires only 40 oxygen atoms rather than 58. Thus, while Hoard[5] was unable to verify Pauling's more open structure on the basis of X-ray diffraction intensity measurements, Keggin[6] obtained an excellent explanation of his diffraction measurements on the basis of a molecule having $T_d$ symmetry with the formulation $H_3PW_{12}O_{40}$. Keggin's work gave the first real knowledge of the structural nature of the heteropoly complexes and marks the starting point of the studies to be reviewed in this article.

The isopolytungstates and isopolymolybdates have long been closely associated with the heteropoly complexes because of their similarities in physical properties and chemical behavior, but they have proved more difficult to study from the structural viewpoint. Metatungstic acid was soon found to form crystals isostructural with the phosphotungstate series, and so it was assumed to have a similar structure, but presumably lacking a nucleate cation. The first clue concerning the structures of some of the numerous other types of isopoly complexes was provided only in the 1950's by Ingvar Lindqvist[7] in a series of structure determinations of crystallized compounds obtained from solutions containing polyanions. These studies, although severely limited mainly by a lack of modern computer facilities, have probably done more to reveal the structural nature of these complexes than all other chemical studies that had been brought to bear up to that time.

A third group of complexes has also been studied to a considerable extent in recent years, namely, the mixed poly complexes. It has been found, for example, that one or two vanadium atoms can be substituted for tungsten in dodecatungstophosphoric acid while Keggin's molecular structure is retained, but frequently the mixing of elements of Groups 5 and 6 in such systems produces what seem to be entirely new complexes. Since no crystal-structure studies have been attempted so far on any of these complexes, we shall not consider them further here.

## II. THE NATURE AND CHEMISTRY OF THE POLY COMPLEXES

Although polynuclear complexes have been known to exist in aqueous solutions of such substances as molybdates, tungstates and vandadates, the determination of what species actually exist, even only in terms of the charge of the complex ion and the number of

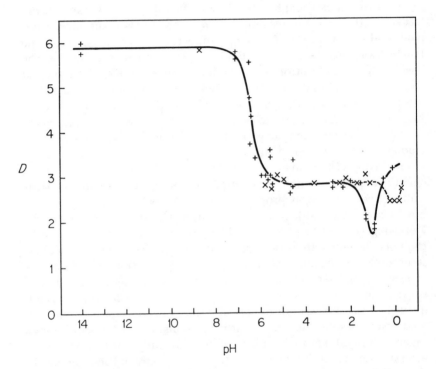

Figure 1. Diffusion rate $D$ (arbitrary units) of molybdenum into molybdenum-free isotonic solution, as a function of acidity. Experimental points of Jander *et al.*:[8] ($+$) for phosphate-free solutions, ($\times$) for solutions with P/Mo = 1/10. For the significance of the broken line see the text.

metal atoms present, has proved to be remarkably difficult. A vast amount of physical-chemical research has yielded very little information about specific species that is unequivocal and not contradicted by some other apparently equally reliable work.

The first direct information concerning the formation and stability of polynuclear oxo complexes in solution was obtained in the 1930's by Jander and his co-workers, by means of diffusion-rate studies. For example, they[8] studied the rate of diffusion of molybdenum in molybdate(VI) solution into supernatant molybdenum-free solution at various pH values. The diffusion rate was found to be strongly dependent on pH, following the step-like variation shown in Figure 1. The clear conclusion is that, if the highest plateau is assumed to correspond to a mononuclear species, certain polynuclear anion species are formed that are stable within definite pH ranges. Further, it was shown that, as the acidity is carried to very high levels, the molybdenum oxide hydroxide precipitated at the isoelectric point (pH 1.2) tends to redissolve with the formation of cationic species. Also, when phosphate was added to the molybdate solutions, the diffusion rate was altered as indicated by the broken line in Figure 1. This region probably corresponds to the familiar phosphomolybdic acid, or, according to proper nomenclature, dodecamolybdophosphoric acid, while the lower region at pH $\sim$ 0.5 may correspond to octadecamolybdodiphosphoric acid.

Another, more modern, general approach to the elucidation of the polynuclear species present in the solutions has been extensively pursued by Sillén[9] and his co-workers, by means of precise emf and analytical measurements of hydrogen ion content of molybdate solutions at varying pH values and molybdenum concentrations. The difference between hydrogen ion activity and analytical hydrogen ion content represents the amount of $H^+$ (or $OH^-$) taken up by the molybdate complexes. Values of $Z$ (ratio of $H^+$ taken up per mole of Mo) as found by Sasaki and Sillén[10] as a function of pH at various Mo concentrations are plotted in Figure 2 (3M-$NaClO_4$ medium, 25°C). While reliable information about the actual size of the polynuclear complexes cannot be obtained from Jander's diffusion curves,[11,12] the shapes of the $Z$/pH curves can be analyzed to find the degree of nuclearity and charge for the main species present in solution, and also the equilibrium constants relating them to one another. The degree of fit of the curves calculated for the annexed equilibria to the experimental points provides excellent support for the correctness of the formulations assigned to the dominant species.

$$p\mathrm{H}^+ + q\mathrm{L}^{2-} \rightleftharpoons \mathrm{H}_p\mathrm{L}_q^{p-2q}; \qquad \mathrm{L} = \mathrm{MoO}_4$$

$$\beta_{p,q} = \frac{[\mathrm{H}_p\mathrm{L}_q^{p-2q}]}{[\mathrm{H}^+]^p[\mathrm{L}^{2-}]^q}$$

| $p$ | $q$ | $\beta_{p,q}$ | $H_pL_q^{p-2q}$ (less $H_2O$) |
|---|---|---|---|
| 1 | 1 | $3.89 \pm 0.09$ | $HMoO_4^-$ |
| 2 | 1 | $7.50 \pm 0.17$ | $H_2MoO_4$ |
| 8 | 7 | $57.74 \pm 0.03$ | $Mo_7O_{24}^{6-}$ |
| 9 | 7 | $62.14 \pm 0.06$ | $HMo_7O_{24}^{5-}$ |
| 10 | 7 | $65.68 \pm 0.06$ | $H_2Mo_7O_{24}^{4-}$ |
| 11 | 7 | $68.21 \pm 0.07$ | $H_3Mo_7O_{24}^{3-}$ |
| 34 | 19 | $196.30 \pm 0.26$ | $Mo_{19}O_{39}^{4-}$ (?) |

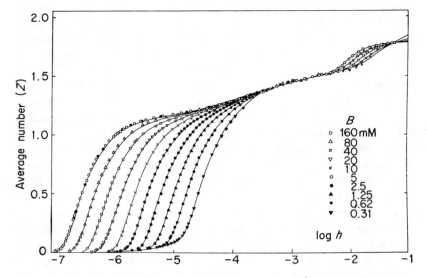

Figure 2. Average number $Z$ of $H^+$ bound per $MoO_4^{2-}$ as a function of acidity (log $h$) and Mo concentration ($B$) (from Sasaki and Sillén[10]). Curves are calculated for equilibria given in text.

Except for $\beta_{12,8}$ and $\beta_{34,19}$, these measurements confirm qualitatively those made by Aveston, Anacker and Johnson,[13] who used ultracentrifuge techniques.

Interactions between polyanions and supporting cations have been largely neglected, but Jahr, Fuchs and Preuss,[14] working with aqueous decavanadate systems, have presented abundant conductimetric evidence that complexes between such ions are formed. The importance of the influence of the cations on the polyanion equilibria has recently been dramatically demonstrated by an emf study of molybdate ion with 1M-Mg(ClO$_4$)$_2$ as supporting electrolyte by Baldwin and Weise.[15] In this case, below pH 6 no heptamolybdate is formed at all, but rather the hexamolybdate $Mo_6O_{20}^{4-}$ predominates, in the presence of some octamolybdate $HMo_8O_{28}^{7-}$.

While equilibrium studies go a long way to identify and characterize important species formed in these complicated solution systems, the number of oxygen atoms present in a given complex must remain undetermined because the number of water molecules involved in the formation of the complex is unknown. Attempts to guess structures under these circumstances have nearly always proved fruitless.

Meanwhile, the number of contributions to the chemistry of these complexes continues to expand rapidly, yielding constantly more significant information as new physical methods are introduced and old methods improved. No attempt to survey this general literature will be made here. Excellent review papers have been published by L. C. W. Baker[11] (general structural chemistry), Sillén[9] (emf studies of isopoly hydrolysis products) and Pope and Dale[16] (polynuclear chemistry of Group 5 elements). A comprehensive survey of the field has been written by P. Souchay.[17]

A matter of prime importance is the identification of a complex species found in a crystal with a species that is supposed to be present in a solution. Usually the identity is more or less assumed, where often more exact tests would be highly desirable. For example, the decavanadate ion found in crystals of its salts has been accepted as being identical (except for protonation) with the decavanadate species found to predominate in acid solution, for the following rather weak reasons: (1) the characteristic orange color is the same for the crystal and the solution; (2) the crystals are very soluble at room temperature; (3) the polyion found in the crystal is so compact that it is easily imagined to remain intact when the crystal dissolves. On the other hand, the fine crystals obtained from the colorless, nearly neutral solutions of vanadate show only chain structures and thus give no clue to the nature of the polyions in the solution. Crystal-structure analysis will certainly contribute enormously to our knowledge of these complex systems in the future; but this approach will be made much more effective if greater care is applied in extracting complex ions from the solutions, as many of these complexes have rather narrow stability ranges in terms of pH and are quite selective in terms of the precipitating cation. The value of the crystal-structure work is increased when special studies are made to ascertain that the solid species is identical with that in solution. Examples of such studies are the application of Raman spectroscopy to molybdates by

Aveston, Anacker and Johnson,[13] and of X-ray diffraction to solutions by Levy, Agron and Danford.[18]

## III. CRYSTAL-STRUCTURE STUDIES OF POLY COMPLEXES

In view of the relative ease with which polyions can be crystallized, and the power of crystal-structure analysis, it is surprising that so few structure determinations have been made in this field. A simple explanation lies in the fact that crystals of the polyion complexes have rather large unit cells and involve the determination of a large number of parameters. Generally there is little difficulty in solving structures of this type, where the unit cell contains a small number of heavy X-ray scatterers, but three-dimensional methods are almost always essential, and modern, high-speed computers are necessary to handle the large amounts of data involved. Thus, reasonably complete structure analysis of poly complexes has not been feasible before the 1960's.

A large proportion of the structures to be reviewed below are consequently incomplete and of preliminary character. Unfortunately, the chemist not trained in structure analysis has a tendency to be overimpressed by the exact appearance of the results reported in such cases and to accept the information as definitive. Even if the reported structure is not actually wrong (see, for example, the conflicting results that have been reported for lithium tungstate, Section VI–D), it is often not understood that many atoms are perhaps only indirectly placed and that interatomic distances given without any estimate of errors may be uncertain by $\pm 0.2$ Å or more. It will be convenient, therefore, to establish some scale of quality of structure determination by which the significance of a given study may be approximately judged. Such a scale of classification is offered in Table 1.

While class AA structures have been appearing more frequently in recent months, class A determinations are still entirely adequate for any chemical interpretations. It is surprising to find that experimental techniques in use thirty years ago (visual estimate of intensities on Weissenberg patterns) are still in common use today, giving structural information that is compatible in accuracy with the current state of theoretical chemistry.

In this Review, standard coordinates have been derived for each molecular structure that has been reported, averaged over the ideal symmetry that is presumed to prevail for the free polyion. Where

Table 1. Quality of crystal-structure determination.

| Class | Range of $R^*$ | Standard error (Å) M–O bonds | Type of data | Remarks |
|---|---|---|---|---|
| AA | < 0.07 | ~ 0.005 | 3-D, counter | Absorption corrections; anisotropic least-squares analysis |
| A | 0.07–0.14 | ~ 0.01 | 3-D, film Mo–$K_\alpha$ rad. | No absorption corrections; isotropic least-squares analysis |
| B | 0.14–0.20 | ~ 0.05 | 3-D, film Cu–$K_\alpha$ rad. | No absorption corrections; Fourier refinement |
| C | 0.20–0.35 | > 0.10 | 3-D, film Cu–$K_\alpha$ | Only heavy-metal positions verified by structure factor calculations |
| C | 0.15–0.20 | > 0.10 | 2-D, film | Fourier or least-squares refinement |
| D | — | — | 3-D or 2-D, film | Heavy metals deduced from Patterson synthesis; no structure factor calculation; positions of light atom from coordination linkages only; structure not proved |

* The conventional reliability index is $R = \Sigma|\Delta F|/\Sigma|F|$, where $F$ is the structure factor derived from diffraction intensities for observed reflections only.

more than one structure determination has been reported, these have been averaged by methods designed to maintain the average bond lengths and angles. Coordinates are given in the Tables in Ångström units, with axes that conform to the molecular symmetry as far as possible. We may note also that generally polyion formulae have been enclosed in brackets [ ] where they correspond to known structural entities, but not when they are only empirical. The stereoscopic figures were generated by the computer program ORTEP, written by C. K. Johnson of Oak Ridge National Laboratories, Oak Ridge, Tennessee.

## IV. HETEROPOLY COMPLEXES

Let us examine first the heteropoly complex structures, since the chemical study of this group is older and fairly well delimited. All heteropoly complexes so far examined involve mainly only molybdenum or tungsten and are characterized by the formation of a cage of corner- and edge-sharing $MoO_6$ or $WO_6$ octahedra enclosing one or two hetero atoms as a nucleus. The only known heteropoly structure not involving Mo or W is a dodecaniobomanganate, which is more naturally considered as an isoniobate coordination complex with manganese(IV). It is added to the list in Table 2, but is treated in an ancillary group at the end of the article (Section VI-C). It is likely, however, that in the future heteropoly complexes will be found in which Group 5 elements and perhaps even other elements play the cage-forming role.

The species whose structures have been directly determined are given in Table 2, together with pertinent data connected with the determination. The magnitude of the structure-analysis problem is suggested by the degree of symmetry (space group) and the number of structure parameters (exclusive of thermal parameters) involved. The quality of the determination is suggested, not only by the class estimate according to Table 1, but also by the number of parameters actually determined.

### A. Dodecamolybdophosphate and Analogous Ions and Acids

The earliest and best known, and therefore the most typical, heteropoly complex is represented by the familiar analytical precipitation agent for phosphate ions, "phosphomolybdic acid." Keggin[6] firmly established its formulation (by analogy with the corresponding tungsten complex which he studied) and showed that by substitution of both the nucleus and shell cations it is a member of a large family of isostructural complexes. The general formula can be written:

$$[T^x M_{12} O_{40}]^{x-8}$$

where T is a tetrahedrally coordinated atom at the nucleus and may represent B, Si, P, $Cr^{III}$, $Mn^{IV}$, $Fe^{III}$, $Co^{II}$, $Co^{III}$, $Cu^{II}$, Zn, Ga or As. Many other substitutions for T are possible and have been suggested by analytical studies, but these are not yet supported by crystallographic measurements. M in the formula is generally $Mo^{VI}$ or $W^{VI}$ but may also be represented partly by $V^V$.

The complexes are formed spontaneously at an appreciable rate, more rapidly when heated, in acid solutions when molybdate or tungstate ions are mixed with the appropriate hetero atom in solution. The solutions and the crystalline salts of the molybdates are yellow and of the tungstates colorless, except where a chromophoric hetero atom masks this color. The strong, free acids can often be separated from strongly acidic solutions, either by ether-extraction

Table 2. Structure determinations of heteropoly complexes.

| No. | Compound | Space group | Parameters total | detd. | Class | Ref., year |
|---|---|---|---|---|---|---|
| 1 | $H_3[PW_{12}O_{40}] \cdot 5H_2O$ | $Pn3m$ | 10 | 10 | B | Keggin,[6] 1933 |
| 2 | $H_3[PW_{12}O_{40}] \cdot 29H_2O$ | $Fd3m$ | 14 | 14 | B | Bradley and Illingworth,[19] 1936 |
| 3 | $(NH_4)_7Na_2[H_2GaW_{11}O_{40}] \cdot 15H_2O$ | $Fm3m$ | 23? | 12 | C | Evans,[20] 1968 |
| 4 | $K_6[Co^{III}W_{12}O_{40}] \cdot 20H_2O$ | $P6_222$ | 67? | 55 | D | Yannoni,[21] 1961 |
| 5 | $K_6[P_2W_{18}O_{62}] \cdot 14H_2O$ | $P\bar{1}$ | 306 | 60 | C | Dawson,[22] 1953 |
| 6 | $K_3[Te^{VI}Mo_6O_{24}] \cdot 7H_2O$ | $Pbca$ | 63 | 9 | D | Evans,[23] 1948 |
| 7 | $(NH_4)_6[Te^{VI}Mo_6O_{24}] \cdot Te(OH)_6 \cdot 7H_2O$ | $A2/a$ | 75 | 75 | A | Evans,[24] 1968 |
| 8 | $Na_4H_6[Ni^{II}W_6O_{24}] \cdot 16H_2O$ | $P\bar{1}$ | 75 | 9 | D | Agarwala,[25,26] 1960 |
| 9 | $Na_3H_6[Cr^{III}Mo_6O_{24}] \cdot 8H_2O$ | $P\bar{1}$ | 63 | 63 | AA | Perloff,[27] 1966 |
| 10 | $Na_3H_6[Cr^{III}Mo_6O_{24}] \cdot 13H_2O$ | $P\bar{1}$ | 141 | 141 | AA | Perloff,[28] 1968 |
| 11 | $(NH_4)_6H_4[Co_2^{III}Mo_{10}O_{38}] \cdot 7H_2O$ | $Pc$ | 189 | 189 | A | Evans and Showell,[29] 1969 |
| 12 | $(NH_4)_6[Mn^{VI}Mo_9O_{32}] \cdot 8H_2O$ | $R32$ | 20 | 3 | D | Waugh, Shoemaker and Pauling,[30] 1954 |
| 13 | $(NH_4)_6H_2[Ce^{IV}Mo_{12}O_{42}] \cdot 6H_2O$ | $R\bar{3}$ | 106 | 106 | A | Dexter and Silverton,[31] 1968 |
| 14 | $Na_{12}[Mn^{IV}(Nb_6O_{19})_2] \cdot 50H_2O$ | $P2_1/n$ | 172 | 172 | A | Flynn and Stucky,[32] 1968 |

11

Table 3. Crystals containing the "Keggin molecule".

In the complex ion $[T^x|M_{12}O_{40}]^{x-8}$, atoms to the left of the bar | are at the nucleus, those to the right are in the cage.

*Cubic, Pn3m; crystal structure type A:*

| No. | Compound | $a$ $\alpha$ | $b$ $\beta$ | $c$ (Å) $\gamma$ | Mol. vol. (Å³) | Ref. |
|---|---|---|---|---|---|---|
| 1 | $H_3[P|W_{12}O_{40}]\cdot 5H_2O$ | 12.165 | | | 900 | 6, 34 |
| 2 | $H_4[Si|W_{12}O_{40}]\cdot 5H_2O$ | 12.13 | | | 892 | 34 |
| 3 | $H_5[B|W_{12}O_{40}]\cdot 5H_2O$ | 12.13 | | | 892 | 34 |
| 4 | $H_8||W_{12}O_{40}|\cdot 5H_2O$ | 12.15 | | | 897 | 34 |
| 5 | $Cs_3[P|W_{12}O_{40}]\cdot 2H_2O$ | 11.854 | | | 833 | 35 |
| 6 | $Cs_3H[Si|W_{12}O_{40}]\cdot 2H_2O$ | 11.801 | | | 821 | 35 |
| 7 | $Cs_3H_2[B|W_{12}O_{40}]\cdot 2H_2O$ | 11.856 | | | 833 | 35 |
| 8 | $Cs_3H_5[|W_{12}O_{40}]\cdot 2H_2O$ | 11.81 | | | 823 | 35 |
| 9 | $Cs_3H_2[Fe^{III}|W_{12}O_{40}]\cdot 2H_2O$ | 11.88 | | | 838 | 36 |
| 10 | $Cs_3H_3[Zn|W_{12}O_{40}]\cdot 2H_2O$ | 11.86 | | | 834 | 37 |
| 11 | $Cs_3H_3[Cu^{II}|W_{12}O_{40}]\cdot 2H_2O$ | 11.85 | | | 832 | 38 |
| 12 | $Cs_3H[Mn^{IV}|W_{12}O_{40}]\cdot 2H_2O$ | 11.80 | | | 821 | 39 |
| 13 | $Cs_3H_2[Cr^{-III}|W_{12}O_{40}]\cdot 2H_2O$ | ? | | | | 40 |
| 14 | $K_3[P|W_{12}O_{40}]\cdot 4H_2O$ | 11.74 | | | 809 | 41 |
| 15 | $(NH_4)_3[P|W_{12}O_{40}]\cdot 4H_2O$ | 11.85 | | | 832 | 41 |
| 16 | $Tl_3[P|W_{12}O_{40}]\cdot 4H_2O$ | 11.89 | | | 840 | 41 |
| 17 | $K_3[As|W_{12}O_{40}]\cdot 4H_2O$ | 11.84 | | | 830 | 41 |
| 18 | $(NH_4)_3[As|W_{12}O_{40}]\cdot 4H_2O$ | 11.94 | | | 851 | 41 |
| 19 | $Tl_3[As|W_{12}O_{40}]\cdot 4H_2O$ | 11.92 | | | 847 | 41 |
| 20 | $K_3[P|Mo_{12}O_{40}]\cdot 4H_2O$ | 11.62 | | | 734 | 41 |
| 21 | $(NH_4)_3[P|Mo_{12}O_{40}]\cdot 4H_2O$ | 11.69 | | | 799 | 41 |
| 22 | $Tl_3[P|Mo_{12}O_{40}]\cdot 4H_2O$ | 11.62 | | | 784 | 41 |

| No. | Formula | | | | | | | | Ref. |
|---|---|---|---|---|---|---|---|---|---|
| 23 | $K_3[As|Mo_{12}O_{40}] \cdot 4H_2O$ | 11.72 | | | | | | 805 | 41 |
| 24 | $(NH_4)_3[As|Mo_{12}O_{40}] \cdot 4H_2O$ | 11.82 | | | | | | 825 | 41 |
| 25 | $Tl_3[As|Mo_{12}O_{40}] \cdot 4H_2O$ | 11.74 | | | | | | 809 | 41 |

*Cubic, Fm3m (or subgroup); crystal structure type D:*

| No. | Formula | | | | | | | | Ref. |
|---|---|---|---|---|---|---|---|---|---|
| 26 | $(NH_4)_7Na_2[Ga|H_2W_{11}O_{40}] \cdot 15H_2O$ | ⎫ | | | | | | ⎫ | ⎫ 20,42,43 |
| 27 | $(NH_4)_{7.3}H_{2.7}[Co^{II}|Co^{II}W_{11}O_{40}] \cdot 17H_2O$ | ⎪ 21.4 | | | | | | ⎪ 1225 | ⎪ |
| 28 | $Rb_{6.5}Co_{0.4}H_{2.7}[Co^{II}|Co^{II}W_{11}O_{40}] \cdot 15H_2O$ | ⎬ to | | | | | | ⎬ to | ⎬ 42,43,44 |
| 29 | $(NH_4)_{6.3}H_{2.7}[Co^{III}|Co^{II}W_{11}O_{40}] \cdot 13H_2O$ | ⎪ 22.5 | | | | | | ⎪ 1424 | ⎪ |
| 30 | $K_{6.5}H_{2.5}[Co^{II}|Co^{II}W_{11}O_{40}] \cdot 13H_2O$ | ⎪ | | | | | | ⎪ | ⎪ |
| 31 | $K_7H_4[Co^{III}|Co^{III}W_{11}O_{40}] \cdot 14H_2O$ | ⎭ | | | | | | ⎭ | ⎭ |
| 32 | $(NH_4)_6H_3[Fe^{II}|Ni^{II}W_{11}O_{40}] \cdot 18H_2O$ | 22.46 | | | | | | 1416 | 45 |
| 33 | $K_6H_3[Fe^{III}|Ni^{II}W_{11}O_{40}] \cdot 13H_2O$ | 21.50 | | | | | | 1242 | 45 |

*Triclinic, PĪ:*

| No. | Formula | a | α | b | β | c | γ | V | Ref. |
|---|---|---|---|---|---|---|---|---|---|
| 34 | $H_3[P|W_{12}O_{40}] \cdot 14H_2O$ | 20.52 | 95°37' | 19.36 | 95°9' | 22.52 | 90°16' | 1108 | 46 |
| 35 | $H_4[Si|W_{12}O_{40}] \cdot 14H_2O$ | 20.51 | 95°36' | 19.42 | 94°43' | 22.44 | 90°56' | 1108 | 46 |
| 36 | $H_5[B|W_{12}O_{40}] \cdot 14H_2O$ | 20.44 | 96°10' | 19.31 | 93°43' | 22.41 | 90°59' | 1097 | 46 |
| 37 | $H_3[P|Mo_{12}O_{40}] \cdot 14H_2O$ | 20.16 | 95°17' | 19.29 | 96°37' | 22.59 | 92°53' | 1085 | 46 |
| 38 | $H_4[Si|Mo_{12}O_{40}] \cdot 14H_2O$ | 20.25 | 95°12' | 19.29 | 96°38' | 22.54 | 92°42' | 1087 | 46 |
| 39 | $Fe^{III}H[Si|W_{12}O_{40}] \cdot 20H_2O$ | 19.15 | 87°55' | 22.55 | 105°31' | 23.97 | 92°25' | 1250 | 47 |

13

## TABLE 3 (continued)

| No. | Compound | a / α | b / β | c (Å) / γ | Mol. vol. (Å³) | Ref. |
|---|---|---|---|---|---|---|
| | | *Hexagonal, P6₂22; crystal structure type C:* | | | | |
| 40 | $K_5[Co^{III}|W_{12}O_{40}]\cdot 20H_2O$ | 19.11 | | 12.54 | 1322 | 11,21,26, 48 |
| | $K_5[Co^{III}|W_{12}O_{40}]\cdot 20H_2O$ | 19.00 | | 12.54 | 1307 | 21 |
| | | *Rhombohedral, R3̄m; crystal structure type B:* | | | | |
| 41 | $H_3[P|W_{12}O_{40}]\cdot 24H_2O$ | 15.63 | | 39.86 | 1405 | 49 |
| 42 | $Fe^{III}H[Si|W_{12}O_{40}]\cdot 24H_2O$ | 15.63 | | 39.88 | 1406 | 47,50 |
| 43 | $Cr^{III}H[Si|W_{12}O_{40}]\cdot 24H_2O$ | 15.63 | | 39.68 | 1399 | 50 |
| 44 | $Li_3H[Si|W_{12}O_{40}]\cdot 24H_2O$ | 15.59 | | 38.98 | 1367 | 49 |
| 45 | $Ni^{II}_{1.5}[P|W_{12}O_{40}]\cdot 24H_2O$ | 15.51 | | 41.19 | 1430 | 51 |
| 46 | $Co^{II}_{1.5}[P|W_{12}O_{40}]\cdot 24H_2O$ | 15.48 | | 41.07 | 1422 | 51 |
| 47 | $Mn^{II}_{1.5}[P|W_{12}O_{40}]\cdot 24H_2O$ | 15.50 | | 41.08 | 1425 | 51 |
| 48 | $Cd_{1.5}[P|W_{12}O_{40}]\cdot 24H_2O$ | 15.51 | | 41.04 | 1427 | 51 |
| 49 | $Ba_{1.5}[P|W_{12}O_{40}]\cdot 24H_2O$ | 15.31 | | 41.16 | 1393 | 51 |
| 50 | $Ba_2[Si|W_{12}O_{40}]\cdot 26H_2O$ | 15.47 | | 41.00 | 1416 | 49 |
| 51 | $Ca_2[Si|W_{12}O_{40}]\cdot 24H_2O$ | 15.26 | | 40.93 | 1376 | 49 |
| 52 | $Li_3H[Si|W_{12}O_{40}]\cdot 26H_2O$ | 15.59 | | 41.18 | 1445 | 49 |
| 53 | $Zn_2[Si|W_{12}O_{40}]\cdot 27H_2O$ | 15.63 | | 41.18 | 1452 | 49 |
| 54 | $Cu_2[Si|W_{12}O_{40}]\cdot 27H_2O$ | 15.67 | | 41.30 | 1464 | 49 |
| 55 | $Th[Si|W_{12}O_{40}]\cdot 27H_2O$ | 16.16 | | 40.28 | 1518 | 49 |
| 56 | $AlH[Si|W_{12}O_{40}]\cdot 28H_2O$ | 15.69 | | 41.43 | 1453 | 50 |
| 57 | $Fe^{III}H[Si|W_{12}O_{40}]\cdot 28H_2O$ | 15.63 | | 41.44 | 1456 | 50 |
| 58 | $Cr^{III}H[Si|W_{12}O_{40}]\cdot 28H_2O$ | 15.61 | | 41.40 | 1456 | 50 |
| 59 | $Th[Si|W_{12}O_{40}]\cdot 30H_2O$ | 16.26 | | 40.28 | 1537 | 49 |

14

*Cubic, Fd3m; crystal structure type B:*

| No | Formula | | | | |
|---|---|---|---|---|---|
| 60 | $H_3[P|W_{12}O_{40}] \cdot 29H_2O$ | 23.35 | | 1591 | 19,49 |
| 61 | $Mg_{1.5}[P|Mo_{12}O_{40}] \cdot 29H_2O$ | 23.11 | | 1543 | 51 |
| 62 | $Zn_{1.5}[P|Mo_{12}O_{40}] \cdot 29H_2O$ | 23.10 | | 1540 | 51 |
| 63 | $Co^{II}_{1.5}[P|Mo_{12}O_{40}] \cdot 29H_2O$ | 23.11 | | 1543 | 51 |
| 64 | $Ni^{II}_{1.5}[P|Mo_{12}O_{40}] \cdot 29H_2O$ | 23.11 | | 1543 | 51 |
| 65 | $Mn^{II}_{1.5}[P|Mo_{12}O_{40}] \cdot 29H_2O$ | 23.11 | | 1543 | 51 |
| 66 | $Cd_{1.5}[P|Mo_{12}O_{40}] \cdot 29H_2O$ | 23.13 | | 1547 | 51 |
| 67 | $Ca_{1.5}[P|Mo_{12}O_{40}] \cdot 29H_2O$ | 23.11 | | 1543 | 52 |
| 68 | $Sr_{1.5}[P|Mo_{12}O_{40}] \cdot 29H_2O$ | 23.10 | | 1540 | 52 |
| 69 | $Ba_{1.5}[P|Mo_{12}O_{40}] \cdot 29H_2O$ | 23.10 | | 1540 | 52 |
| 70 | $Fe^{III}H[Si|W_{12}O_{40}] \cdot 30H_2O$ | 23.15 | | 1551 | 49 |
| 71 | $Nd[P|Mo_{12}O_{40}] \cdot 30H_2O$ | 23.15 | | 1551 | 5 |
| 72 | $Sm[P|Mo_{12}O_{40}] \cdot 30H_2O$ | 23.15 | | 1551 | 5 |
| 73 | $Gd[P|Mo_{12}O_{40}] \cdot 30H_2O$ | 23.15 | | 1551 | 5 |
| 74 | $H_3[P|Mo_{12}O_{40}] \cdot 30H_2O$ | 23.15 | | 1551 | 5 |
| 75 | $Mg_2[Si|Mo_{12}O_{40}] \cdot 31H_2O$ | 23.09 | | 1539 | 5 |
| 76 | $Ni_2[Si|Mo_{12}O_{40}] \cdot 31H_2O$ | 23.05 | | 1531 | 5 |
| 77 | $Be_2[Si|W_{12}O_{40}] \cdot 31H_2O$ | 23.35 | | 1591 | 5 |
| | *Tetragonal, P4/mnc:* | | | | |
| 78 | $H_5[B|W_{12}O_{40}] \cdot 31H_2O$ | 12.46 | 18.42 | 1523 | 53 |
| 79 | $H_4[Si|W_{12}O_{40}] \cdot 31H_2O$ | 13.01 | 18.56 | 1571 | 53 |
| 80 | $H_4[Si|Mo_{12}O_{40}] \cdot 31H_2O$ | 12.93 | 18.45 | 1542 | 52 |
| 81 | $(NH_4)_5[B|W_{12}O_{40}] \cdot 26H_2O$ | 12.83 | 18.44 | 1518 | 53 |
| 82 | $Fe^{III}_{2.5}(OH)_{3.5}[Si|W_{12}O_{40}] \cdot 24.5H_2O$ | 12.22 | 18.86 | 1408 | 54 |

15

or ion-exchange. As the acidity decreases, strong evidence for the presence of an undecamolybdo or undecatungsto hetero complex is found. All these complexes are decomposed into their components at pH $> \sim 8$. The chemical properties of these hetero complex systems are well reviewed by Souchay[17] and by Malaprade.[33]

Although a large number of crystals containing the dodeca complexes have been measured (Table 3), as yet no detailed structure analysis of class A for any representative example has been carried out. Keggin in 1933 first established the configuration and constitution of the complex by a powder diffraction study of $H_3PW_{12}O_{40} \cdot 5H_2O$ (Table 2, no. 1; Table 3, no. 1). Bradley and Illingworth[19] studied more carefully the higher hydrate $H_3PW_{12}O_{40} \cdot 29H_2O$ (Table 2, no. 2; Table 3, no. 60), also using only powder diffraction methods. The high cubic symmetry made it possible to define completely the latter structure in terms of 14 positional parameters, but the experimental evidence consisted of only 19 diffraction intensities. Nevertheless, by applying space limitations, all the oxygen atoms could be approximately located and adjusted so as to obtain a marked improvement in the agreement between observed and calculated intensities (their data give $R = 0.10$ without $H_2O$ molecules, $0.046$ including $H_2O$ molecules).

The structure of the complex (often appropriately referred to as the "Keggin molecule"), as found by Bradley and Illingworth, is described by Figure 3b and Table 4. The structure is further illustrated as linked polyhedra in Figure 3a and as bonded atoms in a stereoscopic view in Figure 4. The reviewer[20] has carried out a three-dimensional refinement of the structure of the compound

$$(NH_4)_7Na_2[H_2GaW_{11}O_{40}] \cdot 15H_2O$$

(Table 2, no. 3; Table 3, no. 26), but, because of the disorder present and the zeolitic nature of the water in the crystal, a degree of refinement better than class C has not been achieved. Nevertheless, excellent Fourier images of the Keggin molecule have been obtained, and the interatomic distances listed in Table 4 are probably known to an accuracy of about $\pm 0.05$ Å. The unshared W–O linkage is probably nearer to 1.72 Å than 1.85 Å, but it is certainly not as short as 1.42 as has been suggested by Baker.[11] The latter value is based on a two-dimensional structure analysis of $K_5[CoW_{12}O_{40}] \cdot 20H_2O$ (Table 2, no. 4; Table 3, no. 40) by Yannoni.[21, 26] By least-squares analysis Yannoni found standard errors of tungsten positions of 0.01 Å and of oxygen positions $\sim 0.15$ Å and that the determined

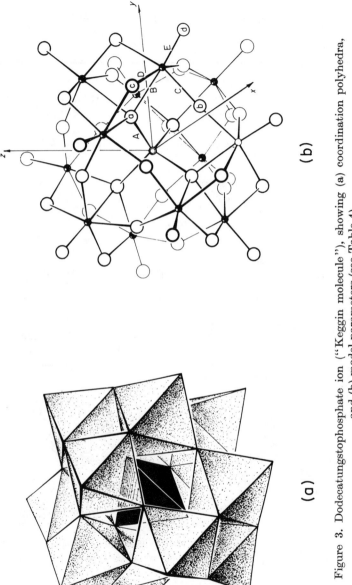

(a)

(b)

Figure 3. Dodecatungstophosphate ion ("Keggin molecule"), showing (a) coordination polyhedra, and (b) model parameters (see Table 4).

17

Table 4. Model dimensions for the "Keggin molecule" ion.

Molecular symmetry $\bar{4}3m$ ($T_d$); see Figure 5.
Lengths in Å, estimated standard error 0.05 Å; derived from data of Bradley
and Illingworth[19] for $[PW_{12}O_4]^{3-}$, and Evans[20] for $[H_2GaW_{11}O_{40}]^{9-}$.

| | *Molecular coordinates:* | | | | | |
| --- | --- | --- | --- | --- | --- | --- |
| | $[PW_{12}O_{40}]^{3-}$ | | | $[H_2GaW_{11}O_{40}]^{9-}$ * | | |
| Atom | $x$ | $y$ | $z$ | $x$ | $y$ | $z$ |
| 1 P (Ga) | 0 | 0 | 0 | 0 | 0 | 0 |
| 12 W | 2.50 | 2.50 | 0.15 | 2.50 | 2.50 | 0.15 |
| 4 $O_a$ | 0.99 | 0.99 | 0.99 | 1.09 | 1.09 | 1.09 |
| 12 $O_b$ | 2.95 | 1.06 | $-1.06$ | 2.85 | 0.97 | $-0.97$ |
| 12 $O_c$ | 3.57 | 1.58 | 1.58 | 3.55 | 1.49 | 1.49 |
| 12 $O_d$ | 3.66 | 3.66 | $-0.41$ | 3.80 | 3.80 | $-0.10$ |

*Bond lengths:*

| | | No. in | Lengths | |
| --- | --- | --- | --- | --- |
| Bonded atoms | Vector | mol. | $[PW_{12}O_{40}]^{3-}$ | $[H_2GaW_{11}O_{40}]^{9-}$ * |
| P(Ga)–$O_a$ | A | 4 | 1.71 | 1.87 |
| W–$O_a$ | B | 12 | 2.29 | 2.18 |
| –$O_b$ | C | 24 | 1.93 | 1.91 |
| –$O_c$ | D | 24 | 1.97 | 1.99 |
| –$O_d$ | E | 12 | 1.84 | 1.72 |

* Disordered over average $T_d$ symmetry.

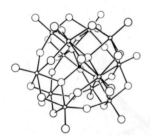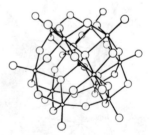

Figure 4. Stereoscopic view of the dodecatungstophosphate ion
("Keggin molecule"), showing bonding.

bond lengths W–$O_d$ (Figure 3b) ranged from 1.43 to 1.72 Å; this
variation, though consistent with the stated standard error, is much
more than would be expected as a result of crystal packing; the
variation found for W–$O_c$ of 1.47 to 2.43 Å suggests that the error of

the determined oxygen positions for this compound is considerably underestimated.

There is no reason to suppose at this stage that the molecule in a free environment in solution has any lower point symmetry than $\overline{4}3m$ ($T_d$). The size of the central $XO_4$ tetrahedron may vary considerably, even to accommodate atoms that normally have higher coordination (for example, $Co^{III}$), but the $W_{12}O_{36}$ cage can expand or contract to compensate without greatly altering the W–O distances.

The existence of a complex with the Keggin structure in solution has been demonstrated directly by an X-ray diffraction study of solutions of dodecatungstosilicic acid by Levy, Agron and Danford.[18] They found that their scattering measurements were closely compatible with the presence of a single molecular species having a cluster of 12 W atoms exactly as found in the Keggin molecule. Expected W–O interactions were detected but could not be measured quantitatively.

Various associations of the Keggin molecule in the solid state have been found, and these seem to be mainly based on a tetrahedral juxtaposition of the molecules with one another, with $W_3O_{13}$ triplets in apposition. The following types are known:

Structure type A: molecular centers lie on a body-centered cubic lattice, with $a \sim 12$ Å (2 molecules per cell). The molecule at $\frac{1}{2}, \frac{1}{2}, \frac{1}{2}$ is turned 90° to its eight neighbors.[6]

Structure type B: face-centered cubic (or pseudocubic) lattice, with $a \sim 23$ Å (8 molecules per cell). The molecules are arranged as the Si atoms are in cristobalite (diamond array). The molecule at $\frac{1}{4}, \frac{1}{4}, \frac{1}{4}$ (and similar sites) is turned 90° to its four neighbors.[19]

Structure type C: hexagonal. The molecules are arranged as the Si atoms are in quartz.[21]

Structure type D: molecular centers lie on a simple cubic lattice with doubled cell with $a \sim 22$ Å (8 molecules per cell). Each molecule is turned 90° to its 6 neighbors.[43]

Structure type A is the most dense, having room for only $5H_2O$ per molecular unit, and structure type B is the most open, allowing $29H_2O$ per molecular unit. Large cations often replace $H_2O$ molecules, sometimes non-stoichiometrically, without changing the structure. In this way, compounds nos. 5–8 (Table 3) are equivalent to nos. 1–4, and no. 81 is equivalent to no. 78. In Table 3, the compounds are arranged approximately in order of increasing hydration. A plot of volume per molecular unit against the number of $H_2O$ molecules is linear and indicates an unhydrated molecular volume for $[TW_{12}O_{40}]$ of 685 $Å^3$ (the specific volume of oxygen is 17.1, consistent with close-packed oxide structures) and a volume per $H_2O$ molecule of 28.8 $Å^3$. Compounds nos. 20–25 indicate that the molecular volume of $[TMo_{12}O_{40}]$ may be about 10% smaller.

The abundance of different hydrates of the same compound (for example, compounds nos. 39, 42, 57, 70) indicate one of the difficulties involved in the preparation of really homogeneous phases for chemical analysis and physical studies. In fact, Kraus[50] has shown that distinctly different hydrates, such as compounds nos. 42 and 57, intergrow in lamellar crystal edifices. For this reason, as well as the increased error due to the low contribution to the formula weight compared with that of tungsten, the exact number of water molecules determined may be uncertain by one or two formula units. In some crystals the water may be zeolitic and vary in a continuous manner (compounds nos. 26–31; see below).

An interesting series of compounds, discovered and first described by Baker and McCutcheon[44] as dodecatungsto complexes of $Co^{II}$ and $Co^{III}$, is now quite well established[43] as an undecatungsto complex series. In this series the molecule has the general formula

$$[T^xX^yW_{11}O_{40}]^{14-x-y}$$

in which T is a tetrahedrally coordinated nucleus atom such as P, Si, $Co^{II}$ or $Co^{III}$, and X is an octahedrally coordinated atom such as $Mn^{II}$, $Fe^{II}$, $Co^{II}$, $Ni^{II}$ or $Cu^{II}$, presumably occupying one of the twelve cage sites normally taken by W. Unpublished three-dimensional crystal structure studies by the reviewer[20] of an undecatungstogallate complex[42,43] $(NH_4)_7Na_2[H_2GaW_{11}O_{40}] \cdot 15H_2O$ (Table 2, no. 3; Table 3, no. 26) show conclusively that the molecule is of the Keggin type. In this case the center is clearly occupied by Ga. The X site is presumably vacant (one unshared oxygen atom then must become $H_2O$) but, unfortunately, the lowered symmetry of the molecule is masked by disorder in a cubic crystal structure of high symmetry. Clearly, the special features of the molecule cannot be discerned until it is obtained in crystals of lower symmetry undisturbed by disorder. The chemistry of this complex system has recently been extensively studied by Weakley and Malik,[55] Tourné and Tourné,[56] and Baker and Figgis.[57]

It has been suggested[43] that the cubic crystals of compound no. 26 actually have a negative thermal expansion coefficient, but it seems equally likely that the shrinkage of the cell parameter with increasing temperature is due to loss of zeolitic water. The reviewer has observed that this reversible change is mainly a function of the water vapor tension in the atmosphere (humidity) and is accompanied by a corresponding change in weight. The relation to humidity has also been observed for the cobalt complexes by Baker and his co-workers (see Rollins[42]). The change in molecular volume in the undecatungstogallate complex[43] between 25° and 35°C corresponds to about $2H_2O$ per formula unit. In the reviewer's electron-density syntheses of the unit-cell contents of this crystal, while the molecule itself is sharply delineated, the intermolecular water molecules and ammonium ions are only partly and very poorly resolved, in line with common experience with other zeolite structure determinations. Zeolitic character is also suggested for compound no. 40 because Yannoni[21] reports two different sets of cell parameters and was unable clearly to locate all of the water molecules and $K^+$ ions.

## B. Octadecatungstodiphosphate Ion

The tungstic acid systems containing phosphate ions that produce the dodecatungstophosphate complexes of the type described in the preceding Section also contain a complex species in which the ions are in the ratio $P/W = 1/9$. If phosphoric acid is present in large excess in the strongly acid solutions, an orange-yellow crystalline product is formed, which is, typically, $K_6[P_2W_{18}O_{62}] \cdot 14H_2O$. So far, such complexes have been described only for tungstates and molybdates with phosphate and arsenate as hetero groups. Cryoscopic measurements by Souchay[58] showed the complex to have a doubled composition (as given above), and this result has been confirmed by a crystal-structure analysis by Dawson.[22]

Dawson found the potassium salt formulated above (Table 2, no. 5) to be triclinic, space group $P\bar{1}$, with a cell content of two formula units. In a prodigious feat of Patterson analysis, he was able to find the arrangement of the 18 W atoms in the asymmetric unit from three two-dimensional projections alone. Oxygen and phosphorus atoms were not located directly, but the configuration of the molecule and its close relation to the dodecatungstophosphate complex were clearly established. By adding 2 P and 62 O atoms in reasonable locations to the W atoms in the structure factor calculations, the reliability index was reduced by 0.03; the final value was $R = 0.18$. No information was obtained about the $K^+$ ions or $H_2O$ molecules.

As shown in Figure 5, the molecular structure of the octadecatungstodiphosphate ion consists of portions of two dodecatungstophosphate molecules (see Figure 3a) joined together by sharing opposite octahedral corners. Each $[PW_{12}O_{40}]^{3-}$ moiety has three $WO_3$ groups removed, and the exposed corners are then joined across a mirror plane producing a $[P_2W_{18}O_{62}]^{6-}$ molecule with point group symmetry $\bar{6}m2$ ($D_{3h}$). The two $PO_4$ groups remain separate and intact within the large tungstate cage.

Souchay[59] and others have recognized the appearance of two distinct crystal forms in the preparations of the ammonium salt, one

Figure 5. Octadecatungstodiphosphate ion, showing coordination polyhedra.

designated $\alpha$ (hexagonal-rhombohedral prisms) and the other $\beta$ (triclinic parallelepipeds). The two forms can be separated and recrystallized until pure, although their chemical and physical properties are closely similar. Souchay[59] has also studied a similar isomerism in potassium dodecatungstosilicate and dodecatungstoboric acid. The structural basis of the observed isomerism of these complexes is unknown, but, as Jahr[60] and O'Daniel[61] have pointed out for the dodeca complex, it is easy to imagine several different ways of constructing cage molecules from $W_3O_{13}$ octahedral triplets by joining the octahedral corners in various ways. Thus, in the molecule shown in Figure 5, the whole upper half could be rotated by 60° before being joined to the lower half, resulting in a molecule with $\bar{3}m$ ($D_{3d}$) symmetry. This and other possible structural isomers have been briefly discussed by Baker and Figgis.[57] An alternative possible arrangement for the dodeca complex is actually found in the tridecaaluminum complex described below (Section VI-C). Without careful crystal-structure studies, it is of little use to speculate further on the structural nature of the isomerism of these complexes.

## C. Hexamolybdo- and Hexatungsto-metallate Complexes

When acid tungstate and molybdate solutions are treated with hetero cations of a group where octahedral coordination rather than tetrahedral is usually found, another series of heteropoly complexes is produced. These have the general formulation

$$[Oc^x M_6 O_{24}]^{12-x}$$

where M may be $Mo^{VI}$ or $W^{VI}$ as before, and Oc may be $Al^{III}$, $Cr^{III}$, $Fe^{III}$, $Co^{II}$, $Co^{III}$, $Ni^{II}$, $Ga^{III}$, $Rh^{III}$, $Te^{VI}$, $I^{VII}$, etc. Of these $[TeMo_6O_{24}]^{6-}$ (Table 2, no. 7) and $[CrMo_6O_{24}]^{9-}$ (Table 2, nos. 9, 10) have been examined in detailed crystal-structure analyses by Evans[24] and by Perloff,[27, 28] and are evidently typical of the group. The structure has also been confirmed for $[NiW_6O_{24}]^{10-}$ (Table 2, no. 8) in an incomplete structure determination by Agarwala.[25, 26]

The molecule consists of seven octahedra condensed by edge-sharing into a flat, trigonal arrangement of symmetry $\bar{3}m$ ($D_{3d}$), shown as polyhedra in Figure 6a and as bonded atoms in a stereoscopic view in Figure 7. The hetero octahedron is thus surrounded by six $MoO_6$ or six $WO_6$ octahedra. This configuration was first proposed by Anderson[62] for the heptamolybdate ion (which, however, subsequently proved to have another structure; see Section V-A) and is thus sometimes referred to as the "Anderson molecule." The model structure is defined in Figure 6b and Table 5. In this case, the interatomic distances (Table 5) in the three independent class A or AA structure determinations are remarkably consistent, agreeing to better than 0.005 Å.

The complexes based on Al, Cr Fe, Co and Ni appear to have six rather firmly bound protons, which appear in the typical compounds nos. 8, 9 and 10 in Table 2. Perloff has found indirectly from the hydrogen-bond arrangement in no. 9 that they are associated with the central $CrO_6$ octahedron, each shared with one Cr and two Mo atoms. These complex ions might thus be written $[Oc^x(OH)_6M_6O_{18}]^{6-x}$. We may also note that Matijević, Kerker, Beyer and Theubert[63] have shown that among these complexes it is possible to replace Mo by W in a stepwise manner, according to the formulation

$$(NH_4)_4H_6[NiMo_{6-n}W_nO_{24}] \cdot 5H_2O,$$

where $n$ ranges from 0 to 6.

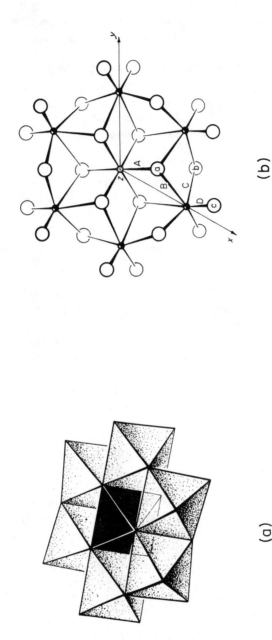

(b)

(a)

Figure 6. Hexamolybdotellurate ion, showing (a) coordination polyhedra and (b) model parameters (see Table 5).

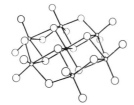 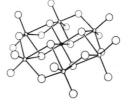

Figure 7. Stereoscopic view of the hexamolybdotellurate ion, showing bonding.

Table 5. Model dimensions for hexamolybdotellurate and hexamolybdochromate(III) ions.

Molecular symmetry $\bar{3}m$ $(D_{3d})$; see Figure 6b.
Lengths in Å, estimated standard error 0.005 Å; derived from data of Evans[24] for $[TeMo_6O_{24}]^{6-}$ and Perloff[27,28] for $[H_6CrMo_6O_{24}]^{3-}$.

*Molecular coordinates* (hexagonal axes):

| Atom | [TeMo₆O₂₄]⁶⁻ | | | [H₆CrMo₆O₂₄]³⁻ | | |
|---|---|---|---|---|---|---|
| | $x$ | $y$ | $z$ | $x$ | $y$ | $z$ |
| 1 Te (Cr) | 0 | 0 | 0 | 0 | 0 | 0 |
| 6 Mo | 3.299 | 0 | 0 | 3.341 | 0 | 0 |
| 6 O$_a$ | 1.890 | 0.945 | 1.037 | 1.963 | 0.981 | 1.006 |
| 6 O$_b$ | 3.766 | 1.883 | −0.944 | 3.776 | 1.888 | −0.905 |
| 12 O$_c$ | 4.691 | 0.734 | 1.218 | 4.740 | 0.717 | 1.210 |

*Bond lengths:*

| Bonded atoms | Vector | No. in mol. | Lengths [TeMo₆O₂₄]⁶⁻ | [H₆CrMo₆O₂₄]³⁻ |
|---|---|---|---|---|
| Te(Cr)–O$_a$ | A | 6 | 1.938 | 1.976 |
| Mo–O$_a$ | B | 12 | 2.299 | 2.295 |
| –O$_b$ | C | 12 | 1.943 | 1.938 |
| –O$_c$ | D | 12 | 1.714 | 1.705 |

## D. Decamolybdodicobaltate(III) Ion

In solutions of the hexamolybdocobaltate(III) complex (emerald-green), a small amount of a 5:1 complex has been known for some time to be present. Tsigdinos[64] found that the hexa complex is almost entirely converted into the 5:1 complex (dark reddish-purple) when the solutions are heated with activated charcoal or Raney nickel, and he showed by sodium sulfate cryoscopy that it is actually a dimer, that is, a 10:2 complex. A complete crystal-structure analysis of an ammonium salt (Table 2, no. 11) by Evans and Showell[29] has shown the compound to be $(NH_4)_6H_4[Co_2Mo_{10}O_{38}] \cdot 7H_2O$. with a structure closely related to that of the hexa complex. This relationship is shown in Figure 8a in terms of polyhedra. Figure 9 shows the molecule as bonded atoms in a stereoscopic view. The model structure is defined in Figure 8b and in Table 6.

Table 6. Model dimensions for decamolybdodicobaltate(III) ion.

Molecular symmetry 222 ($D_2$); see Figure 8b.
Lengths in Å, estimated standard error 0.01 Å; derived from data of Evans and Showell[29] for $[Co_2Mo_{10}O_{38}]^{6-}$.

| | *Molecular coordinates:* | | | |
|---|---|---|---|---|
| Atom | $x$ | $y$ | $z$ |
| 2 Co | 0 | 1.39 | 0 |
| 2 $Mo_I$ | 1.65 | 0.25 | 2.26 |
| 4 $Mo_{II}$ | $-1.47$ | 3.00 | 2.43 |
| 4 $Mo_{III}$ | 0 | 4.69 | 0 |
| 2 $O_a$ | 1.26 | 0 | 0 |
| 4 $O_b$ | $-0.04$ | 1.28 | 1.90 |
| 4 $O_c$ | $-1.30$ | 2.83 | 0.19 |
| 4 $O_d$ | $-2.54$ | 1.41 | 2.14 |
| 4 $O_e$ | $-0.03$ | 4.14 | 1.89 |
| 4 $O_f$ | 1.49 | 0.41 | 3.92 |
| 4 $O_g$ | 2.82 | 1.42 | 1.81 |
| 4 $O_h$ | $-1.29$ | 2.84 | 4.10 |
| 4 $O_i$ | $-2.78$ | 4.10 | 2.40 |
| 4 $O_k$ | $-1.35$ | 5.73 | $-0.01$ |

*Bond lengths* (4 in molecule for each):

| Bonded atoms | Vector | Length | Bonded atoms | Vector | Length |
|---|---|---|---|---|---|
| Co–$O_a$ | A | 1.88 | $Mo_{II}$–$O_b$ | K | 2.30 |
| –$O_b$ | B | 1.90 | –$O_c$ | L | 2.26 |
| –$O_c$ | C | 1.95 | –$O_d$ | M | 1.95 |
| $Mo_I$–$O_a$ | D | 2.31 | –$O_e$ | N | 1.91 |
| –$O_b$ | E | 2.26 | –$O_h$ | P | 1.73 |
| –$O'_b$ | F | 2.01 | –$O_i$ | Q | 1.71 |
| –$O'_d$ | G | 1.89 | $Mo_{III}$–$O_c$ | R | 2.28 |
| –$O_f$ | H | 1.67 | –$O_e$ | S | 1.97 |
| –$O_g$ | J | 1.71 | –$O_k$ | T | 1.71 |

In a manner reminiscent of the construction of the octadeca-tungstodisphosphoric acid molecule (Section IV-B), the decamolyb-dodicobaltate complex ion $[Co_2O_2(OH)_4(Mo_5O_{16})_2]^{6-}$ may be formed by removing $MoO(OH)_4$ from each of two $[Co(OH)_6Mo_6O_{18}]^{3-}$

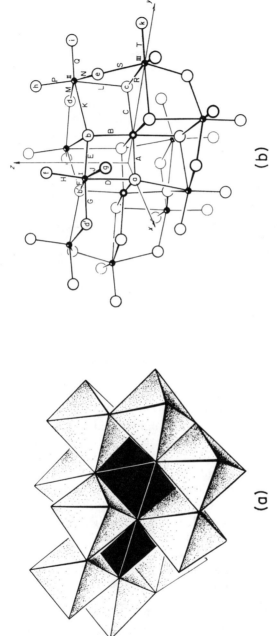

(a)

(b)

Figure 8. Decamolybdodicobaltate(III) ion, showing (a) coordination polyhedra, and (b) model parameters (see Table 6).

27

anions, turning one by 180° with respect to the other, and joining them along octahedron edges so that the two $CoO_4(OH)_2$ octahedra share an edge at the center (see Figure 8a). The resulting molecule has 222 ($D_2$) symmetry, and therefore must be optically active. The crystal studied (Table 2, no. 10) is a raceme; it would be interesting to separate chemically the enantiomers and study the rate of racemization in solution. [However, see p. 59.]

The optical character of the decamolybdodicobaltate(III) anion is reminiscent of another famous polynuclear cobalt(III) complex,

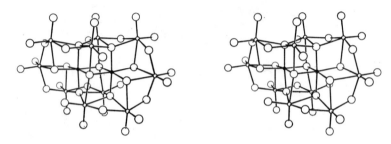

Figure 9. Stereoscopic view of the decamolybdodicobaltate ion, showing bonding.

tris(tetraamminecobalt)hexa-$\mu$-hydroxocobaltic cation, which Werner in 1914 succeeded in resolving into optical isomers, the first resolution of a wholly inorganic substance. The two $Co^{III}$ atoms at the center of the molybdo complex are surprisingly close together (2.78 Å), but the shared oxygen atoms are highly polarized and are only 2.55 Å apart. This geometry may be compared with that found by Prout[65] in another polynuclear cobalt(III) complex, di-$\mu$-hydroxobis[tetra-aminedicobalt(III)] chloride $[(NH_3)_4Co(OH)_2Co(NH_3)_4]Cl_4 \cdot 4H_2O$, in which the Co–OH bond lengths are $1.91 \pm 0.03$ Å and the Co–Co distance is $2.93 \pm 0.01$ Å. It is worth noting that the optic directions in the crystal of the molybdo complex nearly coincide with the three two-fold axes in the molecule and have strikingly different absorption: the direction along the Co–Co axis is olive-yellow; the normal to this direction in a plane containing the shared oxygen atoms is red, and the third direction (vertical in Figure 8) is blue.

### E. Enneamolybdomanganate(IV) Ion

The enneamolybdomanganate(IV) ion represents a third type of heteropoly complex; this has an octahedrally coordinated hetero atom and a ratio of Mo:Mn = 9:1. Hall[66] described the preparation of this complex, $(NH_4)_6[Mn^{IV}Mo_9O_{32}] \cdot 6H_2O$, and also the isomorphous $Ni^{IV}$ complex, in 1907. A very preliminary crystal-structure analysis of the ammonium salt octahydrate (Table 2, no. 12) reported by Waugh, Shoemaker and Pauling,[30] has established the probable configuration of the molecule. As shown in the polyhedral drawing in Figure 10, it may be considered as being built by adding

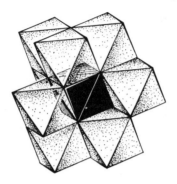

Figure 10. Enneamolybdomanganate(IV) ion, showing coordination polyhedra.

to alternate edges of the central $MnO_6$ octahedron three bent triple-octahedral $Mo_3O_8$ groups, that are themselves joined by two additional oxygen atoms at the top and bottom of the molecule. The triple molybdate groups are tilted with respect to the trigonal axis, giving the molecule a screw-like aspect, with point symmetry 32 ($D_3$). Unfortunately, only hexagonal $hk0$-type intensity data were used to check the heavy-atom arrangement, and the oxygen locations were inferred only by octahedral coordination geometry, so that no dimensional details are yet available for this molecule.

## F. Dodecamolybdocerate(IV) Ion

Baker, Gallagher and McCutcheon,[67] in their study of the dodecamolybdocerate(IV) complex decided, on the basis of the apparent eight-fold basicity and the large size of the hetero atom, that its structure must be different from that of the Keggin molecule. This suggestion has been confirmed by a complete structure determination by Dexter and Silverton[31] of the compound

$$(NH_4)_2H_6[CeMo_{12}O_{42}] \cdot 6H_2O$$

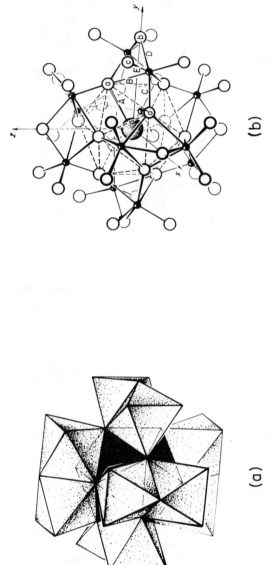

(a)

(b)

Figure 11. Dodecamolydocerate(IV) ion, showing (a) coordination polyhedra, and (b) model parameters (see Table 7).

(Table 2, no. 13). The structure, shown as a polyhedral linkage in Figure 11a and as bonded atoms in a stereoscopic view in Figure 12, is defined by Figure 11b and Table 7. This complex ion has $m3$ ($T_h$) point symmetry.

Figure 12. Stereoscopic view of the dodecamolybdocerate ion, showing bonding. The icosahedral environment of Ce is indicated by light lines.

Table 7. Model dimensions for dodecamolybdocerate(IV) ion.

Molecular symmetry $m3$ ($T_h$); see Figure 11b.
Lenths in Å, estimated standard error 0.01 Å; derived from data of Dexter and Silverton[31] for $[CeMo_{12}O_{42}]^{12-}$.

*Molecular coordinates:*

| Atom | $x$ | $y$ | $z$ |
|------|-----|-----|-----|
| 1 Ce | 0 | 0 | 0 |
| 12 Mo | 1.59 | 3.13 | 0 |
| 12 $O_a$ | 0 | 2.15 | 1.28 |
| 6 $O_b$ | 0 | 4.31 | 0 |
| 24 $O_c$ | 2.42 | 3.78 | 1.34 |

*Bond lengths:*

| Bonded atoms | Vector | No. in mol. | Length |
|--------------|--------|-------------|--------|
| Ce–$O_a$ | A | 12 | 2.51 |
| Mo–$O_a$ | B | 24 | 2.27 |
| –$O_a'$ | C | 12 | 1.94 |
| –$O_b$ | D | 12 | 1.98 |
| –$O_c$ | E | 24 | 1.70 |

Two unusual features are revealed in this structure. First, the nucleus hetero atom has icosahedral twelve-fold coordination, much higher than has been observed in any other heteropoly complex. Secondly, the $Mo_6$ octahedra that form the cage are linked together in pairs by sharing three oxygen atoms. This is the first structure of any molybdenum compound that has been found to contain pairs of $MoO_6$ octahedra sharing a face. A similar complex is apparently formed by thorium as the hetero atom.[67]

## V. ISOPOLY COMPLEXES

The group of polynuclear solution complexes generally termed "isopoly complexes" is more loosely defined than the heteropoly complex field. This is especially true since the penetrating emf studies of Sillén's group[9] have shown that most metal ions in solution with changing hydrogen ion concentration undergo hydrolysis that leads to polynuclear oxo complexes. The limitation imposed here of considering only the polynuclear complexes of Group 5 and 6 elements based on octahedral coordination is thus admittedly arbitrary. Within the stated limitation, all the compounds whose structures have been directly determined are collected in Table 8, which has the same design as Table 2. These structures will be briefly treated in the following Sections.

### A. Heptamolybdate, Octamolybdate and Hexamolybdate Ions

That hydrolyzed molybdate(VI) solutions contain a polyanion of high molecular weight was recognized long ago and was associated with solid crystalline salts termed "paramolybdates." The size of the polyanion could not be determined until Sturdivant[77] measured the unit cell of the ammonium salt $(NH_4)_6[Mo_7O_{24}] \cdot 4H_2O$ (Table 8, no. 2) and showed that the most probable molecular unit contained seven Mo atoms. Lindqvist[78] in 1950 studied the three-dimensional Patterson function of this compound and located the Mo atoms in clusters of seven, from which, by assuming octahedral coordination, he inferred the configuration shown in polyhedral form in Figure 13a. The molecule, which has, surprisingly, somewhat lower symmetry than that proposed by Anderson,[62] has point symmetry $2mm$ $(C_{2v})$. Although Lindqvist was unable, with facilities available at that time, to carry out a complete determination of the crystal structure and its refinement, this has recently been done by Shimao[79] and in more detail by Evans.[24] A structure analysis of the isostructural potassium salt (Table 8, No. 1) has been carried out by Gatehouse and Leverett.[67a] The molecular structure as determined by Evans is shown as bonded atoms in a stereo view in Figure 14, and is defined by Figure 13b and Table 9 (p. 36). The heptamolybdate molecule clearly

Table 8. Structure determinations of isopoly complexes.

| No. | Compound | Space group | Parameters total | detd. | Class | Ref., year |
|---|---|---|---|---|---|---|
| 1 | $K_6[Mo_7O_{24}] \cdot 4H_2O$ | $P2_1/c$ | 123 | 123 | B | Gatehouse and Leverett,[67a] 1968, 1970 |
| 2 | $(NH_4)_6[Mo_7O_{24}] \cdot 4H_2O$ | $P2_1/c$ | 123 | 123 | A | Evans,[24] 1968 |
| 3 | $(NH_4)_6[Mo_8O_{26}] \cdot 5H_2O$ | $P\bar{1}$ | 66 | 12 | C | Lindqvist,[68] 1950 |
| 4 | $Na_{10}H_2[W_{12}O_{42}] \cdot 27H_2O$ | $P\bar{1}$ | 135 | 18 | D | Lindqvist,[69] 1952 |
| 5 | $(NH_4)_{10}H_2[W_{12}O_{42}] \cdot 10H_2O$ | $P\bar{1}$ | 189 | 189 | A | Allman and Weiss,[70], 1969 |
| 6 | $K_2Zn_2[V_{10}O_{28}] \cdot 16H_2O$ | $P\bar{1}$ | 87 | 87 | A | Evans,[71] 1966 |
| 7 | $Ca_3[V_{10}O_{28}] \cdot 17H_2O$ | $I2$ | 85 | 85 | AA | Swallow, Ahmed and Barnes,[72] 1966 |
| 8 | $Na_6[V_{10}O_{28}] \cdot 18H_2O$ | $P\bar{1}$ | 93 | 93 | A | Pullman,[73] 1966 |
| 9 | $Na_7H[Nb_6O_{19}] \cdot 15H_2O$ | $Pmmn$ | 42 | 5 | C–D | Lindqvist,[74] 1953 |
| 10 | $K_8[Ta_6O_{19}] \cdot 16H_2O$ | $I2/c$ | 72 | 9 | C–D | Lindqvist and Aronsson,[75] 1954 |
| 11 | $[(C_4H_9)_4N]_2[W_6O_{19}]$ | $P\bar{1}$ | 102 | 72 | C | Henning and Hüllen,[76] 1969 |

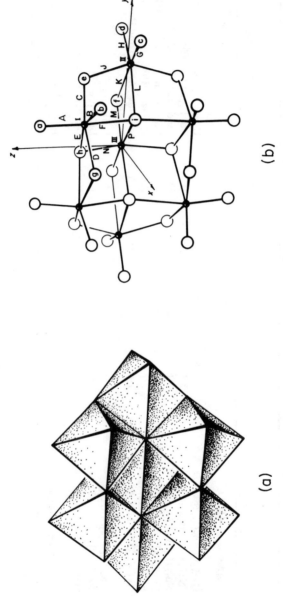

Figure 13. Heptamolybdate ion, showing (a) coordination polyhedra, and (b) model parameters (see Table 9).

34

corresponds to the heptamolybdate species found in solution by Sasaki and Sillén[10] (see Section II) and by Aveston, Anacker and Johnson.[13]

In his studies of various isopoly systems, Lindqvist[68] selected from an "ammonium paramolybdate" sample a crystal that proved

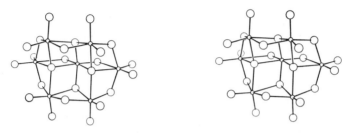

Figure 14. Stereoscopic view of the heptamolybdate ion, showing bonding.

to be $(NH_4)_6[Mo_8O_{26}] \cdot 5H_2O$ (Table 8, no. 3). In a preliminary crystal-structure study he found the molecule ion to be a cluster of eight condensed octahedra, arranged as shown in Figure 15. The molecule has point symmetry $2/m$ $(C_{2h})$. Unfortunately, no direct determination of oxygen atoms or structure refinement has yet been made, so

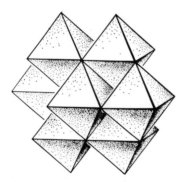

Figure 15. Octamolybdate ion, showing coordination polyhedra.

that the configuration is the only information we have at present. Aveston, Anacker and Johnson[13] (evidently influenced by Lindqvist's work) offered ultracentrifugation and Raman spectroscopic evidence for the existence of octamolybdate in solution, but this

Table 9. Model dimensions for heptamolybdate ion.

Molecular symmetry $2mm$ ($C_{2v}$); see Figure 13b.
Length in Å, estimated standard error 0.01 Å;
derived from data of Evans[24] for $[Mo_7O_{24}]^{6-}$.

*Molecular coordinates* (2-fold axis along $x$):

| Atom | $x$ | $y$ | $z$ |
|------|-----|-----|-----|
| 4 $Mo_I$ | 2.10 | 1.63 | 2.12 |
| 2 $Mo_{II}$ | 0.48 | 3.39 | 0 |
| 1 $Mo_{III}$ | 0 | 0 | 0 |
| 4 $O_a$ | 1.88 | 1.54 | 3.82 |
| 4 $O_b$ | 3.43 | 2.71 | 1.90 |
| 2 $O_c$ | 1.63 | 4.65 | 0 |
| 2 $O_d$ | −1.07 | 4.18 | 0 |
| 4 $O_e$ | 0.60 | 2.86 | 1.84 |
| 2 $O_f$ | −1.07 | 1.38 | 0 |
| 2 $O_g$ | 3.14 | 0 | 1.87 |
| 2 $O_h$ | 0.60 | 0 | 1.81 |
| 2 $O_i$ | 1.62 | 1.57 | 0 |

*Bond lengths:*

| Bonded atoms | Vector | No. in mol. | Length |
|--------------|--------|-------------|--------|
| $Mo_I$–$O_a$ | A | 4 | 1.71 |
| –$O_b$ | B | 4 | 1.73 |
| –$O_e$ | C | 4 | 1.97 |
| –$O_g$ | D | 4 | 1.95 |
| –$O_h$ | E | 4 | 2.18 |
| –$O_i$ | F | 4 | 2.17 |
| $Mo_{II}$–$O_c$ | G | 2 | 1.72 |
| –$O_d$ | H | 2 | 1.74 |
| –$O_e$ | J | 4 | 1.92 |
| –$O_f$ | K | 2 | 2.42 |
| –$O_i$ | L | 2 | 2.16 |
| $Mo_{III}$–$O_f$ | M | 2 | 1.75 |
| –$O_h$ | N | 2 | 1.90 |
| –$O_i$ | P | 2 | 2.26 |

could not be confirmed by the precision emf measurements of Sasaki and Sillén.[10] The latter authors suggest that octamolybdate is formed only at higher temperatures.

As mentioned in Section II, while Sasaki and Sillén[10] found heptamolybdates to predominate in $NaClO_4$ medium, Baldwin and Weise[15] have found that they are absent in $Mg(ClO_4)_2$ medium, where instead hexamolybdate $Mo_6O_{20}^{4-}$ is formed. The relationship (if any) is not clear between this discovery and the findings of Fuchs and Jahr[80] who prepared quarternary ammonium salts of hexamolybdate for which they established the formula $(NR_4)_2[Mo_6O_{19}]$; by infrared spectroscopy they showed this complex to be closely similar in type to the analogous hexatungstate, which they also prepared. Henning and Hüllen[76] have shown by crystal-structure analysis (Table 8, No. 11) that the structure of the $[W_6O_{19}]^{2-}$ ion is the same as that of the hexaniobate and hexatantalate ions (see Sections V-B and V-D and Figure 20).

## B. Metadodecatungstic Acid, Paradodecatungstate and Hexatungstate Ions

The chemistry of hydrolyzed acid tungstate solutions is completely different to, and considerably more complex than that of the analogous molybdate solution system. Initial acidification of alkaline tungstate solutions leads immediately to the formation of a hexatungstate $HW_6O_{21}{}^{5-}$ in the pH range 8–6, but this slowly changes to another species with supposedly the same nuclearity but different constitution. These hexatungstates have been referred to as "paratungstates," and Souchay[81] designates the former "B" and the latter "A." Souchay showed that further acidification yields "$\psi$-metatungstic acid" $H_3W_6O_{21}^{3-}$, but that, when boiled, these solutions afford "true metatungstic acid" $H_6W_{12}O_{42}^{6-}$. From all these systems a bewildering variety of crystalline products has been separated and described, but only one has been subjected to a complete crystal-structure analysis. The chemistry of the polytungstate solution system has been reviewed by Kepert.[82]

From Souchay's "true metatungstic acid" solution, that is, strongly acidic, hot tungstate solution, Keggin[83] obtained crystals of a potassium salt which he found by X-ray powder diffraction to be isomorphous with $K_4SiW_{12}O_{40} \cdot 18H_2O$. He therefore formulated it as a dodecatungstate, $K_4H_4W_{12}O_{40} \cdot 18H_2O$, with the supposition that the Si site in the complex is vacant. Signer and Gross[34] subsequently found that cubic $H_8W_{12}O_{40} \cdot 5H_2O$ (Table 3, no. 4) is completely isomorphous with $H_3PW_{12}O_{40} \cdot 5H_2O$ for which Keggin had solved the structure (Table 2, no. 1; Table 3, no. 1). Further, Santos[35] found $Cs_3H_5W_{12}O_{40} \cdot 2H_2O$ also to be wholly analogous to $Cs_3PW_{12}O_{40} \cdot 2H_2O$ (Table 3, nos. 8 and 5). There seems no doubt that this complex, which we may tentatively designate "metadodecatungstate ion", is a Keggin molecule.

Two of the protons associated with the acid have never been displaced in its salts, and it is generally felt that they are bound in some special way to the molecule. They are so non-labile that it has been strongly argued that they are lodged in the nucleus tetrahedron, replacing in some way the usual hetero atom. Pope and Varga[84] have obtained convincing evidence of the uniqueness of these protons in the form of a very sharp proton magnetic resonance signal of appropriate intensity about 6 ppm downfield from the $H_2O$ signal in solutions of sodium metatungstate, while this resonance is not given by dodecatungstosilicate or dodecatungstocobaltate solutions. The question of these hydrogen atoms has been discussed by Schott and Harzdorff.[85]

Another crystalline product that has been studied by X-ray diffraction is a sodium salt obtained from tungstate solutions acidified to about pH 5, in the "paratungstate" region. Saddington and Cahn[86] found that the triclinic unit cell contained $Na_{10}W_{12}O_{41} \cdot 28H_2O$. Lindqvist[69] carried out a three-dimensional Patterson synthesis of this crystal (Table 8, no. 4), which he interpreted to find the locations of the twelve tungsten atoms. Considering octahedral coordination linkages in the usual way, he postulated a molecular configuration corresponding to the bracketed portion of the formulation

$$Na_{10}H_{10}[W_{12}O_{46}] \cdot 23H_2O.$$

Lipscomb[87] showed hypothetically that a different configuration of octahedra is equally consistent with Lindqvist's tungsten arrangement, but that this is more economical in oxygen sharing and can be formulated as $Na_{10}H_2[W_{12}O_{42}] \cdot 27H_2O$. A recent complete determination by Allmann and Weiss[70] of the crystal structure of the ammonium salt $(NH_4)_{10}H_2[W_{12}O_{42}] \cdot 10H_2O$ (Table 8, No. 5) has wholly confirmed Lindqvist's tungsten atom arrangement with Lipscomb's oxygen configuration.

The molecule as determined by Allmann and Weiss has point symmetry $\bar{1}$ $(C_i)$; it is shown in polyhedral form in Figure 16a and as bonded atoms in stereoscopic view in Figure 17. Its dimensions are defined in Figure 16b and Table 10. Although the symmetry is actually triclinic, the molecule is not far from monoclinic symmetry $(2/m)$. If it were monoclinic, the four atoms $W_{II}$ and $W_{III}$ would form a rectangle with equal diagonals of 5.74 Å; the actual lengths are 5.68 Å and 5.80 Å. Whereas in other polyions the molecular distortions from higher symmetry in the crystal are readily accounted for by packing effects on the exterior of the molecule, in this case it seems likely that an internal effect is present resulting from the empty

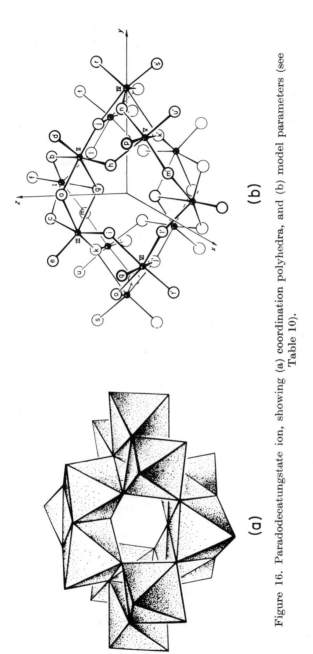

Figure 16. Paradodecatungstate ion, showing (a) coordination polyhedra, and (b) model parameters (see Table 10).

39

cavity at the center. There are two octahedral sites (each of which shares two faces with $WO_6$ octahedra) and two tetrahedral sites. The nature of the molecular distortion appears to be a slight twisting of the top and bottom $W_3O_{13}$ triplets, involving atomic shifts of the order of 0.1–0.2 Å, in a direction that would tend to reduce the space in this central region. Therefore, rather than average the dimensions

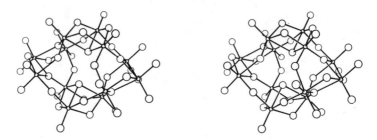

Figure 17. Stereoscopic view of the paradodecatungstate ion, showing bonding.

over the higher monoclinic symmetry, the actual triclinic dimensions as found by Allmann and Weiss[70] are given in Table 10, in such a way that nearly equivalent dimensions can be readily compared.

There is a possibility that one or two protons are lodged somewhere in the center of the molecule, which may also be acting to lower the symmetry of the molecule. The two central $O_g$ atoms are rather close together for a contact not involving a coordination linkage, but the distance 2.79 Å suggests that some form of hydrogen bonding may be involved. Such bonding may be needed to stabilize an otherwise rather loosely assembled molecular structure.

The two dodecatungstate complexes described above are difficult to relate to tungstate solution chemistry. The dominant hexameric species supported by ample physical chemical evidence [42,82] has yet to be found in crystals. On the other hand, Aveston[88] has offered strong ultracentrifugation and Raman spectroscopic evidence for the existence, in the acid solutions, of the dodeca species $W_{12}O_{41}^{10-}$ (or $[H_2W_{12}O_{42}]^{10-}$) and $W_{12}O_{39}^{6-}$ (or $[H_2W_{12}O_{40}]^{6-}$) in addition to the well established $HW_6O_{21}^{5-}$ (Sasaki[89]). The need for much crystal-structure work carefully correlated with the physical studies is apparent.

The preparation[80] and structure analysis[76] of the compound bis(tetra-$n$-butyl)ammonium    hexatungstate    $[(C_4H_9)_4N]_2[W_6O_{19}]$

Table 10. Model dimensions for paradodecatungstate ion.

Molecular symmetry $\bar{1}$ ($C_i$); see Figure 16b.
Lengths in Å, estimated standard error 0.02 Å; derived from data of
Allmann and Weiss[70] for $[H_2W_{12}O_{42}]^{10-}$.
Data in the four right-hand columns are related to those in left-hand
columns by monoclinic pseudosymmetry.

*Molecular coordinates:*

| Atom | $x$ | $y$ | $z$ | Atom | $x$ | $y$ | $z$ |
|------|------|------|------|------|------|------|------|
| $W_I$ | $-2.76$ | 0.08 | 2.34 | | | | |
| $W_{II}$ | 0.18 | 1.70 | 2.27 | $W_{III}$ | 0.11 | $-1.71$ | 2.34 |
| $W_{IV}$ | 0.00 | 4.68 | 0.00 | | | | |
| $W_V$ | 2.82 | 2.91 | $-0.11$ | $W_{VI}$ | 2.82 | $-2.91$ | 0.00 |
| $O_a$ | 0.64 | 0.06 | 3.18 | | | | |
| $O_b$ | $-1.72$ | 1.45 | 3.10 | $O_c$ | $-1.70$ | $-1.35$ | 3.15 |
| $O_d$ | 0.69 | 2.78 | 3.62 | $O_e$ | 0.52 | $-2.88$ | 3.64 |
| $O_f$ | $-4.22$ | 0.11 | 3.33 | | | | |
| $O_g$ | $-0.83$ | 0.03 | 1.12 | | | | |
| $O_h$ | 1.66 | 1.55 | 1.21 | $O_i$ | 1.61 | $-1.58$ | 1.23 |
| $O_j$ | $-0.75$ | 2.89 | 1.03 | $O_k$ | $-0.90$ | $-2.78$ | 1.13 |
| $O_l$ | $-3.27$ | 1.32 | 0.89 | $O_m$ | $-3.34$ | $-1.43$ | 1.03 |
| $O_n$ | 1.67 | 4.15 | 0.81 | $O_o$ | 1.66 | $-4.31$ | 0.88 |
| $O_p$ | 4.15 | 2.88 | 1.11 | $O_q$ | 3.99 | $-3.04$ | 1.31 |
| $O_r$ | $-0.57$ | 5.81 | 1.28 | $O_s$ | 0.51 | $-5.79$ | 1.23 |
| $O_t$ | $-3.40$ | 4.02 | 1.30 | $O_u$ | $-3.40$ | $-4.12$ | 1.31 |

*Bond lengths* (primed atoms are centric equivalents):

| Bonded atoms | Lengths | Bonded atoms | Lengths |
|------|------|------|------|
| $W_I-O_b$ | 1.88 | $W_I-O_c$ | 1.95 |
| $-O_f$ | 1.76 | | |
| $-O_g$ | 2.28 | | |
| $-O_l$ | 1.98 | $-O_m$ | 2.08 |
| $W_{II}-O_a$ | 1.92 | $W_{III}-O_a$ | 2.03 |
| $-O_b$ | 2.09 | $-O_c$ | 2.02 |
| $-O_d$ | 1.81 | $-O_e$ | 1.79 |
| $-O_g$ | 2.26 | $-O_g$ | 2.33 |
| $-O_h$ | 1.83 | $-O_i$ | 1.87 |
| $-O_j$ | 1.95 | $-O_k$ | 1.90 |
| $W_{IV}-O_j$ | 2.20 | $W_{IV}-O_k'$ | 2.38 |
| $-O_n$ | 1.93 | $-O_o'$ | 1.91 |
| $-O_r$ | 1.80 | $-O_s'$ | 1.74 |
| $W_V-O_h$ | 2.22 | $W_{VI}-O_i$ | 2.18 |
| $-O_n$ | 1.93 | $-O_o$ | 2.02 |
| $-O_p$ | 1.80 | $-O_q$ | 1.76 |
| $-O_k'$ | 2.18 | $-O_j'$ | 2.32 |
| $-O_m'$ | 1.82 | $-O_i'$ | 1.88 |
| $-O_u'$ | 1.81 | $-O_t'$ | 1.80 |

(Table 8, no. 11) proves the existence of a hexatungstate polyanion, but it cannot be the same as the hexatungstate ion $HW_6O_{21}^{5-}$, and its relation to the polytungstate system outlined above is unknown. Though they did not determine accurate dimensions for the $[W_6O_{19}]^{2-}$ molecule ion, Henning and Hüllen[76] proved that it is analogous to the hexaniobate and hexatantalate polyions (see Section V-D and Figure 20).

## C. Decavanadate Ion

Orange, acidic vanadate solutions have been shown by precision emf studies by Rossotti and Rossotti[90] to contain mainly the decanuclear species $[V_{10}O_{28}]^{6-}$, $H[V_{10}O_{28}]^{5-}$ and $H_2[V_{10}O_{28}]^{4-}$, a result subsequently confirmed by several other workers using various techniques.[16] Complete three-dimensional crystal-structure determinations of three different compounds containing the decavanadate ion have been carried out, as listed in Table 8, nos. 6, 7 and 8. All three determinations found the same polyanion configuration and are in close agreement concerning the bond-length distribution. Each of these structure determinations has claimed standard deviations for V–O bond-lengths of 0.014 Å or better (0.004 Å for no. 7, 0.008 Å for no. 6, 0.014 Å for no. 8). Comparison of the three sets of results affords an opportunity to see how much real variation one may expect as a result of hydrogen bonding and other effects of the unsymmetrical near surroundings in the crystal structure. All the determined distances and their standard deviations are collected in Table 11 to show the extent of such variations in typical situations.

The decavanadate ion consists of ten highly condensed $VO_6$ octahedra, in a configuration having point symmetry $mmm$ $(D_{2h})$, as shown in polyhedral form in Figure 18a and as bonded atoms in a stereoscopic view in Figure 19. Figure 18b and Table 12 serve to define the molecule, as based on average bond lengths derived from Table 11. (Fig. 18 is on p. 44 and Fig. 19 on p. 45.)

In their careful structure analysis of the calcium salt* (Table 8, no. 7) Swallow, Ahmed and Barnes[72] found that two of the three $Ca^{2+}$ ions are associated with the decavanadate group, each coordinated to a pair of $O_g$ atoms and $5H_2O$, at the top and bottom of the polyion (see Figure 18). They suggested that this feature of the structure may

---

* The natural mineral pascoite. This compound, $K_2Mg_2V_2O_{28}\cdot16H_2O$ known as hummerite (isostructural with the Zn analog, Table 8, No. 6), and an undescribed Na salt are the only isopoly compounds known to occur in Nature. No heteropoly compounds at all have been found in Nature.

Table 11. Comparison of bond lengths in the decavanadate ion in three different crystals.

Lengths in Å, least-squares standard error in parenthesis (last significant figures).

| Bonded atoms (see Figure 18b) | Vector | Determined lengths in $[V_{10}O_{28}]^{6-}$ | | |
|---|---|---|---|---|
| | | Evans[71] | Swallow et al.[72] | Pullman[73] |
| $V_I–O_g$ | A | 1.612(8) | 1.617(3) | 1.620(16) |
| | | 1.603(8) | 1.623(3) | 1.604(17) |
| $V_I–O_a$ | B | 2.246(6) | 2.211(3) | 2.199(15) |
| | | 2.218(6) | 2.222(3) | 2.256(15) |
| $V_I–O_b$ | C | 2.000(6) | 2.009(3) | 2.006(13) |
| | | 1.965(7) | 2.009(3) | 1.969(13) |
| | | 2.014(6) | 2.002(3) | 2.005(13) |
| | | 2.000(6) | 2.002(3) | 2.000(13) |
| $V_I–O_d$ | D | 1.831(7) | 1.803(4) | 1.817(14) |
| | | 1.846(7) | 1.803(4) | 1.839(14) |
| | | 1.839(7) | 1.813(4) | 1.838(15) |
| | | 1.803(7) | 1.813(4) | 1.798(13) |
| $V_{II}–O_f$ | E | 1.602(8) | 1.593(3) | 1.599(16) |
| | | 1.608(9) | 1.593(3) | 1.599(15) |
| $V_{II}–O_a$ | F | 2.317(6) | 2.313(3) | 2.305(14) |
| | | 2.355(7) | 2.313(3) | 2.308(14) |
| $V_{II}–O_c$ | G | 1.837(7) | 1.818(4) | 1.829(14) |
| | | 1.845(7) | 1.818(4) | 1.834(14) |
| $V_{II}–O_d$ | H | 1.895(7) | 1.867(4) | 1.886(14) |
| | | 1.867(7) | 1.874(4) | 1.844(15) |
| | | 1.845(7) | 1.867(4) | 1.886(13) |
| | | 1.907(7) | 1.874(4) | 1.869(14) |
| $V_{II}–O_e$ | J | 2.027(7) | 2.077(3) | 2.033(13) |
| | | 2.077(7) | 2.077(3) | 1.993(13) |
| $V_{I,I}–O_e$ | K | 1.713(7) | 1.681(3) | 1.677(13) |
| | | 1.678(7) | 1.681(3) | 1.712(13) |
| $V_{III}–O_a$ | L | 2.110(7) | 2.135(3) | 2.118(14) |
| | | 2.123(6) | 2.135(3) | 2.111(14) |
| $V_{III}–O_b$ | M | 2.000(6) | 1.903(3) | 1.937(13) |
| | | 1.965(7) | 1.903(3) | 1.937(13) |

indicate the existence in solution of a complex that they formulated as $[(Ca \cdot 5H_2O)_2 V_{10}O_{28}]^{2-}$. This suggestion is reminiscent of the proposals of Jahr, Fuchs and Preuss,[14] based on conductometric studies, that complexes of the type $[Ca(H_2O)_n V_{10}O_{28}]^{4-}$ are formed in solution, although they did not formulate a bicationic complex.

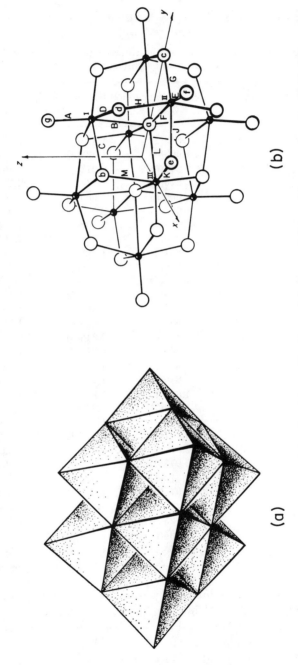

Figure 18. Decavanadate ion, showing (a) coordination polyhedra, and (b) model parameters (see Table 12).

Figure 19. Stereoscopic view of the decavanadate ion, showing bonding.

Table 12. Model dimensions for decavanadate ion.

Molecular symmetry $mmm$ ($D_{2h}$); see Figure 18b. Lengths in Å, estimated standard error 0.01 Å; derived from data for $[V_{10}O_{28}]^{6-}$ of Evans,[71] Pullman,[73] and Swallow, Ahmed and Barnes.[72]

*Molecular coordinates:*

| Atom | $x$ | $y$ | $z$ |
|------|-----|-----|-----|
| 4 $V_I$ | 0 | 1.54 | 2.21 |
| 4 $V_{II}$ | 1.53 | 3.08 | 0 |
| 2 $V_{III}$ | 1.66 | 0 | 0 |
| 2 $O_a$ | 0 | 1.33 | 0 |
| 4 $O_b$ | 1.24 | 0 | 1.88 |
| 2 $O_c$ | 0 | 4.08 | 0 |
| 8 $O_d$ | 1.34 | 2.70 | 1.83 |
| 4 $O_e$ | 2.66 | 1.36 | 0 |
| 4 $O_f$ | 2.70 | 4.16 | 0 |
| 4 $O_g$ | 0 | 1.46 | 3.83 |

*Bond lengths:*

| Bonded atoms | Vector | No. in mol. | Length |
|--------------|--------|-------------|--------|
| $V_I$–$O_g$ | A | 4 | 1.62 |
| –$O_a$ | B | 4 | 2.22 |
| –$O_b$ | C | 8 | 2.00 |
| –$O_d$ | D | 8 | 1.82 |
| $V_{II}$–$O_f$ | E | 4 | 1.60 |
| –$O_a$ | F | 4 | 2.32 |
| –$O_c$ | G | 4 | 1.83 |
| –$O_d$ | H | 8 | 2.06 |
| –$O_e$ | J | 4 | 1.87 |
| $V_{III}$–$O_e$ | K | 4 | 1.69 |
| –$O_a$ | L | 4 | 2.13 |
| –$O_b$ | M | 4 | 1.93 |

## D. Hexaniobate and Hexatantalate Ions

Hydrolyzed niobate and tantalate solutions, according to careful studies by Neumann,[91] Aveston and Johnson,[92] and Nelson and Tobias,[93] contain only one polynuclear species each, namely, hexaniobate $[Nb_6O_{19}]^{6-}$ and hexatantalate $[Ta_6O_{19}]^{6-}$. It has been suggested[16] that these species predominate even in the most alkaline solutions. Two crystals obtained from such solutions were studied in a preliminary way by Lindqvist and Aronsson[74,75] (Table 8, nos. 9 and 10). They were both found to contain regular octahedral clusters of Nb or Ta atoms, with uniform inter-metal-atom distances of 3.3 Å. This arrangement is consistent with a system of six condensed $NbO_6$ or $TaO_6$ octahedra, combined in a molecule with point symmetry $m3m$ ($O_h$), as shown in polyhedral form in Figure 20. Unfor-

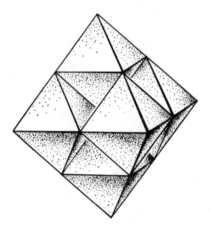

Figure 20. Hexaniobate ion, showing coordination polyhedra.

tunately, no direct determination of oxygen positions in these crystals has yet been made, so that the bond lengths in the undisturbed molecule are unknown; however, these are probably not very different from the distances found in an addition complex of hexaniobate and manganese(IV) ion, $[Mn(Nb_6O_{19})_2]^{12-}$, recently determined in a crystal-structure analysis of the sodium salt (Table 2, no. 14) by Flynn and Stucky.[32] This structure is described in more detail in Section VI-B.

# VI. STRUCTURES OF SOME CLOSELY RELATED POLYIONS

As noted above, the restriction of our concern to octahedral polyion structures of V, Nb, Ta, Mo and W is rather arbitrary. Some molecular structures are known that come near to this borderline but have been excluded. A few of these, however, have such a close bearing on the type of structure under consideration that it seems worthwhile to consider them in this appended Section.

## A. Tridecaaluminyl Cation

In Section IV-B the possibility of isomorphism of the Keggin molecule, first suggested by Jahr,[60] was mentioned. If the $M_3O_{13}$ octahedral triplets that make up the Keggin molecule by corner-sharing (see Fig. 3a) are disassembled, then each rotated 60° around a three-fold axis, they can be rejoined in a new configuration, still with point symmetry $\overline{4}3m$ ($T_d$), but with adjacent octahedra linked by sharing edges, as shown in Figure 21. It has been discovered that a complex of aluminum obtained from partially hydrolyzed solutions[94] has just this structure. Johansson has solved the structures of the cubic $Na[Al_{13}O_4(OH)_{24}(H_2O)_{12}](SO_4)_4 \cdot 13H_2O$ and the corresponding selenate,[95] and also the monoclinic sulfate

$$[Al_{13}O_4(OH)_{25}(H_2O)_{11}](SO_4)_3 \cdot 16H_2O.^{[96]}$$

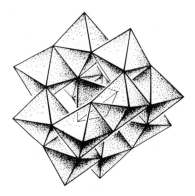

Figure 21. Tridecaaluminyl ion, showing coordination polyhedra.

He refined the last structure in three dimensions to the point where the standard error of the Al–O distances is 0.02 Å (nearly class A). The $[H_{48}Al_{13}O_{40}]^{7+}$ and $[H_{47}Al_{13}O_{40}]^{6+}$ molecule ions were found to be practically identical in the bonding details (within the error of determination) in all three structures. The molecule is shown in poly-hedral form in Figure 21 and as bonded atoms in a stereoscopic view in Figure 22.

We may note in passing that the tridecaaluminyl cation represents a fragment of a spinel-type structure. On the other hand, the structure of the mineral zunyite[97] ($Al_{13}Si_5O_{20}(OH)_{18}Cl$) contains an

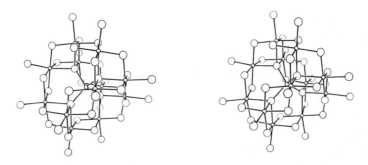

Figure 22. Stereoscopic view of the tridecaaluminyl ion, showing bonding.

$[Al_{13}O_{16}(OH)_{24}]$ configuration (linked to adjacent groups) that is exactly that of the Keggin molecule.

The occurrence of both tetrahedral (nucleus) and octahedral (shell) coordination for aluminum in Johannson's structure is especially note-worthy. Such ambivalence is, of course, well known for aluminum, but it is also known for molybdenum[98] and niobium[99] in oxide structures, and for tungsten in a lithium tungstate[100] (see Section VI-D).

## B. Titanium(IV) Alkoxides

Anhydrous preparations of titanium(IV) alkoxides are known to form polymers $[Ti(OR)_4]_n$ in benzene solutions. Two of these have been isolated in crystalline form and shown by crystal-structure analysis to be tetramers: $[Ti(OEt)_4]_4$ by Ibers[101] and $[Ti(OMe)-(OEt)_3]_4$ by Witters and Caughlan[102] (Me = methyl; Et = ethyl). Both consist of four $TiO_6$ octahedra condensed by edge-sharing to give isopoly $Ti_4O_{16}$ groups, which may be considered as fragments of a layer structure of the $Cd(OH)_2$ type.

Of particular interest in the present context is the study of par-tially hydrolyzed solutions of $Ti(OEt)_4$ in benzene made by

Watenpaugh and Caughlan,[103] in which they discovered an unusual polynuclear titanium complex by crystal-structure analysis of one of the crystalline products. It proved to be heptanuclear, $Ti_7O_4(OEt)_{20}$,* with a $[Ti_7O_{24}]$ configuration exactly like that of heptamolybdate ion, described in Section V-A (Figures 13 and 14). The titanium structure has not been extensively refined (class B), but the oxygen atoms have been sufficiently resolved to show that the $TiO_6$ octahedra are much more regular than they are in heptamolybdate ion.

## C. Bis(hexaniobato)manganate(IV) Ion

Flynn and Stucky[104] have recently studied so-called heteropoly complexes involving niobate with $Mn^{IV}$ and $Ni^{IV}$. The nature of these complexes was revealed in a detailed three-dimensional crystal-structure determination[32] of $Na_{12}[Mn(Nb_6O_{19})_2]\cdot50H_2O$. The molecule is formed by junction of a Mn atom to the face of each of two hexaniobate isopoly anions of the type shown in Figure 20. The Mn atom thus attains octahedral coordination while the $[Nb_6O_{19}]$ groups remain practically unchanged except for local distortions in the vicinity of Mn. The two hexaniobate groups are not linked to each other except through Mn, and therefore do not form a cage around the hetero atom as is usual for the typical heteropoly complex. It seems more natural, therefore, to regard this complex as an addition complex of an isopoly complex with a metal atom rather than as a heteropoly complex.

The Mn atom is joined to the hexaniobate groups along their three-fold axes, to give a large molecule with $\bar{3}m$ ($D_{3d}$) point symmetry, as shown in polyhedral form in Figure 23a, and as bonded atoms in Figure 24 in a stereoscopic view. The molecule is defined by Figure 23b and Table 13 (See pp. 50 and 51).

## D. Lithium Tungstate 4/7-Hydrate

In a hydrothermal reaction Hüllen[100] obtained cubic crystals to which he assigned the formula $7Li_2WO_4\cdot4H_2O$. This product corresponds to that obtained from hot, alkaline solutions by Rosenheim and Reglin.[105] Hüllen's crystal-structure analysis, based on two-dimensional data only (class C), revealed the presence of what may

---

*Watenpaugh and Caughlan[103] give the formula empirically from their incomplete structure analysis as $Ti_7O_{24}(C_2H_5)_{19}$ but, on the assumption that all the titanium is quadrivalent, the rational formula $Ti_7O_4(OC_2H_5)_{20}$ is used here. It is also possible that the compound contains some OH groups.

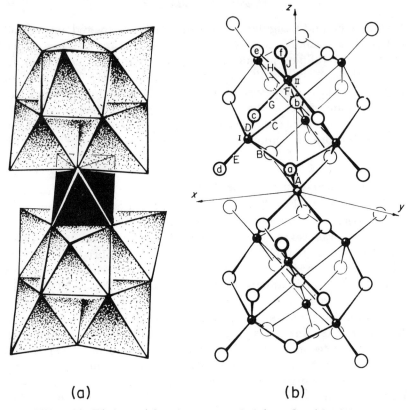

(a)                              (b)

Figure 23. Bis(hexaniobato)manganate(IV) ion, showing, (a) coordination polyhedra, and (b) model parameters (see Table 13).

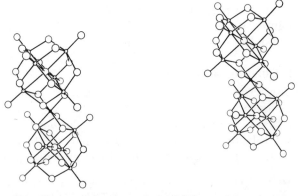

Figure 24. Stereoscopic view of the bis(hexaniobato)manganate ion, showing bonding.

Table 13. Model dimensions for bis(hexaniobato)-
manganate(IV) ion.

Molecular symmetry, $\bar{3}m$ ($D_{3d}$); see Figure 23b.
Lengths in Å; estimated standard error 0.02 Å;
derived from data of Flynn and Stucky[32] for
$[Mn(Nb_6O_{19})_2]^{6-}$.

*Molecular coordinates:*

| Atom | $x$ | $y$ | $z$ |
|---|---|---|---|
| 1 Mn | 0 | 0 | 0 |
| 6 $Nb_I$ | 2.04 | 0 | 2.26 |
| 6 $Nb_{II}$ | 1.93 | 1.93 | 4.93 |
| 6 $O_a$ | 1.44 | 1.44 | 1.20 |
| 2 $O_b$ | 0 | 0 | 3.54 |
| 12 $O_c$ | 3.18 | 1.65 | 3.50 |
| 6 $O_d$ | 3.54 | 0 | 1.33 |
| 6 $O_e$ | 1.63 | 0 | 5.74 |
| 6 $O_f$ | 3.34 | 3.34 | 5.99 |

*Bond lengths:*

| Bonded atoms | Vector | No. in mol. | Length |
|---|---|---|---|
| $Mn–O_a$ | A | 6 | 1.87 |
| $Nb_I–O_a$ | B | 12 | 2.10 |
| $–O_b$ | C | 6 | 2.41 |
| $–O_c$ | D | 12 | 1.92 |
| $–O_d$ | E | 6 | 1.77 |
| $Nb_{II}–O_b$ | F | 6 | 2.38 |
| $–O_c$ | G | 12 | 2.01 |
| $–O_e$ | H | 12 | 1.97 |
| $–O_f$ | J | 6 | 1.76 |

be considered a heteropoly complex ion, tetratungstolithate $[LiW_4O_{16}]^{7-}$. Isolated tetrahedral $WO_4^{2-}$ groups are also present, and the resulting structural formula may be written

$$Li_{13}(WO_4)_3[LiW_4O_{16}] \cdot 4H_2O.$$

The surprising coordination of the hetero lithium atom and the possibility that the supposed heteropoly complex could be an isopoly group have led us to relegate this complex to this appendix group. Nevertheless, its crystal chemistry is very instructive in this connection.

The proposed heteropoly ion is apparent in the center of the unit-cell cube shown in Figure 25. It consists of four condensed $WO_6$

octahedra forming a $W_4O_{16}$ group with point symmetry $\overline{4}3m$ ($T_d$). The lithium atom is presumed to occupy the tetrahedral cavity at the center of the molecule, thus sharing its faces with all four $WO_6$ octahedra. The structure as a whole is based on a cubic close packing of oxygen atoms (spec. vol. of O, 18.0 Å³) strongly bound together by the $W^{6+}$ and $Li^+$ ions into a three-dimensional framework; this

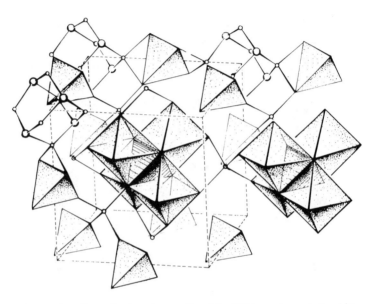

Figure 25. Crystal structure of cubic lithium tungstate 4/7-hydrate, according to Hüllen.[100] Small circles are Li, large circles are $H_2O$; $WO_4$ and $WO_6$ groups are indicated by stipple-shaded tetrahedra and octahedra, respectively. The presumed hetero Li atom is at the center of the line-shaded tetrahedron.

molecule is thus chemically rather different from the ionic molecular type of crystal dealt with above.

Borisov, Klevtsova and Belov[106] have recently made a three-dimensional study (class B) of what is evidently the same structure, independently and apparently unaware of Hüllen's work. Their tungsten arrangement and oxygen coordination agree with Hüllen's data but they find that the central tetrahedron is empty while the lithium atoms are in octahedral instead of tetrahedral coordination, leading to a structure somewhat similar to that of spinel. They assume that their crystals contain no hydrogen atoms, but, instead, Fe(III) replacing Li to the extent of Li:Fe = 11:1, and thus they derive the structural formula $(Li_{11}Fe)(WO_4)_3(W_4O_{16})$. Obviously, the crystal chemistry of this compound is not yet entirely clear.

Rosenheim and Raglin's lithium tungstate compound illustrates a common difficulty that arises in attempts to establish by chemical analysis the constitution of high-molecular-weight complexes of heavy metals such as tungsten and molybdenum.[107] The weight percentage of light hetero atoms or water may vary so little from one formulation to another that it may be difficult to choose the critical atomic ratios correctly. Such was the case, for example, with the undecatungstometallate complexes.[43] In the lithium tungstate compound, it seems possible that there may be 13 Li atoms in the cubic unit cell instead of 14. The charge can be balanced by an extra proton, which would easily find some random location on one of the tungstate oxygen atoms. In this case, the center of the $W_4O_{16}$ group would be unoccupied. Because of the low scattering power of lithium its presence at this point cannot be proved by such a structure determination as this. The analytical data are equally inconclusive, as shown by the following comparison of weight percentages for $xLi_2O \cdot yWO_3 \cdot zH_2O$:

|  | $x:y:z$ | | | |
|---|---|---|---|---|
|  | 7:7:4 | 13:14:9 | Observed[105] | |
| $Li_2O$ | 10.98 | 10.23 | 11.04, | 11.16 |
| $WO_3$ | 85.23 | 85.50 | 84.33, | 84.37 |
| $H_2O$ | 3.78 | 4.27 | 4.90, | 4.97 |

Hüllen[100] finds 3.71 and 3.78 wt. percent for $H_2O$.

## VII. CONCLUDING OBSERVATIONS

The primary chemical question raised by the existence of the family of hetero and isopoly complexes is: what accounts for the critical stability of a particular polynuclear configuration? Little speculation on the question has so far been offered and there seems at present to be no satisfactory theory of structure.

We note first that the complexes obey most of the well-known Pauling rules for the properties of stable oxide structures. Thus, in all cases the outer octahedra of the polyions are highly distorted in such a way that the metal ions are strongly displaced outward, with the result that inner M–O distances are lengthened and outer distances shortened. This distortion is in accord with the repulsive effect of the highly charged cations on one another, and also with the need to balance charges on the inner, multiply linked oxygen atoms. In hexamolybdotellurate ion (see Figure 6 and Table 5), for example, the inner distance $Mo–O_a$, where $O_a$ shares three cations, is the longest (2.30 Å); the lateral distances $Mo–O_b$, where $O_b$ shares two cations, is intermediate (1.94 Å); and the outer distances $Mo–O_c$, where $O_c$ is bonded to only one cation, is the shortest (1.71 Å). Also, the $O_a$ atoms in the central $TeO_6$ octahedron are drawn closer together where they form a shared edge with an adjacent $MoO_6$ octahedron (2.64 Å) than where they do not (2.84 Å). These effects can be seen in all the structures described.

Small as the collection of known structures is, we may look for other more particular common features that might account for stability. Except for the effects noted above, there seems to be no such feature which is not contradicted by at least one notable example. Some of the features that appear to be characteristic have been noted previously by Evans[71] and by Kepert,[108] and are mentioned briefly in the following paragraphs.

1. Certain rather complex structures are critically stable in preference to all other possible structures over rather broad acidity ranges. For example, the decavanadate ion (Section V-C) is stable in solution over a pH range 2–6 to the almost complete exclusion of other species.[90] Such predominant species may be protonated one or more times, and there appears to be no special tendency to eliminate these protons by further condensation within their stability ranges.

2. All the complexes are based on assemblages of $MO_6$ octahedra (M = V, Nb, Ta, Mo or W). These octahedra are highly condensed by sharing edges. The sharing of faces would not ordinarily be expected but was found for the first time in dodecamolybdocerate(IV) ion (Section IV-F).

3. The hetero atom in the heteropoly complexes usually has tetrahedral or octahedral coordination. The known structures are often classified on the basis of these two coordination types. The fact that other hetero atom coordinations are possible was demonstrated recently by the structure determination of dodecamolybdocerate(IV) ion (Section IV-F).

4. Most of the isopoly complexes seem to be based on fragments of a rock-salt structure. The largest is decavanadate ion (Section V-C, Figure 18a) with ten condensed octahedra. Octamolybdate ion

(Section V-A, Figure 15) is similar to decavanadate ion but has two opposite-corner octahedra removed. Heptamolybdate ion (Section V–A, Figure 13a) may be derived from the decavanadate structure by removing three octahedra in a row on one side. Hexaniobate and hexatantalate (Section V–D, Figure 20) may be derived from the decavanadate structure by removing four octahedra at one end. (Hexamolybdocobaltate, decamolybdodicobaltate and enneamolybdomanganate ions may also be considered as fragments of rock-salt type structure.) The tendency to form a rock-salt type structure on hydrolysis is perhaps not surprising, but of course a large charge excess rapidly builds up in the process. For each type of cation a critical barrier is encountered for a degree of nuclearity and configuration that is characteristic for that cation, but how this limit is determined is not known.

The notable exception to this property of isopoly complexes is the isopoly tungstate group. Both the polytungstate complexes (Section V-B) have more complicated configurations, but both resemble the cage-like structures of the heteropoly complexes and may include H as a nucleus atom.

5. A common structural feature of both heteropoly and isopoly complexes is the presence of a bent $MO_2$ group in which the M–O distance is short ($\sim 1.7$ Å) and the angle is close to $104°$. This feature is prominent enough in the vanadates,[71] for example, that Hanic[109] has suggested that it may be considered as a discrete ionic group. All the Mo and W atoms are clearly associated with such V-shaped groups in hexamolybdo- and hexatungsto-metallate complexes (Section IV-C), decamolybdodicobaltate ion (Section IV-D), dodecamolybdocerate ion (Section IV-F), and heptamolybdate ion (Section V-A). In the last of these complexes, even the central $MoO_6$ octahedron is distorted so as to incorporate such a group, although in all the other complexes the associated oxygen atoms are unshared. The $MO_2$ group also appears in many chain and layer molybdate structures; for example, $Ag(MoO_2)PO_4$.[110] The outstanding exception to this trend is the Keggin molecule (Section IV-A). There seems no point in overemphasizing the supposed $MO_2$ grouping, but we may note that the group probably involves directed multiple Mo–O bonds which, when properly incorporated into a complex structure, may contribute to its stability.

6. Although the formation of the poly complexes seems to involve successive, stepwise condensations with changes of acidity, the structures so far revealed do not suggest any mechanism for such

reactions. Only in the cases of dodecatungstophosphate–octadeca-tungstodiphosphate (Section IV-A, B) and hexamolybdocobaltate–decamolybdodicobaltate (Section IV) are obvious structual relationships apparent that might be associated with a conversion equilibrium. Even here, smaller fragments would have to be involved, the existence of which are not known. In other cases, for example, heptamolybdate–octamolybdate, passage from one to the other would require a partial dismantling of the first complex and re-construction to form the second. Thus, a simple chemistry based on the successive building up of larger groups from smaller ones is generally not involved in the formation of iso and heteropoly complexes. This observation is consistent with the sluggishness that is often observed in the attainment of equilibrium in complex solution systems (for example, the polytungstate system, Section V-B).

In this area of chemistry predictions and speculations have had remarkably little success. In view of the size of the complex family involved, we still have rather little good structural information. For a better understanding of the true nature of this system, we must await further development of such information, probably for the most part by the application of crystal-structure analysis, to be closely linked with other physical chemical studies.

## References

1. J. J. Berzelius, *Pogg. Ann.*, **6**, 369, 380 (1826).
2. C. Marignac, *Ann. Chim. Phys.*, [4], **3**, 5 (1864).
3. A. Rosenheim and J. Jaenicke, *Z. anorg. allgem. Chem.*, **100**, 304 (1917).
4. L. Pauling, *J. Amer. Chem. Soc.*, **51**, 2868 (1929).
5. J. L. Hoard, *Z. Krist.*, **84**, 217 (1933).
6. J. F. Keggin, *Nature*, **131**, 968 (1933); **132**, 351 (1933).
7. I. Lindqvist, *Diss. Nova Acta Regiae, Soc. Sci. Upsaliensis*, Ser. IV, Vol. **15**, No. 1 (1950).
8. G. Jander, K. Jahr and W. Henkeshoven, *Z. anorg. allgem. Chem.*, **194**, 383 (1930); G. Jander and H. Witzmann, *ibid.*, **215**, 310 (1933).
9. L. G. Sillén, *Quart. Rev.*, **13**, 146 (1959).
10. Y. Sasaki and L. G. Sillén, *Arkiv Kemi*, **29**, 253 (1968).
11. L. C. W. Baker, *Advances in the Chemistry of Coordination Compounds*, pp. 604–612, Macmillan Co., New York, 1961.
12. L. C. W. Baker and M. T. Pope, *J. Amer. Chem. Soc.*, **82**, 4176 (1960).
13. J. Aveston, E. W. Anacker and J. S. Johnson, *Inorg. Chem.*, **3**, 735 (1964).
14. K. F. Jahr, J. Fuchs and F. Preuss, *Chem. Ber.*, **96**, 556 (1963).
15. W. G. Baldwin and G. Weise, *Arkiv Kemi*, **31**, 419 (1969).
16. M. T. Pope and B. W. Dale, *Quart. Rev.*, **22**, 527 (1968).

17. P. Souchay, *Polyanions et Polycations*, Gautier-Villars, Paris, 1963; *Ions Mineraux Condensés*, Masson et Cie, Paris, 1969.

18. H. A. Levy, P. A. Agron and M. D. Danford, *J. Chem. Phys.*, **30**, 1486 (1959).

19. A. J. Bradley and J. W. Illingworth, *Proc. Roy. Soc.*, *A*, **157**, 113 (1936).

20. H. T. Evans, Jr., unpublished work. The study of $(NH_4)_7Na_2[GaW_{11}O_{39}] \cdot 16H_2O$ as prepared by Rollins[42,43] is based on 664 independent counter-measured data, corrected for absorption. Anisotropic refinement of the Keggin molecule alone (assuming $[GaW_{12}O_{40}]$) gives a reliability index $R = 0.124$.

21. N. F. Yannoni, Doct. Diss., Boston University, 1961; *Diss. Abstr.*, **22**, 1032 (1961).

22. B. Dawson, *Acta Cryst.*, **6**, 113 (1953).

23. H. T. Evans, Jr., *J. Amer. Chem. Soc.*, **70**, 1291 (1948).

24. H. T. Evans, Jr., *J. Amer. Chem. Soc.*, **90**, 3275 (1968).

25. U. C. Agarwala, Doct. Diss., Boston University, 1960; *Diss. Abstr.*, **21**, 749 (1960).

26. K. Eriks, N. F. Yannoni, U. C. Agarwala, V. E. Simmons and L. C. W. Baker, *Acta Cryst.*, **13**, 1139 (1960).

27. A. Perloff, Doct. Diss., Georgetown University, 1966; *Diss. Abstr.*, **27**, 2676 (1967); *Inorg. Chem.*, **9**, 2228 (1970).

28. A. Perloff, unpublished work. The study of $Na_3H_6[CrMo_6O_{24}] \cdot 13H_2O$ was based on 9538 counter-measured data. Anisotropic structure refinement led to $R = 0.041$.

29. H. T. Evans, Jr., and J. S. Showell, *J. Amer. Chem. Soc.*, **91**, 6881 (1969).

30. J. C. T. Waugh, D. P. Shoemaker and L. Pauling, *Acta Cryst.*, **7**, 438 (1954).

31. D. D. Dexter and J. V. Silverton, *J. Amer. Chem. Soc.*, **90**, 3589 (1968).

32. C. M. Flynn, Jr., and G. D. Stucky, *Inorg. Chem.*, **8**, 335 (1969).

33. L. Malaprade, *Nouveau Traité de Chimie Minérale*, Vol. 14, P. Pascal, ed., pp. 903–981, Masson et Cie., Paris, 1959.

34. R. Signer and H. Gross, *Helv. Chim. Acta*, **17**, 1076 (1934).

35. J. A. Santos, *Proc. Roy. Soc.*, *A*, **150**, 309 (1935).

36. D. H. Brown and J. A. Mair, *J. Chem. Soc.*, **1962**, 1512.

37. D. H. Brown and J. A. Mair, *J. Chem. Soc.*, **1958**, 2597.

38. D. H. Brown and J. A. Mair, *J. Chem. Soc.*, **1962**, 3946.

39. D. H. Brown, *J. Chem. Soc.*, **1962**, 4408.

40. D. H. Brown and J. A. Mair, *J. Chem. Soc.*, **1962**, 3322.

41. A. Ferrari and O. Nenni, *Gazz. Chim. Ital.*, **69**, 301 (1939).

42. O. W. Rollins, Doct. Diss., Georgetown University, 1965; *Diss. Abstr.*, **26**, 7024 (1966).

43. L. C. W. Baker, V. S. Baker, K. Eriks, M. T. Pope, M. Shibata, O. W. Rollins, J. H. Fang and L. L. Koh, *J. Amer. Chem. Soc.*, **88**, 2329 (1966).

44. L. C. W. Baker and T. P. McCutcheon, *J. Amer. Chem. Soc.*, **78**, 4503 (1956).

45. R. Ripan and M. Puscasin, *Z. anorg. allgem. Chem.*, **358**, 82 (1968).

46. O. Kraus, *Z. Krist.*, **100**, 394 (1939).

47. O. Kraus, *Z. Krist.*, **96**, 330 (1937).

48. K. Eriks, N. F. Yannoni, U. C. Agarwala, V. E. Simmons and L. C. W. Baker, *Acta Cryst.*, **13**, 1139 (1960).
49. O. Kraus, *Z. Krist.*, **94**, 256 (1936).
50. O. Kraus, *Z. Krist.*, **93**, 379 (1936).
51. A. Ferrari, L. Cavalca, M. Nardelli, A. Selegari and M. Cingi, *Gazz. Chim. Ital.*, **79**, 61 (1949).
52. A. Ferrari, L. Cavalca, M. Nardelli, O. Scaglioni and E. Tognoni, *Gazz. Chim. Ital.*, **80**, 352 (1950).
53. O. Kraus, *Z. Krist.*, **91**, 402 (1935).
54. O. Kraus, *Naturwissenschaften*, **27**, 740 (1939).
55. T. J. R. Weakley and S. A. Malik, *J. Inorg. Nucl. Chem.*, **29**, 2935 (1967); S. A. Malik and T. J. R. Weakley, *J. Chem. Soc., A*, **1968**, 2647.
56. C. Tourné and G. Tourné, *Bull. Soc. Chim. France*, **1969**, 1124.
57. L. C. W. Baker and J. S. Figgis, *J. Amer. Chem. Soc.*, **92**, 3794 (1970).
58. P. Souchay, *Ann. Chim. (Paris)*, [12], **2**, 203 (1947).
59. P. Souchay, *Bull. Soc. Chim. France*, **1951**, 365.
60. K. F. Jahr, *Naturwissenschaften*, **29**, 505 (1941).
61. H. O'Daniel, *Z. Krist.*, **104**, 225 (1942).
62. J. S. Anderson, *Nature*, **140**, 850 (1937).
63. E. Matijević, M. Kerker, H. Beyer and F. Theubert, *Inorg. Chem.*, **2**, 581 (1963).
64. G. Tsigdinos, Doct. Diss., Boston University, 1961; *Diss. Abstr.*, **22**, 732 (1961).
65. C. K. J. Prout, *J. Chem. Soc.*, **1962**, 4429.
66. R. D. Hall, *J. Amer. Chem. Soc.*, **29**, 692 (1907).
67. L. C. W. Baker, G. A. Gallagher and T. P. McCutcheon, *J. Amer. Chem. Soc.*, **75**, 2493 (1953).
67a. B. M. Gatehouse and P. Leverett, *Chem. Commun.*, **1968**, 901; P. Leverett, Doct. Diss., Monash University, Australia, 1970.
68. I. Lindqvist, *Arkiv Kemi*, **2**, 349 (1950).
69. I. Lindqvist, *Acta Cryst.*, **5**, 667 (1952).
70. R. Allmann and G. Weiss, *Acta Cryst.*, **A25**, S106 (1969); G. Weiss, *Z. anorg. allg. Chem.*, **368**, 279 (1969).
71. H. T. Evans, Jr., *Inorg. Chem.*, **5**, 967 (1966).
72. A. G. Swallow, F. R. Ahmed and W. H. Barnes, *Acta Cryst.*, **21**, 397 (1966).
73. N. Pullman, Doct. Diss., Rutgers University, 1966; *Diss. Abstr.*, **28B**, 140 (1967).
74. I. Lindqvist, *Arkiv Kemi*, **5**, 247 (1953).
75. I. Lindqvist and B. Aronsson, *Arkiv Kemi*, **7**, 49 (1954).
76. G. Henning and A. Hüllen, *Z. Krist.*, **130**, 162 (1969).
77. J. H. Sturdivant, *J. Amer. Chem. Soc.*, **59**, 630 (1937).
78. I. Lindqvist, *Arkiv Kemi*, **2**, 325 (1950); *Acta Cryst.*, **3**, 159 (1950).
79. E. Shimao, *Bull. Chem. Soc. Japan*, **40**, 1609 (1967).
80. J. Fuchs and K. F. Jahr, *Z. Naturforsch.*, **23b**, 1380 (1968).
81. P. Souchay, *Ann. Chim. (Paris)*, **18**, 61, 169 (1943).
82. D. L. Kepert, *Progress in Inorganic Chemistry*, vol. 4, pp. 199–274, F. A. Cotton, ed., Interscience Publ., New York, 1962.
83. J. F. Keggin, *Nature*, **131**, 908 (1933).

84. M. T. Pope and G. M. Varga, *Chem. Commun.*, **1966**, 653.
85. G. Schott and C. Harzdorf, *Z. anorg. allgem. Chem.*, **288**, 15 (1956).
86. K. Saddlington and R. W. Cahn, *J. Chem. Soc.*, **1950**, 3526.
87. W. N. Lipscomb, *Inorg. Chem.*, **4**, 132 (1965).
88. J. Aveston, *Inorg. Chem.*, **3**, 981 (1964).
89. Y. Sasaki, *Acta Chem. Scand.*, **15**, 175 (1961).
90. F. J. C. Rossotti and H. Rossotti, *Acta Chem. Scand.*, **10**, 957 (1956).
91. G. Neumann, *Acta Chem. Scand.*, **18**, 278 (1964).
92. J. Aveston and J. S. Johnson, *Inorg. Chem.*, **3**, 1051 (1954).
93. W. H. Nelson and S. Tobias, *Inorg. Chem.*, **2**, 985 (1963).
94. G. Johansson, G. Lundgren, L. G. Sillén and R. Söderqvist, *Acta Chem. Scand.*, **14**, 769 (1960).
95. G. Johansson, *Acta Chem. Scand.*, **14**, 771 (1960).
96. G. Johansson, *Arkiv Kemi*, **20**, 321 (1963).
97. L. Pauling, *Z. Krist.*, **84**, 442 (1933); W. B. Kamb, *Acta Cryst.*, **13**, 15 (1960).
98. L. Kihlborg, *Arkiv Kemi*, **21**, 471 (1963).
99. B. M. Gatehouse and A. D. Wadsley, *Acta Cryst.*, **17**, 1545 (1964).
100. A. Hüllen, *Naturwissenschaften*, **51**, 508 (1964); *Ber. Bunsenges. Phys. Chem.*, **70**, 598 (1966).
101. J. A. Ibers, *Nature*, **197**, 686 (1963).
102. R. D. Witters and C. N. Caughlan, *Nature*, **205**, 1312 (1965).
103. K. Watenpaugh and C. N. Caughlan, *Chem. Commun.*, **1967**, 76.
104. C. M. Flynn, Jr., and G. D. Stucky, *Inorg. Chem.*, **8**, 332 (1969).
105. A. Rosenheim and W. Reglin, *Z. anorg. allg. Chem.*, **120**, 115 (1922).
106. S. V. Borisov, R. F. Klevtsova and N. V. Belov, *Kristallographia*, **13**, 980 (1968) (in Russian); *Soviet Physics Cryst.*, **13**, 852 (1969) (in English).
107. A. Rosenheim and J. Jaenicke, *Z. anorg. allg. Chem.*, **101**, 215 (1917).
108. D. L. Kepert, *Inorg. Chem.*, **8**, 1556 (1969).
109. F. Hanic, *Chem. Zvesti*, **12**, 579 (1958).
110. P. Kierkegaard and S. Holmen, *Arkiv Kemi*, **23**, 213 (1963).

[Added in proof] See page 28. The crystal of the decamolybdodicobaltate studied (Table 2, No. 10) was a raceme; the stereoisomers have more recently been resolved chemically with tris(tetraammine)cobalt cation by T. Ada, J. Hidaka, and Y. Shimura (*Bull. Chem. Soc. Japan*, **43**, 2654 (1970)).

# Conformational Equilibria in the Gas Phase

O. BASTIANSEN and H. M. SEIP, Department of
Chemistry, The University of Oslo, Oslo 3, Norway

and JAMES E. BOGGS, Department of Chemistry,
The University of Texas, Austin, Texas 78712,
U.S.A.

## I. INTRODUCTION

During the last few decades conformational analysis has been acknowledged as one of the most prominent fields of modern chemistry. Since the work on the present article was started, the field has gained further prestige and has even been made known to the layman through the 1969 Nobel Prize in Chemistry, awarded to Odd Hassel at the University of Oslo and to Derek H. R. Barton of the Imperial College of Science and Technology (London) "for developing and applying the principles of conformation in chemistry." The paper by Eliel about Hassel and Barton published in this connection[1] is in fact a short history of conformational analysis.

This now flourishing field emerged from a series of scattered observations, some made already at the end of the 19th century,[2] but most of them 30–40 years ago. As so often in science, some of the basic ideas had apparently been clearly understood very early by single individuals, but formulations that reached the literature prematurely were largely overlooked. The diffuse conception of "free rotation" about single bonds had been introduced to explain the lack of isomerism in ethane derivatives. However, up to 1930 no direct physical method had been applied to demonstrate the lack of free rotation about a single bond. That year three groups independently offered an idea which was ingenious in its simplicity:[3-5] if there were entirely free rotation about a single bond, the meso form and the optically active forms of a compound ought to have the same dipole moments. The observed difference in dipole moments for such compounds had to be interpreted as due to restriction to free rotation; and in the same year Debye[6] and Wierl,[7] using X-ray and electron-diffraction, respectively, also produced experimental evidence for the lack of free rotation about the C–C bond in ethane derivatives.

During the following twenty years a number of physical chemists became active in the field, using optical-rotation measurements, spectroscopy, X-ray crystallography, electron diffraction, thermo-dynamics, and theoretical calculations. This activity resulted in a basic understanding of the underlying principles by around 1950. This part of the history of the conformational analysis has been extensively reviewed in a book by Mizushima,[8] and the book by Eliel, Allinger, Angyal, and Morrison[9] includes more recent results.

It was some time before chemists in general accustomed themselves to the fact that for a compound which, according to all physical and chemical criteria, was a pure compound two or more geometrically different molecular species might coexist in the liquid or in the gaseous phase. In the early fifties these ideas became generally accepted, thanks first to Barton[10] who introduced them into organic chemistry. Since then conformational analysis has had an ever increasing importance, attracting attention from researchers dealing with problems ranging from quantum-mechanical calculation to bio-chemistry.

Though even the most recent literature is not quite consistent as to the semantics of conformational analysis, a rather coherent nomenclature seems to have developed, due above all to the comprehensive work of Eliel. The present authors try to adopt this nomenclature and try also to avoid involving themselves in the semantic controversies that still seem to exist. The term conformer is thus used in this article as synonymous with conformational isomer: it thus denotes a molecular species that differs from another species of the same compound by rotation about one or several single bonds provided that the molecular arrangement corresponds to a minimum in the potential energy function of the molecule, the geometrical structure parameters being taken as independent variables. This definition may serve practical chemical purposes quite well, but certainly leaves ambiguity when a more rigorous theoretical treatment is aimed at. Even consideration of the cyclohexane conformational problem leads to difficulties. The cyclohexane conversion cannot take place exclusively by torsional motion; valence angles must also change. Further, reference to minima in the potential function is only meaningful when the minimum is well defined and surrounded by appreciable barriers. Cases are known where two minima are close together and separated by such a small barrier that the first vibrational level is above the barrier maximum (see Section VI-B-1). In such cases the whole double minimum complex may well be referred to as a conformer.

The definition of conformation has been limited to rotation about single bonds for practical rather than for logical reasons. One may argue that the *cis–trans*-isomerism of ethylene derivatives is essentially the same phenomenon as the *anti-gauche* "conformerism" of ethane derivatives. There exist in both cases two types of molecular species, each corresponding to a minimum in a torsional-dependent energy function. In both cases the energy function has a barrier between the minima. The difference is thus only of quantitative nature, the barrier being considerably higher for ethylene derivatives than for ethane derivatives (approximately 60 and 3 kcal/mole, respectively, for the hydrocarbons themselves), leading to the practical significant difference of separability. If the barrier is large enough, the two molecular species will exist as two stable compounds on the time scale of ordinary laboratory manipulations, whereas a low barrier between the two molecular species may make separation impossible. Which barriers are considered low and which are considered high is given by the thermal energy and connected to activation energies and reaction rates. The difference in barrier heights of ethylene and ethane derivatives is ascribed to the difference in the character of the double and the single bonds. The barrier height also depends upon the kind of substituent linked to the carbon atoms forming the bond in question: if large enough groups are introduced in an ethane derivative, the barrier may be so high that one would rather describe the phenomenon as isomerism than conformerism. A molecule such as 1,2-dichlorotetraphenylethane can, for example, be separated into *anti* and *gauche* isomers.[11]

Most studies in conformational analysis have been carried out with the liquid phase, either a pure liquid or a solution. However, some of the early studies contributing to the discovery of the field were carried out with the gas phase. For many measurements the liquid is the most convenient phase and has by many researchers been considered as the "natural phase" for conformational studies. Wood and Woo,[12] criticizing the electron-diffraction method, refer to the fact that this method is usually limited to the gas phase, "which is not of great interest to most organic chemists." It is a major point of the present authors that the gas phase *is* the "natural phase" for studying conformational problems. It is only in the gas phase that intramolecular forces are exclusively responsible for the conformational choice. In cases where two or more conformers coexist in the gas phase, usually only one exists in the crystal, and this may even not be the one that is the most stable in the vapour. The torsional angles may also be quite different in the gas and in the crystal. For example, biphenyl and some of its derivatives, exhibiting an angle between the phenyl planes of approximately 45° in the gas phase, are planar in the crystal (see Section V-B-4).

The conformational equilibrium in a pure liquid or in a solution is qualitatively more similar to that of a gas. However, the equilibrium in a gas and in a liquid cannot be directly compared. In a pure liquid or in a solution, intermolecular forces, also involving the solvent molecules, compete with the intramolecular forces, often leading to a compromise that may be difficult to account for theoretically. It has been shown that a polar solvent usually favours the conformer with the highest dipole moment. It has even been shown that a conformer that is the more stable as gas and in some solvents may be the less stable in other solvents (see Section VI-C).

However interesting solvent effects may be, the competition between inter- and intra-molecular forces can hardly be quantitatively described before the free-molecule behaviour is understood.

## II. FUNDAMENTAL PRINCIPLES OF CONFORMATIONAL ANALYSIS

In a molecule exhibiting conformational phenomena at least one mode of torsional motion must be present. In addition other modes of internal motion also necessarily exist. This non-torsional motion is often referred to as the "framework vibration" and is customarily assumed to be separable from the torsional motion. To the approximation of this assumption the torsional motion of a molecule with one degree of torsional freedom may be described by a potential function of only one independent variable naturally chosen as the torsional or dihedral angle ($\varphi$). From lack of a comprehensive and quantitative theory for barriers to internal rotation, the potential function of a one-dimensional case is usually expanded in a trigonometric series. In many cases the potential is expressed as

$$V(\varphi) = \tfrac{1}{2} \sum_n V_n(1 - \cos n\varphi) \qquad \ldots (1)$$

Figure 1 (A—E) offers a few typical examples of potentials that can be described by equation (1). The pure term of $n = 3$ (curve C) is, for example, generally used to describe the potential of the ethane molecule. The staggered form corresponds to $\varphi = 0°$ and $\pm 120°$, and the eclipsed form to $\varphi = \pm 60°$ and $\pm 180°$. By adding further terms with $n = 1$ and/or $n = 2$ qualitative description of the potentials for symmetric ethane derivatives may be obtained (curves D and E). If we consider, for example, ethane derivatives of the type $CH_2X–CH_2X$, $\varphi = 0$ corresponds to the more stable *anti* conformer (often referred to as the *trans* conformer) and $\varphi$ near $\pm 120°$ corresponds to the *gauche* conformer. Curve D applies when the *anti* conformer is the more stable, curve E when the *gauche* conformer is the more stable. In

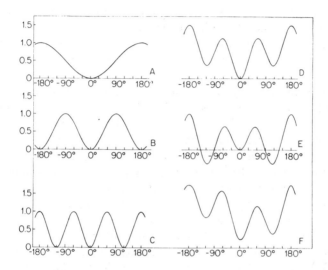

Figure 1. Examples of trigonometric functions used to represent potential curves in one-dimensional cases. A, B, and C correspond to $V(\varphi) = \frac{1}{2}(1 - \cos n\varphi)$ for $n = 1$, $n = 2$, and $n = 3$, respectively. Curve D is $V(\varphi) = \frac{1}{4}(1 - \cos \varphi) + \frac{1}{2}(1 - \cos 3\varphi)$; and $E$ is $V(\varphi) = -\frac{1}{4}(1 - \cos 2\varphi) + \frac{1}{2}(1 - \cos 3\varphi)$. Curve F includes also a sine term, *i.e.*, $V(\varphi) = \frac{1}{4}(1 - \cos \varphi) + \frac{1}{2}(1 - \cos 3\varphi) + \frac{1}{4}(1 - \sin \varphi)$.

general there are two *gauche* conformers ($\varphi = 120°$ and $\varphi = -120°$); in symmetric cases these are mirror images of each other. The experimental techniques used in conformational analysis are often inadequate for distinction between these two molecular species.

If the potential function lacks symmetry around $\varphi = 0$, the curve may be represented by adding sine terms of the kind $\frac{1}{2}V_n'(1 - \sin n\varphi)$ to equation (1). In curve F such a sine term is added to curve D. A molecule of the type $CH_2X–CHXY$ may be qualitatively accounted for by such a curve. The semantics in such cases is less obvious. It would be natural to refer to the conformer with the two X atoms in *anti* position as the *anti* conformer, but one would have to distinguish between two *gauche* conformers (see Section V-B-1). The term *skew* conformer has been introduced in such cases in addition to the *gauche* conformer. Other terminology is used and is certainly practical, particularly in larger organic molecules with many torsional degrees of freedom. For our purpose, when molecules of only moderate size are studied, the situation is less complicated. We therefore do not feel it necessary to elaborate more rigorous semantics. For a molecule with three degrees of torsional freedom, as, for example, $CH_2Br$–$CHBr$–$CHBr$–$CH_2Br$, a special system has been introduced,[13] and similar systems are necessary for truly large molecules (see Section V-C).

In principle any one-dimensional torsional barrier may be expressed in trigonometric series of $\varphi$. The constants $V_n$ are usually considered as parameters to be determined experimentally. In some cases the use of a trigonometric series may be impractical because of a too slow convergence, but in most cases it seems sufficient to include only a few terms of the series (see Section V-B-2). The symmetry of the molecule under investigation often indicates which term of the series is the most important and which terms may be neglected. As mentioned, the single term of $n = 3$ is generally considered to be a good representation of the barrier in ethane. Other terms that may be considered for ethane would be those with $n = 6$ and $n = 9$. Benzaldehyde may serve as another example (Section V-B-3). Here the term with $n = 2$ is likely to be predominant, and the terms with $n = 1$ and $n = 3$ may be neglected. Introduction of a *para*-substituent in the benzaldehyde molecule does not change this, but *meta*- or *ortho*-substitution necessitates inclusion of additional terms.

The $V_n$ parameters are introduced for practical reasons as coefficients in a trigonometric series; they may not have any particularly clear physical meaning. Many attempts have been and are still being made to attribute the barriers to free rotation to various "effects" that are intuitively more easily acceptable than the $V_n$ parameters. These "effects" should, in the ideal case, add up to the total potential function. Such effects are (1) steric repulsion between non-bonded atoms, (2) London-force attraction, (3) steric interaction involving lone pairs, (4) electrostatic interaction, (5) interaction between bonds, (6) dipole interaction, (7) resonance effect, and (8) intramolecular hydrogen bonding. Unfortunately, it seems impossible to assign simple empirical rules from which potential curves can be deduced by adding such effects. Even with a limited number of compounds belonging to a rather homogeneous group, no convincing method has been devised to give a real quantitative description of the potential. On the other hand, arguments based on such effects have proved very useful for qualitative descriptions and have also in several cases been successfully applied in semiquantitative reasoning, particularly when similar molecules are compared. As will be described in Section IV-C, the quantum-mechanical calculations of potential curves may be divided into two kinds, the semiempirical calculations and the calculations *ab initio*, the latter based on a minimum of *a priori* assumptions. Calculations *ab initio* have produced interesting and convincing results during the last few years, although only simple and small molecules have been studied so far. The development of *ab initio* calculations is coupled to the development of computer technology, and for a real break through one may have to wait for the next generation of computers.

Even in cases with only one torsional degree of freedom the use of potentials varying only with the torsional angle is an approximation that may not always be satisfactory; for example, both in semiempirical and *ab initio* calculations of the barrier in ethane, improvements are obtained by optimalizing all the structural parameters. In cases with several torsional degrees of freedom the situation may be exceedingly complex; the potential is then to be described in a many-dimensional space; several minima may exist, and the barrier of particular interest between two minima should be the lowest possible one.

The goal of experimental conformational analysis in gases may be disputed. Many investigators consider the conformational problems to be solved when the various coexisting conformers are detected, their structure determined, and the energy difference between the conformers measured. This means that the positions of the minima of the potential have been determined as well as their relative energy values. In the present article we have tried to consider the complete potential, although the necessary data exist only in few cases. If the complete potential function can be found, not only are the classical conformational problems solved, but information is also available for determining the probability of molecular forms between the actual conformers. Unfortunately there is at present no single experimental method that can give all the information necessary for determination of the complete potential function, which must therefore be deduced from individual items of information obtained by various methods. One method may contribute measured values of the energy difference between conformers, another may give the torsional barrier, and still others may give the torsional angles corresponding to the minima of the potential. In general, microwave studies give information about the barrier heights. The microwave lines are so well-defined and so accurately measurable that the reproducibility of barrier determination is very high, say about 1%. On the other hand, uncertainties in the molecular structure may introduce considerably greater error. It should be noted that errors assigned to microwave barrier data rarely include structure parameter uncertainty and are therefore often illusory. Far-infrared data may also give barrier heights: recent development in this field seems promising and the method may be of considerable importance in the future. Infrared and Raman intensity comparison gives the energy difference between conformers. The same can be obtained from electron diffraction. The latter method may also give torsion angles. A series of studies of liquids, particularly of solutions, based on infrared, Raman, nmr, and dipole-moment measurements, has been used to derive free-molecule barriers and energy differences. However, the uncertainty involved in correcting for intramolecular forces makes these conclusions often rather questionable. Some of the most important experimental methods for gas work are treated in Section III.

In several methods the estimated energy difference between conformers is based on measurements of molecular ratios. For two co-existing conformers we have for the molecular ratio

$$\frac{N_1}{N_2} = n \frac{f_1}{f_2} \exp\left(-\Delta E/RT\right) \qquad \ldots(2)$$

where $n$ is the relative weight factor of conformer 1 relative to conformer 2 given by molecular symmetry. $\Delta E$ is the energy difference between the two conformers. $f_1$ and $f_2$ are the vibrational and rotational partition functions for the two conformers given as

$$f = \frac{(I_A I_B I_C)^{1/2}}{\prod_i [1 - \exp\left(-h\nu_i/kT\right)]} \qquad \ldots(3)$$

$I_A$, $I_B$, and $I_C$ are the momenta of inertia, and $\nu_i$ represents a normal frequency. For many molecules the frequencies are not all known for the various conformers, and the temperature-dependent ratio $\omega = f_1/f_2$ cannot be determined very accurately. The ratio is usually close to unity and is often neglected in semiquantitative calculations.

A typical example is provided by 1,2-dichloroethane for which $\Delta E$ (the energy difference between *gauche* and *anti* forms) is approximately 1 kcal/mole. The error introduced by putting $\omega = 1$ instead of the actual value is only about 0.03 kcal/mole at the temperatures used for electron-diffraction studies.

Unfortunately the energy difference between conformers is in practice a rather ill-defined quantity. A natural way of defining it would perhaps be by the difference between the potential values at minima in the potential curve. Often the difference between the lowest vibrational level of the two conformers is referred to.

In the present article we refer to energy differences as $\Delta E$ values. Often enthalpy differences ($\Delta H$) are cited, and these should be very similar in the gas phase. Free-energy differences ($\Delta G$) are most directly useful when thermodynamic equilibria are discussed. Although the conversion from one to the other should, in principle, be well understood for the gas phase, in practice the accuracy of the conversion may suffer from uncertainty in partition-function calculations.

## III. EXPERIMENTAL METHODS

### A. Gas Electron Diffraction

The electron-diffraction method has played an important part in the development of conformational analysis. The method is well suited for determining fairly accurately the molecular structure of the most abundant conformers and also for estimating their relative abundances. It was used at an early date to show that the prevailing conformers of cyclohexane[14] and *cis*-decalin[15, 16] were made from chair rings, and it also provided an early direct proof that two different molecular species of the same compound may coexist in the gas phase.[17]

The electron-diffraction method is based on measuring the intensity of electrons scattered from a gas jet injected into a high vacuum.[18-20] The scattering picture consists of a series of concentric rings superimposed upon a steeply descending background, falling off from the diffraction centre towards higher scattering angles. A photographic representation of the diffraction picture is so dominated by the steep background that a photometer trace shows very little else than a monotonic fall of intensity. The background itself does not contribute information concerning the molecular structure; it is mainly determined by the charge distribution in the atoms of which the molecule is built. In order to obtain a useful photometer trace of the diffraction pattern, it is necessary to screen the photographic plate by a rotating device referred to as a sector. The sector is placed immediately above the photographic plate with its axis of rotation coinciding with the incoming beam and is so shaped that it compensates for the steep fall of the background. Sectored pictures can easily be recorded by a microphotometer, and the background can be subtracted. By straightforward modifications the so-called modified experimental molecular intensity function is obtained. Theoretically this function may be expressed to an approximation sufficient for most structure work as follows:

$$I(s) = \text{const.} \sum_{i \neq j} \sum g_{ij/kl}(s) \exp\left[-\tfrac{1}{2}u_{ij}^2 s^2\right] \frac{\sin r_{ij}s}{r_{ij}} \qquad \ldots(4)$$

As the independent variable, $s = (4\pi/\lambda)\sin\theta$, is used instead of the diffraction angle, $2\theta$; $\lambda$ is the electron-beam wavelength. The geometrical parameters of the molecule are expressed by the internuclear distances $r_{ij}$, and the summation is taken over all such distances. The values $u_{ij}$ are the vibrational amplitudes expressed as root-mean-square deviations from the equilibrium distances. The function $g_{ij/kl}(s)$ depends on the complex scattering amplitudes

$$f(s) = |f(s)| \exp\left[i\eta(s)\right] \qquad \ldots(5)$$

for the involved atoms, *i.e.*:

$$g_{ij/kl}(s) = \frac{|f_i|\,|f_j|}{|f_k|\,|f_l|} \cos \Delta\eta_{ij} \qquad \ldots(6)$$

The phase factor $\Delta\eta_{ij}$, which does not appear in a "classical" treatment of the scattering process,[21, 22] is zero if the $i$th and the $j$th atoms are of the same kind, *i.e.*, the cosine factor is independent of $s$ and equal to unity. If the $i$th and $j$th atoms are near neighbours in the Periodic Table the cosine term falls off very slowly, having an effect somewhat similar to the exponential term in equation (4). If the $i$th and the $j$th atoms have very different atomic numbers, the cosine term may reach zero at an $s$ value where the intensity date is easily measurable. As an example of a case where the cosine factor is of the greatest importance, Figure 2 shows the experimental and two theoretical intensity curves for $UF_6$.[22, 23] Curve B, which agrees quite well with the experimental one (A), was calculated according to equation (4) with complex scattering amplitudes, while curve C was calculated with real scattering amplitudes. The curves correspond to an electron wavelength of about 0.0645 Å, *i.e.*, an accelerating voltage of about 36 kV. The $g$ function for the UF bonds is zero for $s$ near 10 Å$^{-1}$ ($\Delta\eta = \pi/2$). A second zero point, corresponding to $\Delta\eta = 3\pi/2$,

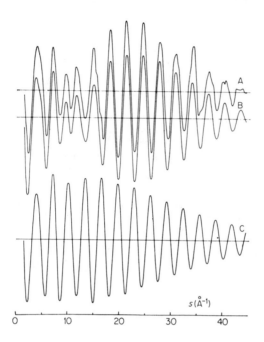

Figure 2. Experimental (A) and theoretical (B, C) intensity curves for UF$_6$. The theoretical curves were calculated according to equation (4). Curve B was obtained by using complex scattering amplitudes [cf. equations (5) and (6)]. Curve C was obtained by using real scattering amplitudes.

seems to be near $s = 50$ Å$^{-1}$, although an increasing intensity for still larger $s$ values could not be identified because of noise.

The factor $g(s)$ may be obtained both from experiment and from theory. For most electron-diffraction structure work the factor may be taken with sufficient accuracy from existing tables or be deduced by straightforward computer programmes.

Equation (4) shows that the molecular intensity function is a sum of damped sine terms, each modified by a function that varies slowly and often not very much over the observed $s$ range. The sum contains one term from each individual internuclear distance. For symmetry reasons, molecules often have groups of identical distances. The summation in equation (4) may therefore be taken over all types of distances rather than over atom pairs.

An electron-diffraction structure determination is in principle based on comparing experimental and theoretical molecular intensity curves or experimental and theoretical curves derived from the intensities. The structure parameters are adjusted until the best fit is obtained. A least-square procedure is most often used for this purpose. The expression "structure parameter" is used in this connection in the widest sense. It is not limited to parameters describing the geometry of the molecule, but includes parameters describing intramolecular motion.

Another curve that is much used in electron-diffraction study is the Fourier transform of the molecular intensity function, the radial distribution curve. It is customarily defined by

$$\sigma(r)/r = \int_{s_1}^{s_2} I(s) \exp\left(-ks^2\right) \sin rs \; ds \qquad \ldots(7)$$

The integration limits $s_1$ and $s_2$ are defined by the experimentally available $s$ range. Often $s_1$ is set equal to zero by introducing theoretical values in the innermost part of the intensity curve. The constant $k$ is an artificial damping constant that helps to suppress error ripples in the radial distribution curve.

The radial distribution curve has the advantage of being intuitively more intelligible than the intensity curve. While the intensity curve is built up by one modified sine term for each type of internuclear distance, the radial distribution curve has a peak for each internuclear distance. The peak is rather narrow and is centred around the $r$ value corresponding to the equilibrium distance. The peaks are in most cases approximately Gaussian in shape, and the dispersion parameter of each peak corresponds (if $k = 0$) to the $u_{ij}$ value of equation (4). In the $g_{ij/kl}(s)$ function (6) the apparently arbitrary values $|f_k|$ and $|f_l|$ in the denominator are chosen to make the shape of the peaks in the radial distribution curve as near Gaussian as possible. The atoms $k$ and $l$ are usually chosen as representative atom pairs of the molecule under investigation. For each such atom pair, the deviation from Gaussian shape is usually quite small.

For illustration a simple and easily interpretable radial distribution curve is given in Figure 3, including both the theoretical and the experimental version of a radial distribution curve.[24] The molecule in question is hexafluorobenzene, which is a rigid molecule with

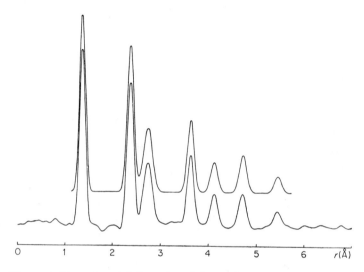

Figure 3. Experimental (lower) and theoretical (upper) radial distribution curves for hexafluorobenzene.

high symmetry. The 10 distances show up as only 7 peaks because of overlap. For example the C–C bond distance of 1.394 Å and the C–F bond distance of 1.327 Å lead to one unresolved peak. But in spite of this unfortunate overlap the two geometric parameters may easily be deduced from the seven peaks, and the vibrational amplitudes may also be obtained. Figure 4 shows that the U–F distances in $UF_6$ give a double peak in the radial distribution curve, though they are identical.[22, 23] The reason is the large values of $\Delta\eta$.

In cases when two or more conformers coexist, each conformer contributes peaks to the radial distribution curve according to its relative abundance. A particularly illustrative example of the application of electron diffraction to conformational analysis is the study of *trans*-1,4-dichlorocyclohexane and the corresponding bromo compound (see Section VI-C). The procedure is best demonstrated by Figure 5 which refers to the chloro compound. The lower full curve is the experimental radial distribution curve. The two upper curves are

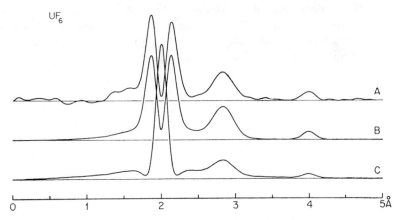

Figure 4. Experimental (A) and theoretical (B, C) radial distribution curves for $UF_6$ calculated by Fourier inversion of the corresponding intensity curves in Figure 2.

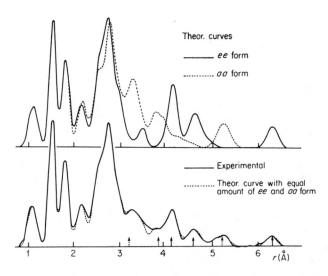

Figure 5. Theoretical and experimental radial distribution curves for 1,4-*trans*-dichlorocyclohexane.

the theoretically calculated curves for the two pure conformers **that** are to be expected. The experimental curve fits neither of these but contains features characteristic for both. The dashed lower curve, which is partly covered by the experimental one, is a theoretical curve based on a 50% mixture of the two conformers.

The electron-diffraction method is particularly successful when the energy difference between the two conformers is rather small, corresponding to nearly equal contribution of the various conformers as in the already applied example. If one conformer is strongly dominant,

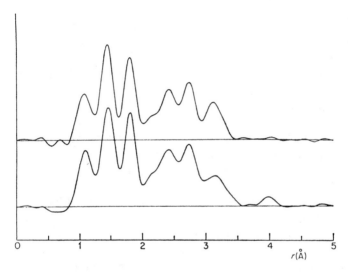

Figure 6. Experimental radial distribution curves for 2-chloro-ethanol corresponding to 310°K (upper curve) and 473°K (lower curve).

the minor conformer is hard to recognize and even harder to study quantitatively. In such cases it helps to carry out the experiment at the highest possible temperature in order to increase the relative population of the minor conformer. Figure 6 demonstrates such an example. The two curves are both experimental radial distribution curves of ethylene chlorohydrin (2-chloroethanol) based upon experiments carried out at 310°K and 473°K, respectively. For ethylene chlorohydrin the *gauche* conformer predominates. At 310°K the *anti* peak is just observable but is nearly lost in the noise. At 473°K the *anti* peak stands out clearly. The increase in the *anti* contribution is, of course, accompanied by a corresponding decrease in the area under the *gauche* peak (see Section V-C).

Unfortunately the temperature range of electron-diffraction experiments is rather limited. The lower limit is set by the temperature necessary to obtain a large enough pressure for the experiment (approximately 10 mm), and the higher limit is set by the temperature at which the compound starts to decompose. The upper temperature limit is often too low to populate the less stable conformer sufficiently for accurate study. Further, the temperature measurement is uncertain and the nozzle temperature may not be exactly the same as that corresponding to the internal motion.[25, 26]

The two chosen examples of conformational analysis are characteristic of cases where there exists a barrier of at least 2–3 kcal/mole between the conformers. In such cases the probability of intermediate molecular forms is so small that their contribution cannot be detected. The analysis is then reduced to the study of two or a few structurally different coexisting species. If the barrier is as low as, say, a few tenths of a kcal/mole, the situation is different. Then the radial distribution curve is most adequately described by assuming a mixture of an infinite number of molecular species with different torsional angles, each contributing according to a certain probability distribution. If this probability distribution can be obtained, the potential function can be deduced. However, this kind of study is severely hampered by theoretical and practical difficulties. The theoretical approach is to assume a separation of the internal motion into two types. One type is a torsional motion of an otherwise rigid model, the other one is vibration without torsional motion. Or differently expressed, one assumes a torsional motion superimposed on a framework vibration. It appears that the framework vibration may vary considerably with the torsion angle.[27, 28] It is in most cases very difficult and perhaps even impossible to determine the framework vibration and the potential barrier simultaneously with any degree of accuracy. However, if the framework vibration can be calculated with some confidence from vibrational spectra, the electron-diffraction studies may give quantitative information about the complete potential function.

In ethane and many related molecules the root-mean-square amplitude for the framework vibration ($u_{\mathrm{fr}}$) may be expressed as:[28]

$$u_{\mathrm{fr}} = (\alpha + \beta \cos \varphi + \gamma \cos^2 \varphi)^{1/2}/R(\varphi) \qquad \ldots (8)$$

where $\alpha$, $\beta$, and $\gamma$ are constants and $R$ is the internuclear distance.

1,4-Dibromo-2-butyne, $BrH_2C—C\equiv C—CH_2Br$, may serve as an illustration (cf. Section V-B-1). If free rotation is assumed, each equal torsion-angle interval has the same probability, but of course if equal distance intervals are used, as in the radial distribution curve, the probability distribution is quite different. The Br–Br contribution to the radial distribution curve of 1,4-dibromo-2-butyne is given in Figures 7. The curve in Figure 7a corresponds to constant framework

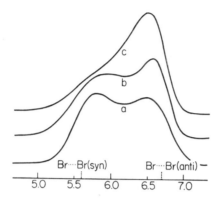

Figure 7. The Br$\cdots$Br contribution to the radial distribution curve of 1,4-dibromo-2-butyne calculated for: (a) constant framework vibration ($u_{fr} = 0.1966$ Å) and free rotation, (b) $u_{fr}$ given by equation (8) with $\alpha = 1.5922$ Å$^4$, $\beta = 0.9765$ Å$^4$, and $\gamma = -0.0165$ Å$^4$ and free rotation, and (c) $u_{fr}$ as in (b), $V(\varphi)$ as given in Section V-B-1.

vibration for the Br$\cdots$Br distance, while the curve in 7b was obtained using the expression (8) with $\alpha = 1.5922$ Å$^4$, $\beta = 0.9765$ Å$^4$, and $\gamma = -0.0165$ Å$^4$. With these constants $u_{fr} = 0.285$ Å for the *syn*-form and 0.116 Å for the *anti*-form. The Br–Br distances corresponding to rigid *syn*- and *anti*-forms are indicated on the Figures. If a barrier of some kind is introduced, this will modify the shape of the curve (Figure 7c; cf. Section V-B-1).

No standard procedure has been established for conformational analysis, particularly for that involving molecules with low torsional barriers. Most studies seem to be based on radial distribution curves, but least-squares calculation based on intensity data has also been applied with some success, e.g., for perfluorobiphenyl[29] and ferrocene.[30] For this kind of work it is necessary to describe framework vibration and torsional potentials by a limited number of parameters.

An attempt is then made to refine these parameters together with the more classical structure parameters.

Carbon suboxide may be mentioned as a final example of the use of electron-diffraction data in the determination of a potential function.[25, 31, 32] It has been realized for some time that the force constant for alteration of the CCC angle must be very low. The potential given in Figure 8 seems to be consistent with available electron-diffraction

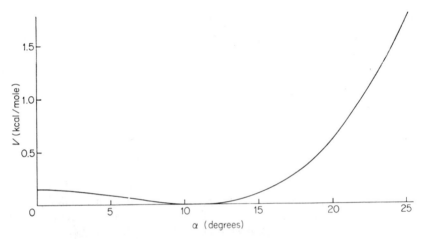

Figure 8. The proposed potential function for the CCC "bending" in $C_3O_2$ ($2\alpha = 180° - \angle CCC$).

data recorded at three temperatures as well as with spectroscopic data.[31] The accuracy is not sufficiently high to rule out a potential function with zero barrier for $\alpha = 0°$. Tanimoto, Kuchitsu, and Morino[32] have obtained a potential function in quite good agreement with that in Figure 8.

According to experience accumulated by many researchers, the electron-diffraction method can contribute significantly to conformational analysis of gases. But the method has its limitations. It is well suited to detect the various conformers present, to characterize them by providing fairly accurate information about geometrical structure, to describe the relative abundances of the various conformers, and, accordingly, to determine the energy differences between them. It does not seem well suited for determining large barriers, but it is capable of contributing, at present rather modestly, to the determination of low barriers. There is considerable promise of future improvement in this direction.

## B. Microwave Spectroscopy

Microwave spectroscopy is primarily the study of spectra resulting from dipole transitions between quantized rotational states of molecules. It is limited to polar molecules that can be obtained in the gas phase at a pressure of at least a few microns at some suitable temperature.

The rotational behaviour of a rigid body can be described in terms of three moments of inertia, $I_a$, $I_b$, and $I_c$, about the principal axes. To the extent that a real molecule can be approximated as a rigid body, these three parameters, together with known dipole selection rules, completely determine the rotational spectrum of the substance. For linear or symmetric-top molecules, the spectral frequencies can be obtained from closed-form expressions, but in the general case of asymmetric-top molecules approximation methods must be used in the computations. With the aid of reasonably small computer programmes, such calculations can readily be made to the eight-figure accuracy required for comparison with experiment. Details of the mathematical methods used are available in many textbooks.[33-35] The analysis of a microwave spectrum, then, begins with the accurate determination of the frequencies of the observed absorption lines, the assignment of quantum numbers characterizing the rotational transition to at least some of the observed lines, and the determination of the moments of inertia that give the best fit to the data.

Since the microwave spectrum is determined by three moments of inertia, not more than three independent structural parameters can be directly determined from the spectrum of one molecular species. The number may be less, since for a linear molecule, a symmetric top, or a planar molecule fewer than three independent, non-zero moments of inertia are obtained. To obtain additional information, use is made of isotopic substitution, since equilibrium internuclear distances are essentially unchanged by the substitution of one isotope for another while there is an appreciable change in the moments of inertia. Thus, data must be gathered on a sufficient number of isotopic species of the same molecule to obtain as many independent moments of inertia as there are independent bond distances and angles to be determined. In many studies this is not done, and some of the structural parameters are assumed from other information or by analogy with other molecules, the spectral data then being used to determine the remaining parameters.

The resolution obtainable in microwave spectroscopy is extremely high, so that it is easy to measure line frequencies to an accuracy better than one part in $10^7$. Even the shifts in line frequency caused by isotopic substitution can be read to such high accuracy that measurement errors for the shifts are not usually the limiting factor in the accuracy with which structural parameters can be determined. More important is the fact that molecules are not rigid bodies and, in particular, the observation that, while isotopic substitution leaves equilibrium internuclear distances unchanged, it has a noticeable effect on vibrational ground-state internuclear distances. In principle, sufficient information is available from the microwave spectrum to correct for internal motions. In practice, this requires such a large amount of work that it has actually been carried out only for molecules containing three or four atoms. It is customary to assume that ground-state internuclear distances are the same for all isotopic species and accept the limits on accuracy which such an assumption entails.

For a vibrating molecule, the structure obtained by fitting all the structural parameters to measured moments of inertia of a suitable number of isotopically substituted species is often referred to as the $r_o$ structure. The name is somewhat deceptive, since the parameters obtained are not simply related to the interatomic distances in the vibrational ground state of the molecule, although the deviations are generally not large. The measured average moments of inertia are proportional to $\langle r^2 \rangle$. Since $\langle r^2 \rangle$ is not in general equal to $\langle r \rangle^2$, the measured moment of inertia is not identical with the equilibrium value, even for a strictly harmonic vibration. Anharmonicity in the vibration, of course, can make the difference even larger. Furthermore, as mentioned above, ground-state interatomic distances do change with isotopic substitution.

Kraitchman[36] has obtained solutions for the structural parameters in terms of the equilibrium moments of inertia that are of great value if isotopic substitution can be made at the positions of atoms of particular interest. The absolute values of the coordinates of any atom at which substitution has been made can be calculated by means of simple algebraic expressions. Thus a particular bond length in a complicated molecule can be determined from the analysis of the spectrum of only three isotopic species: the normal species, one with substitution at one end of the bond of interest, and one with substitution at the other end of the bond. The method loses in accuracy if the bond lies too near a principal axis. If isotopic substitution can be made for every atom in the molecule, the entire structure can be determined in this manner. Such a structure is known as an $r_s$ structure. The relationship between equilibrium, $r_o$, and $r_s$ structures has been discussed by Costain.[37] In general, the $r_s$ structure is nearer the equilibrium structure than is the $r_o$ structure. As a rough rule, the differences may be expected to be on the order of a few thousandths of an Ångstrøm. Structures obtained from electron-diffraction measurements, often known as $r_g$ structures, are still different and exact comparison has been carried out in only rather few cases.[38]

A molecule in an excited vibrational state has, in general, moments of inertia different from those of the ground-state molecule. Thus, the complete microwave spectrum of a substance shows a set of absorption lines corresponding to the allowed rotational transitions of the molecules that are in the ground vibrational state and additional similar sets, displaced in frequency, for those molecules that are in excited vibrational states. At ordinary temperatures, only the lower-lying vibrational states are appreciably populated, but the vibrational satellite pattern surrounding each of the observed rotational lines may be quite complex. In many cases, however, it is possible to pick out sequences of satellite lines corresponding to molecules in successively higher excited states of a particular vibrational mode. A typical example is shown in Figure 9 which diagrams the $2_{12} \rightarrow 3_{13}$ rotational transition of vinylene carbonate (1,3-dioxolone) in the ground vibrational state and in the first four excited states of the ring-puckering vibrational mode.[39] The relative intensities reflect the reduced populations of higher vibrational states, the usual steady decrease being modified in this case by the nuclear spin weight factor which alternates between 3 and 1 depending on the symmetry of the vibrational wave function. Additional weak absorption lines are observed in the spectral region covered by Figure 9, but the lines corresponding to the sequence shown can be identified by their characteristic Stark effect, their uniform progression in frequency, their relative intensity, and the ability to obtain a good fit for the moments of inertia of each excited state when using these frequencies and those observed for other rotational transitions in the same vibrational state.

If the potential function governing a vibrational mode of a molecule has two or more energy minima, the molecule may have two or more conformations. To understand the effect on the microwave spectrum, we may consider two different cases: molecules having identical moments of inertia at the different vibrational minima, and molecules having different moments of inertia in the two forms.

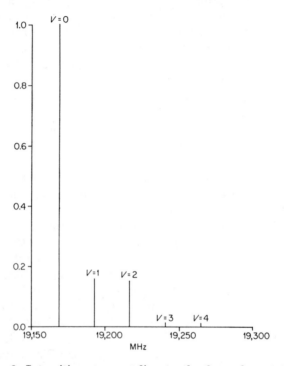

Figure 9. Intensities corresponding to the $2_{12} \rightarrow 2_{13}$ rotational transition of vinylene carbonate (1,3-dioxolone) in the ground vibrational state and in the first four excited states of the ring-puckering vibrational mode.

The classical examples of the first group are the inversion motion of ammonia and the internal rotation of ethane. The potential curve for the inversion motion of ammonia is shown in Figure 10. The central maximum corresponds to the coplanar arrangement of the atoms. The two equivalent minima correspond to pyramidal structures with the nitrogen atom on either side of the plane made by the hydrogen atoms. Quantum-mechanical tunnelling through the barrier causes the energy levels to split into doublets, as shown. Ammonia is nearly unique in that the magnitude in the inversion splitting is such that transitions between the ground-state inversion doublet fall in the microwave frequency region. In most other cases, the doublets produced by tunnelling through a potential barrier are so closely spaced compared with rotational frequencies that the effect appears as a doubling of the rotational absorption line.

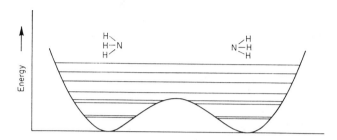

Figure 10. The potential curve for the inversion of ammonia.

For ethane, rotation of one methyl group by 120° produces an equivalent geometry and the potential energy curve has three minima (see Figure 1, C). Ethane itself has no microwave spectrum, since it has no dipole moment, but a similar situation exists in many other molecules such as $CH_3CH_2F$. Tunnelling through the barrier again results in splitting of the rotational absorption lines, the magnitude of the splitting depending on the molecular geometry and the height of the barrier to internal rotation. For barriers higher than a few kilocalories per mole, the splitting is generally too small to be seen in the ground state. If the magnitude of the splitting of the rotational lines is measured and the geometry of the molecule is accurately known, the barrier height can often be computed with great accuracy. If splitting of the rotational transitions is not observable either in the ground or in an excited state, the barrier height may still be determined, although with less accuracy, by measuring the relative intensities of the ground- and the excited-state transitions. These relative intensities reflect the relative populations of the various states, which in turn give the energy differences. From the energy-level spacings, information on the shape of the potential function can be deduced.

The case which has been most extensively treated is the common one in which the molecule can be considered to consist of two parts rotating with respect to each other, one part being a symmetric top with a three-fold axis of symmetry. Such molecules would include $CH_3OH$, $CH_3CH_2Cl$, $(CH_3)_3COH$, $CF_3PH_2$, etc. The mathematical formulation of the problem is treated in detail elsewhere,[33-35] but the result is a highly accurate prediction of the spectrum, the only assumptions being a knowledge of the molecular geometry, the barrier to internal rotation, and the dipole-moment components. These quantities can, of course, be obtained by seeking the best fit between the observed and the calculated spectra. In practical application, the quality of the agreement between the observed and the calculated spectra is not limited by the accuracy of either the measurement or the calculations, but rather by limitations of the physical model chosen; *e.g.*, assumptions that the molecule is rigid except for the internal rotation motion. Much recent work has been directed towards more general treatment of the problem, including efforts to consider the case in which both rotating portions of the molecule are asymmetric and to include interactions between internal rotation and other low-frequency internal motions. In spite of these limitations, the accuracy of potential functions derived from microwave spectroscopy in favourable cases is very satisfactory and is better than that obtainable from any other experimental method.

If the minima in a potential function correspond to different molecular geometries, the sample may simply be thought of as a mixture of two different substances. For example, Hirota[40] has shown that propyl fluoride, $CH_3CH_2CH_2F$, has two stable conformers with the terminal methyl group either *anti* or *gauche* to the fluorine atom. The potential function is shown in Figure 11. The observed microwave spectrum consists of the set of lines expected for a molecule having the moments of inertia of the *gauche* form superimposed on the set of lines arising from the *anti* form. The independence of the two spectra can, of course, be modified if the barrier separating the two conformers is sufficiently low for tunnelling to produce detectable splitting of the observed transitions.

The parameters determining the shape of the potential function for a molecule such as propyl fluoride are obtained from several different types of observation on the microwave spectrum. The molecular geometries of the separate conformers can be obtained from observations on the ground vibrational state spectra of the different species present, provided that enough isotopic forms are studied or that enough of the geometrical parameters can be assumed from related molecules. The relative stabilities of the different conformers and the energy difference between them can be determined by measuring the relative intensities of corresponding transitions in the spectra of the two conformers. The height of the barrier separating two conformers can be determined from the relative intensities of transitions involving molecules (a) in excited states of the torsional motion and (b) in the ground state, or from splittings of the lines due to tunnelling (if observed). Analysis of the spectrum of each such molecule is an individual matter, and the amount of information that can be obtained varies greatly from case to case.

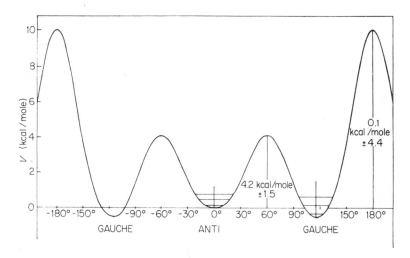

Figure 11. The potential function for the rotation of the $CH_2F$ group in propyl fluoride.

## C. Vibrational Spectroscopy

A considerable contribution to the understanding of molecular conformation has come from gas-phase infrared studies. If the conversion of one conformer into another can be thought of as occurring along one normal vibrational coordinate of the molecule, and if the vibrational transitions associated with that mode can be observed, the motion can be analysed and the barrier height determined. Even if this cannot be done, the infrared spectrum may indicate the coexistence of two or more molecular forms and the molecular symmetry of these forms may be determinable.

Several difficulties accompany efforts to observe transitions associated with the vibrational mode connecting the two conformers, which is usually an internal rotation or ring-puckering mode. First, for many interesting molecules such transitions are infrared-inactive or very weak. In a few cases, the frequencies of inactive internal rotation modes have been deduced from combination bands, although such a procedure is fraught with the usual danger of misassignment. Also, since the potential barrier between conformers is relatively low, the corresponding transitions fall in the far-infrared region where work has been difficult until recent years. The low frequency of the fundamental means that many excited states are appreciably populated, so that many hot bands arising from transitions from vibrationally excited states must be expected. If these are sufficiently separated in frequency from the fundamental they can provide a rich source of additional information, but if they overlap they merely serve to broaden the observed band and hinder interpretation. Few molecules of conformational interest are simple enough to have re-solvable rotational structure in the infrared band, so that this poten-tial source of information is usually denied the investigator. In some cases, however, it has been found useful to use approximate struc-tural data and calculate the expected shape of the band envelope.

A measurement of the fundamental torsional frequency of a hindered internal rotor leads to a determination of the barrier height only if certain approximations are made. If the potential function can be considered to approximate a harmonic oscillator in the vicinity of the minimum, the torque constant is given by $k = 4\pi^2 a^2 I_1 I_2 / I$, where $I_1$ and $I_2$ are the moments of inertia of the two vibrating parts of the molecule, each considered to be rigid. Furthermore, if the complete potential function can be considered to be given by $V = (V_n/2)(1 - \cos n\varphi)$, then $V_n = 2k/n^2$, or $V_n = 8\pi^2 a^2 I_1 I_2 / n^2 I$. While these approximations limit the accuracy of barrier-height deter-minations from infrared spectroscopy, in practice they are often not as damaging as lack of the accurate structural information ($I$, $I_1$, and $I_2$) that cannot be obtained from infrared studies alone. In the more general case for which the barrier potential has a more complex shape, use of hot-band transitions starting in excited vibrational states must be used to determine the additional terms in the Fourier expansion of the barrier. More general expressions for the reduced mass of the rotor are also available for the case in which both portions of the molecule are asymmetric tops.[41,42]

When several non-overlapping bands can be observed, correspond-ing to successively higher transitions ($v = 0 \rightarrow 1$, $1 \rightarrow 2$, $2 \rightarrow 3$, etc.) of the same vibrational mode, quite detailed information can be obtained. This has been accomplished most often for ring-puckering modes of small rings. The vibrational mode observed is most commonly the ring-puckering mode itself in the far-infrared, but similar results can come from the observation of difference vibrations. Ueda and Shimanouchi,[43] for example, examined cyclopentene which has a

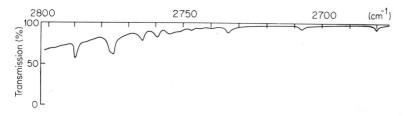

Figure 12. The progression of difference bands between the C–H stretch and the ring-puckering vibration for cyclopentene.

C–H symmetric stretch vibration at 2861.03 cm$^{-1}$ and observed a progression of absorption bands between 2680 and 2800 cm$^{-1}$, as shown in Figure 12. These were interpreted as difference bands between the C–H stretch and the ring-puckering vibration. From the frequency difference the separation between the ring-puckering vibrational levels could be deduced, leading to the pattern shown in Figure 13. A potential curve could then be obtained to fit the observed

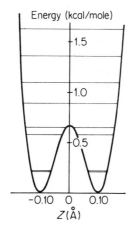

Figure 13. The ring-puckering potential and vibrational levels for cyclopentene derived from the progression of bands shown in Figure 12.

energy-level pattern. For cyclopentene, the result confirmed the conclusion reached earlier by microwave spectroscopy[44, 45] that the molecule is non-planar.

In the relatively few cases in which rotational structure can be resolved in infrared torsional bands, the analysis proceeds in a manner similar to that used in interpreting the microwave spectrum. In fact, the molecular dynamics of vibration–rotation interaction were originally examined in a series of papers by Dennison and his co-workers[46-48] in connection with the infrared spectrum of methanol.

The infrared technique has proved to be particularly valuable for the investigation of conformational problems involving large barriers between the stable conformers. In principle, nuclear magnetic resonance could also be of use in these molecules, but such work has proved very difficult except in condensed phases. Determination of high potential barriers from microwave studies requires measurement of the relative intensities of rotational transitions of molecules in different vibrational states and subsequent evaluation of the energy spacings between the vibrational levels. In favourable cases, these spacings can be determined with greater accuracy by direct observation of the vibrational transition in the infrared spectrum. It is possible that recent developments leading to more accurate intensity measurements in microwave spectroscopy may alter the relative merits of the two methods. In any case, thorough and accurate studies of a given molecule often appear to require the application of several experimental techniques. In a particular case, the infrared spectrum may provide the best value of the vibrational energy level pattern, but it cannot provide the necessary structural data. For a high-barrier case, microwave spectroscopy may provide a more convincing assignment of the torsional oscillation, but may be unable to measure the energy spacings as accurately. The microwave spectrum may provide an excellent determination of some of the structural parameters, but may be unable to determine certain others with sufficient accuracy. Electron diffraction may locate the heavy atoms with great accuracy, but it cannot determine accurately the positions of hydrogen atoms. No set pattern can apply to the study of all molecules, and information must come from the source best able to supply it.

Frequencies of molecular vibrations may also be obtained by the techniques of Raman spectroscopy. For many purposes this method has clear advantages since, to a good approximation, only fundamental transitions are seen, without the sum and difference bands that can complicate an infrared absorption spectrum. Until recent years only limited application of the technique was made to gas-phase spectroscopy, but this situation has been changed by the advent of high-powered lasers which give good scattering intensities even for gases at moderate pressure. Ion lasers filled with argon or krypton give a choice of several frequencies for the exciting radiation, and resolution in the measurement of the frequency difference between the incident and the scattered radiation is comparable to that of direct infrared spectroscopy. The Raman technique is particularly convenient for studying low-frequency vibrations since it avoids the temperamental detectors required in the far-infrared region.

## IV. THEORETICAL CALCULATIONS OF MOLECULAR GEOMETRY AND ENERGIES

### A. Introduction

We do not intend to review all the calculation methods applied to predict molecular geometry, relative energies, and physical properties. However, since theoretical calculations have become an important tool in chemistry, we feel that some of the more important methods and results ought to be mentioned.

In Section IV-B the widely used method treating the molecule as a classical mechanical system is described. The method was first discussed in detail by Westheimer[49] and later modified by Hendrickson,[50-52] Wiberg,[53] and others.[54-63]

Methods based on quantum mechanics are discussed in Section IV-C. During the last decades a great variety of such calculations, for example, of potential curves, has been carried out with various degrees of success, most of them being semiempirical. The most ambitious attempts are based on pure quantum-mechanical calculations, *i.e.*, on the Schrödinger equation for special distributions of the nuclei of the molecule. The current development of such *ab initio* calculations seems rather promising, particularly in view of the dramatic improvement in speed and capacity of modern computers (for a review see ref. 64). On the other hand, the *ab initio* calculations do not provide the intuitive understanding that chemists seek in order to predict barriers in more complicated molecules. Many organic chemists have felt that, even if *ab initio* calculation were to produce better potentials, much of the useful classical thinking would be lost, that classical chemical "effects" would be drowned in quantum mechanics. In contrast to this view it is interesting that the present development of *ab initio* calculation seems rather to lead to a revival of the use of additive effects than the abolition of them. By systematic computer experimentation one may separate the various contributors to the total potential. These contributors may, however, not be identical with the classical effects but may serve the same purpose for predicting barriers in more complex molecules.[65]

The semiempirical methods may be considered as approximations to the *ab initio* calculations and are therefore discussed after the latter.

## B. Calculations Based on a Classical Molecular Model

The calculations are based on the assumption that the energy may be expressed as a sum of terms

$$E = E^f + E^a + E^t + E^{nb} \qquad \ldots (9)$$

each term having a special physical meaning.

A "natural" bond length $(r_i^0)$ is ascribed to each bond type. If the actual bond distance differs from this value, the contribution to the energy is given by:

$$E_i^f = \tfrac{1}{2}k_i^f(r_i - r_i^0)^2 \qquad \ldots (10)$$

$E^f$ in equation (9) is the sum of all terms of this type. In many cases it is of little importance for conformational analysis and may be neglected.

A "natural" value $(\Theta_i^0)$ is also assumed for each bond angle. The bond-angle strain (Baeyer strain) is then usually calculated by:

$$E_i^a = \tfrac{1}{2}k_i^a(\Theta_i - \Theta_i^0)^2 \qquad \ldots (11)$$

There is some evidence that this expression gives too large energy contributions. A modified expression may be used,[54] or the constant $k_i^a$ may be taken somewhat smaller than found from spectroscopic data.

For each bond angle in the molecule a term of type (11) should be included. However, when, say, a methylene group is considered, a composite force constant may be used for the CCC angle, and the contributions from the HCC and HCH angles may be neglected.[51]

$E^t$ is the sum of torsional strain (Pitzer strain) energy contributions. For many bond types it may be assumed that:

$$E_i^t = \frac{V_i^0}{2}(1 - \cos n\varphi) \qquad \ldots (12)$$

where $\varphi$ is the torsional angle around the bond (see Section II).

The term $E^{nb}$ comprises all non-bonded interactions. In hydrocarbons and molecules with just one hetero atom it is probably satisfactory to include only the van der Waals' interactions. However, experience with molecules containing even one hetero atom is limited.[56, 61] In molecules with two or more polar groups the term $E^{nb}$ should include dipole–dipole interactions[56] or electrostatic interactions between point charges on the atoms[63] as well as van der Waals interactions. The former of these approaches suffers from the invalidity of the expression for dipole–dipole interactions at short distances.

The method outlined in this Section has been applied most successfully to saturated hydrocarbons.[50–54, 57] The procedures applied differ in the choice of constants in the energy expression and in the technique for finding the energy minimum. A somewhat different approach using a Urey–Bradley field rather than a valence force field has been applied by Bartell and his co-workers.[58] The force constants and the constants in the expressions for the non-bonded interactions used in their calculations are quite different from the values used by most other investigators; however, it seems possible to obtain reasonably good agreement with experimental data with various sets of constants.

The agreement with experimental data is somewhat poorer for unsaturated hydrocarbons.[55, 59, 60] The main difficulty probably lies in finding a reasonable potential function for torsion about a double bond. It is sometimes assumed that this energy ($E_\pi^t$) may be given as:

$$E_\pi^t = \frac{V_\pi}{2} (1 - \cos 2\varphi) \qquad \ldots(13)$$

where $V_\pi$ is the difference in bond energy between a double bond and a single bond, *i.e.*, for carbon–carbon bonds about 60 kcal/mole. However, Allinger *et al.*[55] applied

$$E_\pi^t = V_\pi(1 - \cos \varphi) \qquad \ldots(14)$$

and Favini *et al.*[59, 60] applied

$$E_\pi^t = V_\pi(1 - 0.7418 \cos \varphi + 0.2582 \cos^2 \varphi) \qquad \ldots(15)$$

In equations (14) and (15) $|\varphi| \leqslant 90°$.

Estimates of $E_\pi^t$ are usually based on the change in the overlap integral between the pure $p$ orbitals of two $sp^2$ hybridized carbon atoms. However, a pyramidal arrangement of the bonds at each atom will also result in a twist around the double bond. Furthermore, the orbitals on the C atoms do not necessarily point directly towards the adjacent atoms; *i.e.*, the single bonds may be bent. Measurements of heats of hydrogenation indicate that considerable twist about a double bond may occur with fairly small increase in energy, probably less than given by equation (14) (see ref. 66, p. 58).

Results obtained by the method described in this Section are discussed below, especially in Section VI-B.

### C. Quantum-mechanical Calculations

#### 1. *Calculations* ab initio

The first determination of a barrier to internal rotation *ab initio* was made in 1963 by Pitzer and Lipscomb[67] for ethane. The value obtained, 3.3 kcal/mole, is in quite good agreement with the experimental value of 2.93 kcal/mole.[68]

*Ab initio* calculations on molecules follow the same pattern as calculations of electron distribution in atoms, where the procedure may be divided into three steps, as follows:

1. Hartree–Fock calculation. This part consists of the solution of a set of differential equations to obtain the one-electron wave functions (orbitals). Each electron is exposed to the field of the average distribution of all the other electrons, and the calculation is repeated until self-consistency has been achieved (SCF calculation).

2. Corrections for electron–electron correlation.

3. Corrections for relativistic effects.

The first part has to be somewhat different for molecules. The molecular orbitals (MO) are restricted to be linear combinations of atomic orbitals (LCAO).[69] The calculation by Pitzer and Lipscomb[67] was, for example, carried out using as basis set only $1s$ orbitals on hydrogen and $1s$, $2s$, and $2p$ orbitals on carbon. The calculation procedure for molecules may thus be divided into the following steps, where 1a and 1b together are often, somewhat imprecisely, called the Hartree–Fock approximation:

1a. MO's are constructed by linear combinations of atomic orbitals (LCAO-MO).

1b. The coefficients in the linear expansions are adjusted until self-consistency is obtained.

2. Correlation corrections.

3. Relativistic corrections.

The corrections 2 and 3 have been neglected in all the published calculations of barriers to internal rotation. This might appear as a serious omission but fortunately it seems that fairly accurate barriers may be obtained within the Hartree–Fock scheme; this would imply that the corrections for correlation and relativistic effects are nearly the same for all points on the potential curve. A calculation by Lévy and Moireau[70] of the barrier in ethane, including a partial correlation correction, supports this conclusion, since the correction amounted to only about 0.1 kcal/mole. The correction for relativistic effects is believed to be much smaller.

Even if the corrections 2 and 3 are neglected, a compromise must be made between accuracy and computing time. As mentioned, Pitzer and Lipscomb obtained quite good results for ethane with a very limited basis set, and most calculations on hydrocarbons with corresponding basis sets have given similar accuracy. The data from a number of calculations on ethane have been compared by Pitzer.[71]

The results may be improved by including more atomic orbitals, for example, $d$ orbitals on carbon, and by optimizing the molecular geometry. This is of particular importance if various points on the potential function correspond to drastically different geometries. Veillard[72] has shown that the barrier in ethane can be improved in this way and has obtained interesting differences in the C–C bond lengths and HCH angles between staggered and eclipsed forms. Some of the results are given in Table 1. These values were obtained by six calculations for the staggered and six for the eclipsed form, starting with the experimental geometry and varying the C–C bond length and the HCH angles.

Table 1. Calculated[72] and experimental values of barriers, bond lengths, and bond angles for ethane.

|  |  | Calc. | Exptl. |
|---|---|---|---|
| Barrier (kcal/mole) |  | 3.07 | 2.93 |
| C–C bond length (Å) | staggered | 1.551 | 1.534 |
|  | eclipsed | 1.570 |  |
| ∠HCH | staggered | 107.3° | 109.3° |
|  | eclipsed | 107.0° |  |

Calculation of barriers around bonds between carbon and another atom seems to require larger basis sets than are necessary for carbon–carbon bonds. A calculation by Fink and Allen[65] for methanol gave a barrier of 1.06 kcal/mole, in excellent agreement with the experimental value of 1.07 kcal/mole.

As an example of a more complicated problem, some results obtained by Veillard[73] for hydrogen peroxide are given in Table 2. The potential function has been determined experimentally by Hunt *et al.*[74] (see Section V-B-1). It is particularly interesting that the calculation gives a small barrier at the *anti* position, as found experimentally.

Table 2. Calculated[73] and experimental values for hydrogen peroxide. $\Delta r$ is the difference in the O–O bond lengths in the *syn* and *anti* forms, and $\Delta \angle$OOH the corresponding difference in the angle.

| | | Calc. | Exptl. | Ref. |
|---|---|---|---|---|
| Barrier (kcal/mole) | *syn* ($\varphi = 0$) | 10.9 | 7.0 | 74 |
| | *anti* | 0.6 | 1.1 | 74 |
| Torsional angle, $\varphi$ | | 123° | $\begin{cases} 111° \\ 120.0°, 116.1°^a \end{cases}$ | 74 <br> 75 |
| $\Delta r$ (Å) | | 0.025 | | |
| $\Delta \angle$OOH | | 5.35° | | |

*a* For the lower and the upper state of the ground-state doublet, respectively.

Two *ab initio* calculations of the energy difference between the twist and the chair form of cyclohexane may also be mentioned. Hoyland[76] obtained about 6 kcal/mole with a minimum basis set, while a slightly higher value (7.3 kcal/mole) was obtained by Veillard.[77]

## 2. *Semiempirical methods*

A number of approximations to the method described in Section IV-C-1 have been developed to reduce the computing time. The common feature of these semiempirical methods is the use of experimental information to adjust selected parameters. Various extended Hückel calculations have been applied[78-80] to barrier problems. Hoffmann[78] found, for example, a barrier of 4.0 kcal/mole in ethane by this method. Special schemes have also been developed for $\pi$-electron systems.[81-83]

Recent approximations to the self-consistent molecular orbital method have been proposed, for example, by Pople *et al.*[84-87] and by Baird and Dewar.[88] Davidson *et al.*[89] have recently given a critical examination of two of these methods, namely, the CNDO[84, 85] or rather CNDO/2[86] (Complete Neglect of Differential Overlap), and the NDDO method[84] (No Diatomic Differential Overlap). The conclusion is that, except for the prediction of bond lengths, the results obtained by the CNDO method parallel experimental trends somewhat better than the NDDO method, though the latter introduces less drastic approximations. As an illustration, Table 3 gives a comparison of experimental and calculated barriers for various compounds. The CNDO results are all somewhat lower than the experimental values, while the NDDO results deviate less systematically from the observed barriers.

Table 3. Calculated (CNDO and NDDO methods) and experimental barriers to internal rotation. (References are in parentheses.)

| Compound | CNDO | NDDO | Exptl. |
|---|---|---|---|
| Ethane | 2.32 (89) | 2.39 (89) | 2.93 (68) |
| Propane | 2.20 (89) | 2.61 (89) | 3.6 (92) |
| Isobutane | 2.64 (89) | 2.01 (89) | 3.9 (94) |
| Propene | 1.00 (89) | 1.68 (89) | 2.00 (90,91) |
| cis-2-Butene | 0.44 (89) | 2.54 (89) | 0.748 (93) |
| Methanol | 0.67 (85) | | 1.07 (48) |
| Methylamine | 1.21 (85) | | 1.98 (95) |

# V. CONFORMATIONAL PROBLEMS IN OPEN-CHAIN SYSTEMS

## A. Introduction

Molecules with one degree of freedom for internal rotation and molecules with several degrees of freedom are discussed separately in Section V-B and V-C, respectively. However, this distinction is not followed very strictly, since it is frequently convenient to compare molecules belonging to different Sections. A further subdivision has been made for molecules with one degree of torsional freedom. Again some compounds do not clearly belong to any specific group. Hydrogen peroxide and hydrogen disulphide as well as 2-butyne and its analogues, for example, are discussed in connection with ethane and related compounds (Section V-B-1). The cyclopropyl group is considered analogous to the vinyl group; bicyclopropyl is thus discussed in the same Section as 1,3-butadiene (V-B-3). Compounds containing benzoyl or related groups are also discussed in one Section.

The greatest amount of detailed experimental information is available for molecules in which there is one mode of interval rotation with a relatively low barrier compared with other possible internal motions of the molecule. In all but a few of the most recent studies,[96, 97] the assumption has been made that the internal rotation is separable from other vibrational modes with no interaction terms, so that the problem is reduced to a model of a rigid molecule except for the single internal motion considered. Such studies are limited to quite simple molecules, but they may lead to an understanding of the principles involved, so that qualitative extrapolation to larger molecules is possible.

As explained in Section II, the potential function is usually expressed as a Fourier series. For the rotation of a methyl group the terms corresponding to $n < 3$ must be zero because of symmetry. If, for symmetry reasons, the first non-zero term corresponds to $n = 6$, a very low barrier is usually found. The $V_6$ term for toluene[98] is only 14 cal/mole, while 6 cal/mole and 74 cal/mole have been reported for $CH_3NO_2$[99] and $CF_3NO_2$,[100] respectively.

It is interesting that the $V_6$ terms for both 2-methylpyridine[101] and 4-methylpyridine[102] are almost the same as for toluene. However, a considerable $V_3$ term (258 cal/mole) is found for 2-methyl-pyridine.

Tables of barriers to internal rotation are given in refs. 103 and 104.

## B. Molecules with One Degree of Freedom for Internal Rotation

### 1. *Ethane and related molecules*

As discussed in Section II, the equilibrium conformation of ethane is the staggered form. The same result has been obtained for a number of singly substituted ethane derivatives such as $CH_3CH_2F$,[105] $CH_3CH_2Cl$,[106] and $CH_3CH_2Br$.[107] For these molecules, and for many others involving $sp^3$–$sp^3$ hybridization, the potential function is dominated by a three-fold cosine term. The barrier to internal rotation in ethane is $2.928 \pm 0.025$ kcal/mole. The $V_6$ and higher terms are negligibly small.[68] The barriers in many related molecules range from about 3.0 to 3.5 kcal/mole. With increasing substitution, particularly by more bulky atoms, the barrier rises. For example, the rotational barrier in $CF_3CF_3$ was found[108] from thermodynamic data to be 4.35 kcal/mole. This value probably suffers from errors in the assumed structural parameters and has since been amended[109] to 3.92 kcal/mole. The barrier in $CCl_3CCl_3$ has been reported as 11 kcal/mole[27] and 17.5 kcal/mole.[110] In the latter molecule it is clear that steric effects predominate.

One type of molecule that played a particularly important part in the understanding of the conformational equilibrium of ethane derivatives is $CH_2XCH_2X$, where X is halogen. For 1,2-dichloro- and 1,2-dibromo-ethane the *anti* conformers predominate in the gaseous state, with some contribution from the *gauche* conformers. Energy differences between the *gauche* and the *anti* conformers given in Mizushima's book[8] range from 1.0 to 1.3 kcal/mole for 1,2-dichloroethane and from 1.4 to 1.8 kcal/mole for 1,2-dibromoethane. These values are mainly based on infrared and Raman studies. The energy difference for 1,2-diiodoethane has been estimated from nmr spectra of solutions to be 2.45 kcal/mole in the vapour phase.[111] Electron-diffraction studies of 1,2-dihaloethanes[112, 113, 19] essentially verify the spectroscopic data and yield estimated energy differences of 0.95 ± 0.2 kcal/mole and 1.6 ± 0.3 kcal/mole for the chloro and bromo compound, respectively. The uncertainties in these investigations are larger than would be expected if the study were repeated with present-day methods. First, the temperature can now be varied over a larger range and measured more accurately. Secondly, the calculation procedures now in use are more powerful than those used earlier. The inherent uncertainty caused by the factor $\omega$ (see Section II) is unlikely to affect the results seriously, as the estimated error due to this uncertainty is only a small fraction of the total uncertainty. The $\omega$ values used presently have been recalculated by using the most recent experimental vibrational frequencies.

The electron-diffraction data give a torsional angle ($\varphi$) for the *gauche* conformer of 64° for the chloro compound and 74° for the bromo compound.

In contrast to the chloro, bromo, and iodo compounds, 1,2-difluoroethane exhibits a *gauche* predominance. The electron-diffraction data[113, 19] leave no doubt about this, which is also demonstrated by infrared and Raman studies carried out in the gas phase as well as in other states of aggregation.[114] However, the agreement between the results is not complete, in that the electron-diffraction data suggest that the *gauche* contribution exceeds 67% (corresponding to energy equality of the two conformers) whereas the spectroscopic data show no measurable temperature effect on the integrated extinction of any band and therefore lead to zero energy difference. The energy difference suggested by the electron-diffraction data is about 1 kcal/mole, where the uncertainty range is rather large but does not cover 0 kcal/mole. (The spectroscopic studies show that the *gauche* form predominates in the liquid.)

The *anti* preference in 1,2-dichloro- and 1,2-dibromo-ethane is naturally explained by halogen–halogen repulsion, though the barrier against rotation cannot be attributed to non-bonded interaction alone. For ethane itself, hydrogen–hydrogen repulsion can account for only a small fraction of the total barrier. However, it may be argued that the changes in the potential going from ethane to an ethane derivative may be caused primarily by changes in non-bonded atom interaction. The *gauche* form of 1,2-dichloroethane would then be destabilized by the too short $Cl \cdots Cl$ distance. This would also explain the observed deviation in torsional angle from the "ideal" *gauche* position. The barrier between the *anti* and the *gauche* conformers is usually considered to be about the same as in ethane, *i.e.*, near 3 kcal/mole. Repulsion effects would be expected to make it slightly larger than in ethane since the passage through the form with halogen and hydrogen eclipsing should be less favourable than the eclipsing of mere hydrogen atoms. The barriers in $CH_3CH_2F$ and $CH_3CH_2Cl$ are accordingly somewhat higher than in ethane; *i.e.*, 3.3 and 3.7 kcal/mole, respectively.[115] The barrier that has to be overcome when a 1,2-dihaloethane molecule passes through the *syn* position is expected to be considerably higher as it involves eclipsing of two halogen atoms. Bernstein[116] reported the barrier between the two conformers in 1,2-dichloroethane to be about 1.7 kcal/mole smaller than the barrier corresponding to the *syn* position.

If the *anti* preference in the chloro, bromo, and iodo compounds is to be explained by halogen–halogen repulsion it would be natural to explain the *gauche* preference in 1,2-difluoroethane by assuming a fluorine–fluorine attraction near the torsional angle corresponding to the *gauche* position. It is known that halogen–halogen attraction of the London force kind may be decisive for conformational choice. A typical example is that of the 2,2′-dihalobiphenyls (Section V-B-4).

*n*-Butane is also found as a mixture of *anti* and *gauche* conformations with a relatively small ( < 1.0 kcal/mole) energy difference, as shown from thermodynamic data by Pitzer[117] and more recently from electron-diffraction data by Bonham and Bartell.[118]

Infrared studies of the rotational isomerism of meso- and racemic 2,3-dichlorobutane[119] and the corresponding bromo compounds[120] have been carried out (see Figure 14). The *anti* forms of the meso compounds are, as expected, considerably more stable than the *gauche* forms (energy differences about 1.7 and 2.2 kcal/mole, respectively). The racemic compounds are most stable when the halogen atoms are in the *anti* position. The energy difference between the two other forms is small, the rotamer with both groups in the *gauche* position probably being the less stable in both cases.

As mentioned in Section IV-C-1, hydrogen peroxide has been extensively studied both experimentally and theoretically. Analysis of the far-infrared spectrum resulted in a three-term expansion for the potential with the $V_1$ and $V_2$ terms dominating.[74] Further results

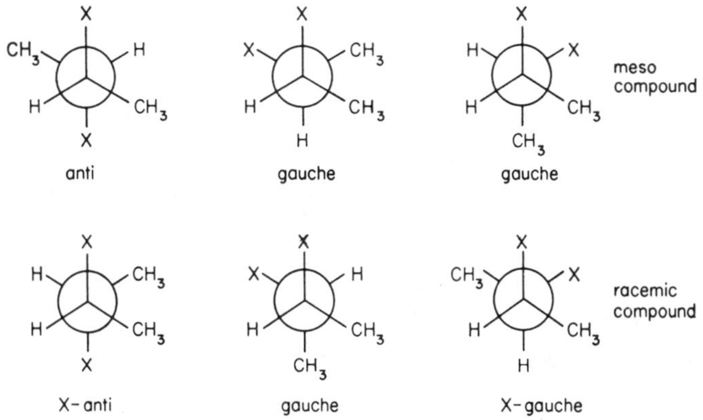

Figure 14. The possible conformers of *meso-* and racemic 2,3-dihalobutanes.

are given in Table 2. The conformation of hydrogen peroxide in solids is considerably influenced by intermolecular forces. Thus the *anti* conformer has been found in, for example, $Na_2C_2O_4 \cdot H_2O_2$[121] and $Li_2C_2O_4 \cdot H_2O_2$.[122] The torsional angle in $H_2S_2$ is $90.6°$[123] and thus considerably smaller than in $H_2O_2$ (see Table 2). The barriers in $H_2S_2$ are, according to *ab initio* calculations, 6.0 ($\varphi = 180°$) and 9.3 kcal/mole.[124] The much higher *anti* barrier than in $H_2O_2$ is in agreement with experimental evidence.

The barrier to internal rotation in 2-butyne, $CH_3—C\equiv C—CH_3$, and its derivatives is much lower than in ethane. Upper limits for the barrier in 2-butyne itself have been given as 0.10 kcal/mole[125] and 0.03 kcal/mole.[126] The upper limit for the barrier in $CH_3—C\equiv C—SiH_3$ has been estimated to be only 0.003 kcal/mole.[127] From infrared and Raman spectra of hexafluoro-2-butyne it has been concluded[128] that the effective point group for the vibrational potential function for this molecule is $D_{3d}$, which should not be strictly true if the barrier is very low.[129, 130] A recent electron-diffraction investigation[131a] showed that the rotation is nearly free.

A barrier less than 0.1 kcal/mole has been reported for 1,4-dichloro-2-butyne from dipole-moment measurements[132] and electron-diffraction data.[133] An electron-diffraction investigation of 1,4-dibromo-2-butyne also led to essentially free rotation.[134] A more recent investigation[131b] showed that, although the assumption of free rotation gives quite good agreement between experimental and theoretical curves, even better agreement is obtained for the potential

$$V(\varphi) = \tfrac{1}{2} \cdot 0.5[(1 + \cos \varphi) + (1 + \cos 2\varphi) + (1 - \cos 3\varphi)]$$

which has a minimum near $\varphi = 120°$ ($\varphi = 0$ corresponds to *syn*). In view of the difficulties (Section III-A) in determining the potential function this result can only be regarded as an indication of a minimum closer to the *anti* than to the *syn* position. The problem should be investigated further by other methods.

### 2. *Propene and related molecules*

In contrast to the staggered *gauche* or *anti* conformers observed with simple ethane derivatives involving $sp^3$–$sp^3$ hybridization, it has been observed that a double bond is generally eclipsed by a single bond. This is the case in propene,[135] as well as in $CH_3NO$[136] and $CH_3CHO$,[137] for which the three equivalent equilibrium conformations have the double bond eclipsed by a C–H bond of the methyl group. If the double bond is regarded as two bent single bonds, as advocated by Pauling,[138] the staggering of the bonds adjacent to the central single bond is preserved.

Particularly extensive studies have been made on compounds of the type $CH_3COX$, where X is F,[139] Cl,[140] Br,[141] I,[142] or CN.[143] For all of these molecules the carbonyl group eclipses a methyl-hydrogen atom, and the barriers to internal rotation fall in the range 1.0–1.3 kcal/mole.

Much higher barriers are, of course, found if the carbonyl or nitroso group is conjugated with another double-bonded system, as in $C_6H_5NO$,[144, 145] $C_6H_5CHO$,[146, 147] or $CH_2{=}CH{-}CHO$[148, 149] (see Section V-B-3). Interesting conformational questions arise for substituted propenes. The barriers for both *trans*- and *cis*-isomers are given in Table 4 for some molecules. Because of the double bond, the barrier separating the two isomers is very large. The barrier restricting rotation of a methyl group is strikingly different in the *cis*- and the *trans*-isomers. The equilibrium conformation of the methyl group

Table 4. Barriers (in kcal/mole) to internal rotation for methyl groups in propene and derivatives thereof. (References are in parentheses.)

|  | *cis* | | *trans* | |
|---|---|---|---|---|
| $CH_3CH=CH_2$ | | 2.00 | (90,91) | |
| $CH_3CH=CHCH_3$ | 0.748 | (93) | 1.950 | (150) |
| $CH_3CH=CHCH_3$ | 0.73 | (151) | — | |
| $CH_3CH=CHF$ | 1.057 | (40) | 2.20 | (152) |
| $CH_3CH=CHCl$ | 0.620 | (153) | 2.170 | (154) |
| $CH_3CH=CHBr$ | — | | 2.120 | (155) |
| $CH_3CH=CHCN$ | 1.40 | (156) | >2.10 | (156) |
| $CH_3CH=CHCH=CH_2$ | 0.741 | (157) | 1.805 | (157) |

was not determined in all these investigations, but it seems very likely that one of the methyl-hydrogen atoms eclipses the double bond. Dauben and Pitzer[150] attributed the difference in the $CH_3$ barrier in *cis*- and *trans*-2-butene to steric repulsion between hydrogens in the two methyl groups in the *cis*-isomer. Similar steric interactions in the other molecules may explain the lowering of the barriers in the *cis*-forms. However, since equilibration studies of $CH_3CH=CHF$ and $CH_3CH=CHCl$ indicate that the *cis*-isomers are more stable than the *trans*-isomers, other effects besides steric interactions may be of importance.

Stone, Srivastava, and Flygare[158] have argued that the changes in the barrier to methyl rotation as halogen atoms replace hydrogen in propene may be additive, and they used the measured differences between the barriers for the six monosubstituted fluoro- and chloropropenes and propene itself to predict the barriers of the various polysubstituted compounds. They have measured a barrier of 2.565 kcal/mole for *Z*-1-chloro-2-fluoropropene, in good agreement with the

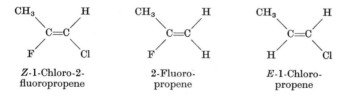

Z-1-Chloro-2-fluoropropene     2-Fluoropropene     E-1-Chloropropene

predicted value of 2.61 kcal/mole obtained by addition of the 2.0 kcal/mole barrier in propene[91] and increments of +0.44 kcal/mole for fluorine and +0.17 kcal/mole for chlorine, these increments being obtained from the barrier[159] of 2.44 kcal/mole of 2-fluoropropene and the barrier[154] of 2.17 kcal/mole of *E*-1-chloropropene.

Figure 15 shows the potential function for internal rotation in a propene derivative, where the rotor does not have three-fold sym-

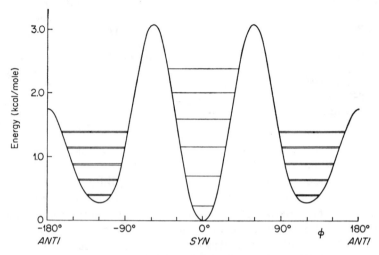

Figure 15. Potential function for internal rotation in
$CH_2{=}CH{-}CH_2F$.

metry, *e.g.*, 3-fluoropropene.[160] The lowest energy is obtained for the *syn* form (**1**) ($\varphi = 0°$). However, the *gauche* form with $\varphi = 120°$ is almost equally stable. In this case six terms were used in the Fourier expansion of the potential. The values obtained were (in $cm^{-1}$): $V_1 = -247$, $V_2 = 185$, $V_3 = 857$, $V_4 = 188$, $V_5 = 7$, and $V_6 = -93$. The $V_3$ term is thus the largest, but except for $V_5$ the other terms are far from negligible.

2-Methyl-1-butene (**2**) is also slightly more stable in the conformation where the methyl group eclipses the double bond, than in the *gauche* form.[161]

## 3. 1,3-Butadiene and related compounds

In this Section we start with a discussion of the conformations of molecules containing the group $=\overset{|}{C}-\overset{|}{C}=$ and then proceed to related problems. As shown in Table 5, the majority of these molecules are most stable in the *anti* conformation. The $\sigma,\pi$ picture of the double bond leads to the prediction of minima in *anti* and *syn* conformers for molecules related to 1,3-butadiene. (In this Section the *anti* form

Table 5. 1,3-Butadiene and some related molecules. The observed conformations are given; the most stable is given first.

| Compound | Conformation(s) | Ref. |
|---|---|---|
| $CH_2{=}CH{-}CH{=}CH_2$; 1,3-butadiene | *anti, gauche?* | 162–178 |
| $CH_2{=}C(CH_3){-}CH{=}CH_2$; isoprene | *anti* | 179,180 |
| $CH_2{=}C(CH_3){-}C(CH_3){=}CH_2$; | *anti* | 181 |
| 2,3-dimethyl-1,3-butadiene | | |
| $CH_3CH{=}C(CH_3){-}C(CH_3){=}CHCH_3$; | | |
| 3,4-dimethyl-2,4-hexadiene | | |
|    $\begin{cases} cis,cis \\ cis,trans \\ trans,trans \end{cases}$ | *anti* (distorted) | 182 |
| | *gauche* (114°) | 182 |
| | *gauche* (113°) | 182 |
| $CH_2{=}CF{-}CH{=}CH_2$; fluoroprene | *anti* | 183 |
| $CF_2{=}CH{-}CH{=}CH_2$; | *anti* | 184 |
| 1,1-difluoro-1,3-butadiene | | |
| $CF_2{=}CF{-}CF{=}CF_2$; | *gauche* (133°) | 185 |
| hexafluoro-1,3-butadiene | | |
| $CCl_2{=}CCl{-}CCl{=}CCl_2$; | *gauche* | 186 |
| hexachloro-1,3-butadiene | | |
| $CH_2{=}CH{-}CH{=}CH{-}CH{=}CH_2$; | | |
| 1,3,5-hexatriene$\begin{cases} cis\ (\mathbf{4}) \\ trans\ (\mathbf{5}) \end{cases}$ | *anti-anti* (distorted) | 187 |
| | *anti-anti* | 188 |
| $CH_2{=}CH{-}CH{=}O$; acrolein | *anti* | 148,149, 174,189 |
| $CH_3CH{=}CH{-}CH{=}O$; crotonaldehyde | *anti* | 190,191 |
| $CH_2{=}CH{-}C(OH){=}O$; acrylic acid | *syn,anti* ($\Delta E = 360 \pm$ 100 cal/mole) | 192 |
| $CH_2{=}CH{-}CF{=}O$; acryloyl fluoride | *anti,syn* ($\Delta E = 90 \pm$ 100 cal/mole) | 193–195 |
| $CH_2{=}CH{-}CCl{=}O$; acryloyl chloride | *anti,syn* ($\Delta E \sim 0.6$ kcal/ mole) | 196 |
| $CH_2{=}C(CH_3){-}CCl{=}O$; methacryloyl chloride | *anti,syn?* | 196 |

Table 5 (continued)

| Compound | Conformation(s) | Ref. |
|---|---|---|
| $CH_3CH=CH—CCl=O$; crotonoyl chloride | *anti,syn* | 196 |
| $(C_4H_3O)—CH=O$; 2-furaldehyde (**6**) | *syn,anti* | 147, 197–199 |
| $(C_4H_3S)—CH=O$; 2-thiophenecarbaldehyde (**7**) | *syn* | 199 |
| $O=CH—CH=O$; glyoxal | *anti* | 174,200 |
| $O=CCl—CH=CH—CCl=O$; fumaroyl chloride (**3**) | mixture (probably *anti-anti*, *syn-anti*, and *syn-syn*) | 196 |
| $(C_3H_5)—CH=CH_2$; vinylcyclopropane (**8**) | *anti,gauche* | 201,202 |
| $(C_3H_5)—CH=O$; cyclopropanecarbaldehyde (**9**) | *syn,anti* ($\Delta E \sim 0$) | 203 |
| $(C_3H_5)—CCl=O$; cyclopropanecarbonyl chloride | *syn,anti* | 204,205 |
| $(C_3H_5)—C(CH_3)=O$; cyclopropyl methyl ketone | *syn,anti* | 204 |
| $(C_2NH_4)—C(CH_3)=O$; N-acetylethylenimine | *gauche* (100°) | 206 |
| $(C_3H_5)—(C_3H_5)$; -bicyclopropane (**11**) | *anti,gauche* (140°–145°) | 207,208 |
| $(C_2NH_4)—(C_2NH_4)$; 1,1′-biaziridine (**12**) | *anti* | 209,210 |

corresponds to $\varphi = 0$, the *syn* form to $\varphi = 180°$.) Quantum-mechanical calculations by Mulliken[162] favoured the *anti* form of butadiene by about 2.5 kcal/mole over the *syn* form. Pauling's model of the double bonds as two bent single bonds[138] would be consistent with an essentially three-fold barrier to rotation. The *anti* form corresponds to staggering of the bonds and should therefore be more stable than the *syn* form, which corresponds to eclipsing of the bonds. However, according to this picture the non-planar *gauche* form with $\varphi$ about 120° should also correspond to an energy minimum.

The potential function for rotation around the C–C single bonds in these molecules may probably be written to a good approximation as:[163]

$$V(\varphi) = \frac{V_1}{2}(1 - \cos\varphi) + \frac{V_2}{2}(1 - \cos 2\varphi)$$
$$+ \frac{V_3}{2}(1 - \cos 3\varphi) \quad \dots(16)$$

All the terms have minima in the *anti* conformation ($\varphi = 0°$) (cf. Figure 1). A subsidiary minimum is obtained for the *syn* conformation only if $V_2$ is the dominating term, *i.e.*, if $4V_2 > V_1 + 9V_3$.

Much attention has been paid to 1,3-butadiene itself. The earliest structure determination was carried out by Schomacher and Pauling.[164] Their results were not conclusive, but they found evidence for an essentially coplanar *anti* form. They also indicated the possibility of a small amount of a *syn* conformer. A series of other early papers also discussed the *syn* and *anti* conformations, though there was no unanimity as to which conformer was the more stable.[165-170] Aston *et al.*[171] estimated the potential for rotation about the central bond from calorimetric data. Their potential curve has minima corresponding to the two planar conformers; the energy difference was given as 2.3 kcal/mole, and the barrier between them as 4.9 kcal/mole, measured from the *anti* form which was taken as the more stable.

1,3-Butadiene was studied by electron diffraction once more in 1948 (ref. 13, pp. 56, 80). The results of the study may be summarized as follows: (1) The planar *anti* form predominates. (2) No evidence was found for a *syn* form. (3) A *gauche*-like conformer seems to coexist with the *anti* form. The existence of a *gauche* form of 1,3-butadiene is to be expected if the double bonds are considered in Pauling's bent-bond picture,[138] but could also arise as a compromise between resonance energy and non-bonded repulsion.

The *gauche* form was not conclusively established and, since later electron-diffraction studies[172-174] and a high-resolution infrared and Raman investigation[175] did not reveal this conformer, it was abandoned as an important contributor to the conformational equilibrium. There are, however, reasons to revive the *gauche*-conformation hypothesis. First, some derivatives of 1,3-butadiene (hexafluoro-butadiene, hexachlorobutadiene, *cis,trans*- and *trans,trans*-3,4-di-methyl-2,4-hexadiene; cf. Table 5) are most stable in *gauche*-like conformations in the gas phase. There is nmr evidence for the existence of similar conformations of some 1,1,3-trihalobutadienes.[211, 212] The experimental evidence may perhaps not justify a generalization, but it seems as if a fairly small steric repulsion in the *anti* conformation may be released by a small rotation about the central bond (*cis,cis*-3,4-dimethyl-2,4-hexadiene). However, if the strain in the *syn* conformation were large, rotation might well produce a *gauche* form with a torsional angle $\varphi$ about 120°. Secondly, the possibility of a coexisting *gauche* conformer in 1,3-butadiene vapour is supported by a recent semiempirical quantum-mechanical calculation by Dewar and Harget.[176] Their potential function contains a minimum about 30° away from the *syn* position. This value is considerably smaller than that suggested in the early electron-diffraction work (about 60°), but neither of the two values can be regarded as reliable. The barrier height and the energy difference were both found to be somewhat higher than the values of Aston *et al*. Additional recent evidence for the existence of a *gauche* form is given by nmr spectroscopy.[177]

The reason why later electron-diffraction experiments gave no evidence for a *gauche* form may be that these studies were carried out at the lowest possible temperature in order to optimize the accuracy in the structural data obtained. The older values were recorded at higher temperatures. If the energy difference between the *gauche* and the *anti* form is about 2.5 kcal/mole, the *gauche* contribution at 0°C should be only about 2% and would not be detectable. At present, electron-diffraction data recorded at about 200°C are being analysed.[178] There seems to be some evidence for a *gauche* conformer, though with a smaller angle than indicated by the older study.

The *anti* conformation is also found for acrolein, $CH_2\!\!=\!\!CH\!\!-\!\!CH\!\!=\!\!O$ (see Table 5), and many related compounds. However, acrylic acid seems to be slightly more stable in the *syn* than in the *anti* conformation. Acryloyl fluoride, $CH_2\!\!=\!\!CH\!\!-\!\!CF\!\!=\!\!O$, is about equally stable in these two conformations, in contrast to fluoroprene for which only the *anti* form has been observed. The energy difference between *syn* and *anti* conformers is probably small also for acryloyl chloride and crotonoyl chloride. There is evidence that fumaroyl chloride (**3**) exists as a mixture of *anti-anti* (**3a**), *syn-syn* (**3b**), and *syn-anti* (**3c**) conformers. However, *cis-* (**4**) and *trans*-hexatriene (**5**) are both considerably more stable in the *anti-anti* than in any other conformation. The extra strain in the *cis*-isomer seems to be released by a modest increase in the central $C\!\!-\!\!C\!\!=\!\!C$ valence angle and possibly also by a small twist ($\sim 10°$) about the central double bond.

2-Furaldehyde (**6**) exists as a mixture of two rotamers as first shown by Allen and Bernstein.[197] The potential function has been studied by Mönnig *et al*.[198, 199] and by Miller *et al*.[147] Both groups found the *syn* conformer (**6a**) (the double bonds in the *syn* position) to be more stable than the *anti* form (**6b**), but the reported energy differences are not in very good agreement. The $V_3$ term in the potential function was included by Mönnig *et al*. but was assumed to be zero by Miller *et al*. This difference may partly explain the discrepancy, but there is also a considerable disagreement in the observed frequencies.

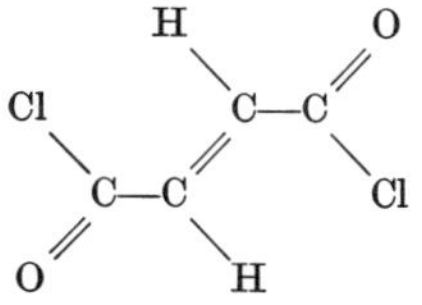

(3a) 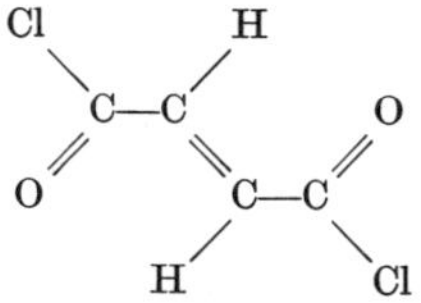          (3b) 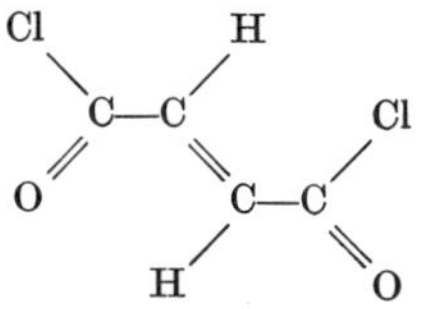          (3c)

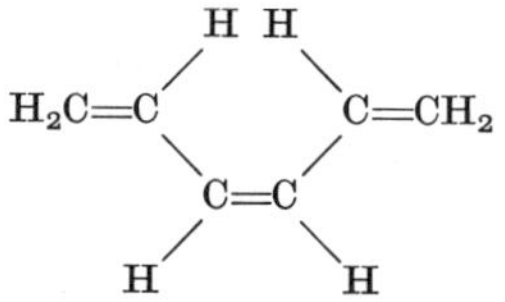

(4) 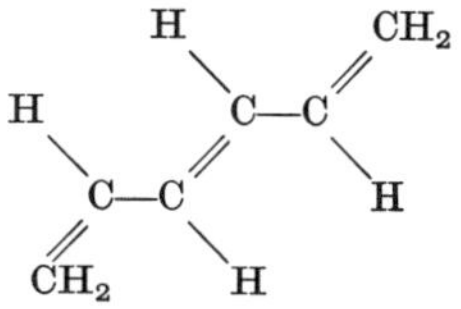                    (5)

Also 2-thiophenecarbaldehyde seems to be most stable in the *syn* form (7).

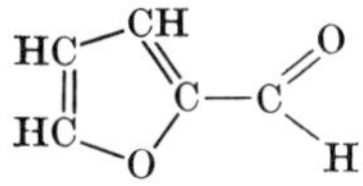

(6a) 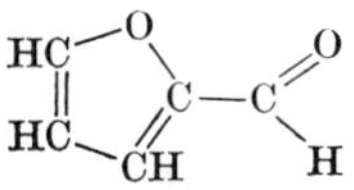              (6b) 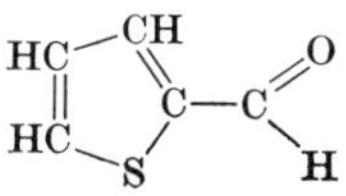                (7)

Several theoretical treatments of cyclopropane,[213-215] especially that of Walsh, suggest an analogy between the vinyl and the cyclopropyl group. Vinylcyclopropane (8) is, as expected, most stable in the *anti* conformation. There is evidence for a second conformer which does not seem to be a *syn* form, but rather a *gauche* form with $\varphi$ in the range 110–120°,[201, 202] although non-bonded repulsion in the *syn* form is not expected to be very great. In contrast to this result, cyclopropanecarbaldehyde (9) is found to be about equally stable in the *syn* and the *anti* form. Thus the $V_2$ term in the potential function seems to dominate in (9), while the $V_3$ term is more important both in (8) and in (10).[203, 216]

Bicyclopropane (11) is also most stable in the *anti* form. However, the potential function seems unusually flat in this region.[207, 208] A less stable conformer with $\varphi$ in the range 130–160° has also been detected. A spectroscopic study[209] as well as preliminary electron-diffraction results[210] show that 1,1′-biaziridine (12) exists almost exclusively in the *anti* form. The preference for the *anti* form in (12) may be steric in origin, since the geometry of the molecule causes quite large non-bonded repulsions in a *gauche* conformation.

In the introduction to this Section the two pictures of a double bond ($\sigma$, $\pi$ or two bent single bonds) were mentioned. The experimental results are not entirely in favour of either picture. There

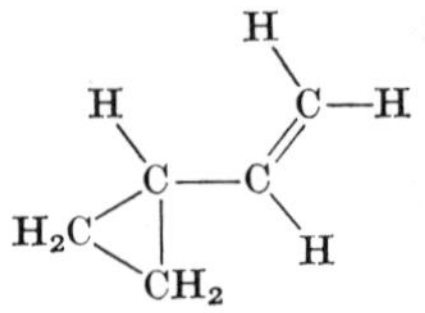

(8) 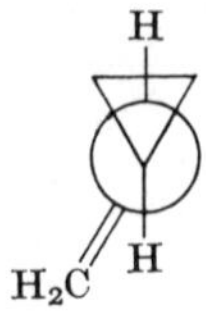       anti 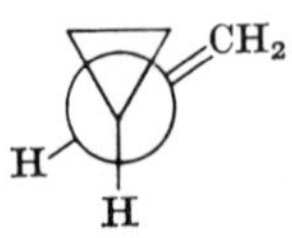       gauche

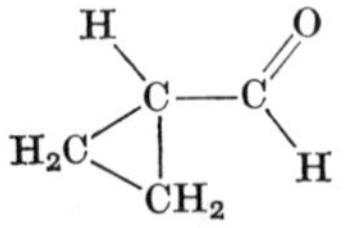

(9) 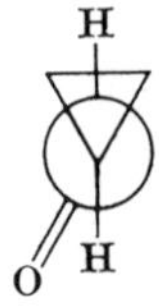       anti 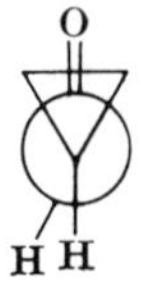       syn

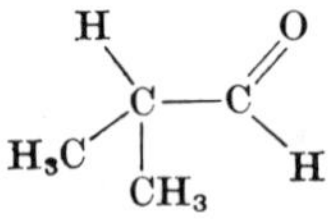

(10)

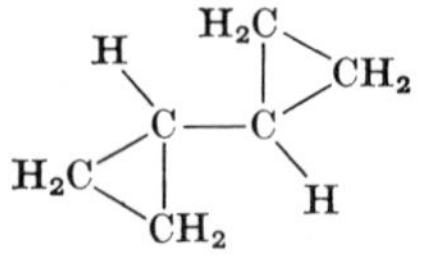

(11) 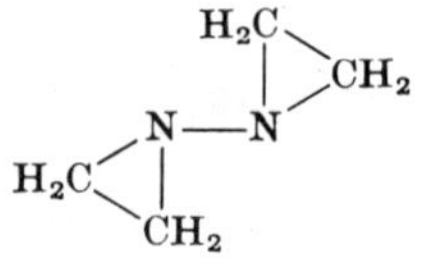                    (12)

seems to be a difference between carbonyl and vinyl groups in this respect, since *syn* conformations have been observed in some molecules containing the former group but not the latter. Part of the difference may be steric, though other effects are likely also to be of importance. Perhaps some interaction involving oxygen lone pairs is involved. There is very little evidence for the existence of *syn* conformations of any of the other molecules listed in Table 5, but *gauche* forms have been found.

An interesting far-infrared investigation of benzaldehydes and related compounds has been carried out by Miller *et al.*,[147] who assigned torsional frequencies to the molecules. Equation (16) was applied for the potential function, and further simplifications were obtained by expanding the cosine functions in power series.

In benzaldehyde and *para*-substituted derivatives equation (16) reduces to:

$$V = \frac{V_2}{2}(1 - \cos 2\varphi) \approx V_2\varphi^2 \qquad \ldots(17)$$

provided that the substituent is symmetrical. $V_2$ may thus be found from the torsional frequency. Table 6 gives some results. There is

Table 6. $V_2$ (kcal/mole) in benzaldehyde and *para*-substituted derivatives.[147]

| | |
|---|---|
| Benzaldehyde | 4.66 (cf. ref. 146) |
| *p*-Fluorobenzaldehyde | 3.58 |
| *p*-Chlorobenzaldehyde | 2.81 |
| *p*-Bromobenzaldehyde | 2.37 |
| *p*-Tolualdehyde | 3.47 |

considerable variation in $V_2$; the barriers decrease in the order $H > F \approx CH_3 > Cl > Br$, but the reason for this is not clear.

The observed barrier in benzaldehyde (4.7 kcal/mole) is in relatively good agreement with semiempirical calculations by Forsén and Skancke.[217]

In asymmetrically substituted benzaldehydes and related compounds Miller *et al.* make the additional assumption that $V_3$ is small. Two torsional frequencies are found for most of these molecules, presumably corresponding to oxygen-*syn* (**13a**) and oxygen-*anti* (**13b**)

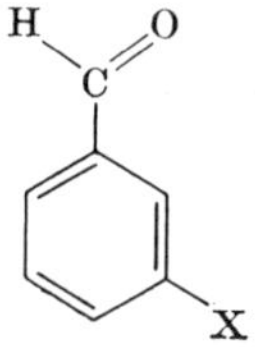

(13a) 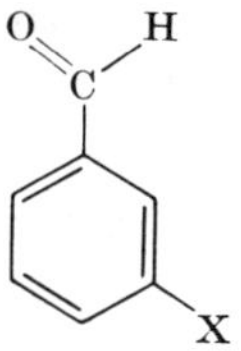    (13b)

to the substituent. Table 7 gives some results. For the *meta*-compounds, the *syn* conformer is somewhat more stable than the *anti*, while the opposite holds for the *ortho*-compounds, probably for steric reasons. The energy difference between the two forms was found to be small for the *ortho*-substituted fluoro and chloro compounds. For the corresponding bromo and methyl compounds only one torsional frequency was observed. If steric interaction destabilizes the *syn* form, a greater energy difference is expected for these two compounds than for the fluoro and chloro derivatives. However, the differences between F and Cl on one hand and Br and $CH_3$ on the other seem surprisingly large.

Table 7. $V_1$ and $V_2$ (kcal/mole) in some *meta-* and *ortho-*substituted benz-aldehydes.[147] $V_{max}$ is the barrier to rotation.

| Substituent in benzaldehyde | $V_1$ | $V_2$ | Stable rotamer | Relative population | $V_{max}$ |
|---|---|---|---|---|---|
| *m*–F | 1.37 | 4.14 | O-*syn* | 10 | 4.85 |
| *m*–Cl | 0.66 | 4.02 | O-*syn* | 2.5 | 4.38 |
| *m*–Br | 1.44 | 4.09 | O-*syn* | 7 | 4.77 |
| *m*-CH$_3$ | 1.71 | 4.18 | O-*syn* | 11 | 4.99 |
| *o*–F | 0.28 | 3.85 | O-*anti* | 1.6 | 4.32 |
| *o*–Cl | 0.70 | 7.07 | O-*anti* | 2.6 | 4.21 |

## 4. Biphenyl and related compounds

Biphenyl and its derivatives played an important part in the development of our understanding of conformational phenomena. The existence of optical isomerism in biphenyl derivatives with bulky *ortho*-substituents[218] offered the first demonstration of the effect of steric hindrance to rotation about a single bond.[219, 220] Smaller groups in the *ortho*-position also cause rotational barriers. In the case of biphenyl itself the barrier must be rather low, since the molecule undergoes conformational changes by phase transition.[13, 221] Biphenyls without *ortho*-substituents,[222–225] including biphenyl itself,[226] appear often to be planar in the solid state, while the angle between the rings ranges from 42° to 54° in the gas phase.[221, 227] It is interesting that the inter-ring angle in crystalline 4,4′-dimethylbiphenyl (**14**) is close to 40°.[228]

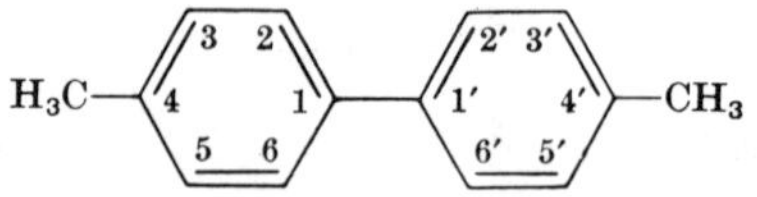

(**14**) 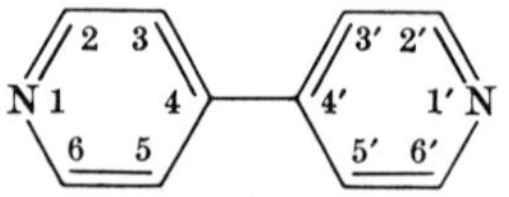          (**15**)

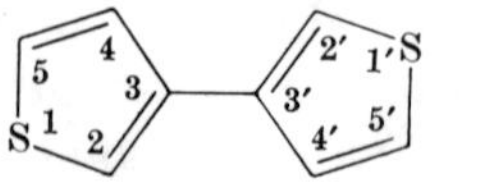

(**16**)

Similar conformational differences exist for molecules such as 1,3,5-triphenylbenzene[229, 230] and hexaphenylbenzene,[231, 232] although these molecules are not actually planar in the solid state.

Hexaphenylbenzene is an interesting case. In the vapour the peripheral rings are orthogonal to the central ring, with a rather large torsional amplitude of at least 10° to either side; this molecular arrangement is probably a result of statistics rather than of potential energy; from general knowledge of biphenyl-like molecules a propeller-shaped conformer ought to be energetically the most stable one, a conformation also found in the solid. On the other hand this conformation is statistically unfavourable since, if only one of the six rings is brought out of order, the propeller conformer cannot be realized.

The height of the barrier to rotation in biphenyl itself is not known experimentally. Theoretical calculations [233, 234, 176] lead to a flat minimum at about 40° from planarity and to barrier heights around 2–4 kcal/mole. The barrier through the planar form is probably slightly higher than the barrier through the orthogonal form.

The conformational behaviour of biphenyls without *ortho*-substituents and analogues thereof should be very similar, at least in the gas phase. For example, 4,4'-bipyridine (**15**) assumes essentially the same conformation as biphenyl, though the angle of twist is slightly smaller (approximately 37° from planarity).[227] Another analogous molecule also having only hydrogen atoms in the positions adjacent to the bridge bond is 3,3'-bithiophene (**16**) where the angle of twist is found to be about 30°.[235] In contrast to the molecules discussed above, 3,3'-bithiophene has two conformers, one with an angle of twist about 30° away from the *syn* position and one with an angle of twist about 30° away from the *anti* position. According to electron-diffraction findings the latter conformer is the more stable by approximately 0.3 kcal/mole. Results of theoretical calculations [236] are in good agreement with the experimental results. The calculated barrier has two flat minima at $\varphi = 30°$ and 150°, respectively ($\varphi$ being set equal to zero at the *syn* position). The 150° minimum lies about 0.7 kcal/mole lower than that at 30°. The barrier height from the lowest minimum is approximately 2 kcal/mole.

It seems accordingly clear that biphenyls without *ortho*-substituents and their analogues behave in the same way in the gas phase. The deviation from planarity may, at least qualitatively, be described by a combined effect of non-bonded interaction and conjugation.

To get further information it is natural to study the effect of *ortho*-substitution. This should, of course, introduce more steric hindrance and increase the angle of twist. The non-bonded interaction may, however, be eased by replacing a CH group in the 2-position by a single atom such as nitrogen or sulphur. This might be loosely described as replacing hydrogen by a lone pair of electrons. For a molecule such as 2,2'-bipyridine one might expect a virtually planar *anti* conformer, arguing that the conjugation effect would probably overcome the interaction between nitrogen lone pairs and hydrogen atoms. Electron-diffraction results do not, however, favour a rigid planar *anti* conformer as the main contributor.[227] There seems to be no strongly preferred conformational choice in the vapour, but the torsional motion is considerably more free than in biphenyl or in 4,4'-bipyridine. The potential curve seems rather complex and has no doubt a maximum at the *syn* position, probably a shallow minimum at 20–40° away from this position, and probably also a very shallow minimum at or near the *anti* position. An analogous molecule also studied by electron diffraction is 2,2'-bithiophene.[237] In this case also it is clear that a rigid planar *anti* conformer does not predominate. The prevailing angle of twist seems to be around 34°, but the electron-diffraction work suggests nearly free rotation over a large angular interval. Theoretical calculations indicate nearly free rotation with a very shallow minimum at approximately 60°.[236] This apparent discrepancy between experiment and theory probably only reflects the fact that 2,2'-bithiophene as well as 2,2'-bipyridine have rotational barriers considerably lower than those in biphenyls with free *ortho*-positions and their analogues. This means, in the language used above, that the repulsion between a lone pair and a hydrogen atom is smaller than the hydrogen–hydrogen repulsion, a finding that is not too surprising. On the other hand, hydrogen–lone pair interaction seems not entirely negligible for the barrier.

Turning now to the genuine *ortho*-substituted biphenyls it would be natural to consider first molecules with only one such substituent. Unfortunately, only 2-fluorobiphenyl has been studied in the gas phase by electron diffraction.[238] The angle of twist is 49°, only slightly and not significantly larger than that in biphenyl itself (42°).

More extensive studies have been carried out on the 2,2'-dihalo-biphenyls: 2,2'-difluoro-, -dichloro-, -dibromo-, and -diiodo-biphenyl have been studied by electron diffraction in the gas phase.[238, 239] As expected, the deviations from planarity are larger than those just discussed. The angles of twist are listed in Table 8 for all the 2,2'-dihalobiphenyls as well as for perfluorobiphenyl. Except for the last compound these results are based on rather old measurements and should be treated with caution. In the earlier studies the twist angle was calculated from the maximum in the halogen–halogen distance,

Table 8. Angles of twist and halogen–halogen distances in 2,2'-dihalobiphenyls. ($\varphi = 0$ corresponds to the *syn* position.) Pauling's van der Waals distances are included as well as the difference between these data and the experimental halogen–halogen distances. (Reference numbers are in parentheses.)

| Biphenyl substituents | Angle of twist, $\varphi$ | | X–X (Å) | van der Waals distance (Å)[138] | Diff. (Å) |
|---|---|---|---|---|---|
| | obs. | calc. | | | |
| 2,2'-Difluoro | 60° (238) | 42°, 143° (29) | 2.85 | 2.70 | −0.15 |
| 2,2'-Dichloro | 74° (239) | 72°, 120° (29) | 3.46 | 3.60 | +0.14 |
| 2,2'-Dibromo | 75° (239) | 82°, 112° (29) | 3.62 | 3.90 | +0.28 |
| 2,2'-Diiodo | 79° (239) | 93° (29) | 3.82 | 4.30 | +0.48 |
| Perfluoro | 70° (29) | 46° (29) | 3.12 | 2.70 | −0.42 |

but in more recent investigations the reported angle corresponds to the maximum in the probability distribution, $P(\varphi)$, or to the mean value of $\varphi$. The difference between the two last estimates seems to be small for perfluorobiphenyl[29] but may be larger in the other compounds.

It is interesting that the prevailing conformer in all 2,2'-dihalobiphenyls is found to be one with the two halogen atoms on the same side (*i.e.*, closer to the *syn* than to the *anti* position). Table 8 contains also the halogen–halogen distances for comparison with Pauling's van der Waals distances.[138] The observed conformational preference is probably due to London forces between halogen atoms, although the observed halogen–halogen distances are shorter than the van der Waals distances, except in the fluoro-compounds. It may be noted that the packing F···F distance in perfluorobiphenyl is even larger than in 2,2'-difluorobiphenyl.[29]

Rough calculations of the potential functions for torsion around the bridge bond have been carried out for some halobiphenyls by combining conjugation energy and van der Waals energy.[29, 240] For 2,2-diiodobiphenyl a calculated minimum was obtained for $\varphi$ close to 90°, in reasonable agreement with the observed value. For the other 2,2'-dihalobiphenyls the calculated potential curves showed a minimum for $\varphi > 90°$, *i.e.*, closer to the *anti* than to the *syn* form, and a subsidiary minimum for $\varphi < 90°$ (Table 8). In contrast to the observations, the calculations therefore suggest that the conformer with $\varphi > 90°$ should be the more stable. Further, the agreement between the experimental twist angles and the positions of the calculated minima for $\varphi < 90°$ is not particularly good. However, the calculated minima are very flat, and small changes in the parameters may shift the position of the minimum considerably. The observed torsional angles in the fluoro compounds are considerably larger than the calculated values, probably because electrostatic interactions were neglected.

High-temperature electron-diffraction experiments are now being carried out to search for a possible second conformer.

The results for the 2,2′-dihalobiphenyls seem to indicate that the non-bonded interactions are somewhat more important, and the conjugation energy perhaps less important, for the twist angles than was assumed in the calculations. Calculations using the CNDO method are now being carried out.[241] [See, however, p. 165.]

As stated above, experimental rotational barriers are not available for biphenyl derivatives. Several attempts have been made to obtain information about the potential from electron diffraction. As mentioned in Section III-A, this method is not well suited for barrier determinations, but it can under special conditions give quantitative information about the shape of the potential curve near the minima. 3,3′-Dibromo- and 3,5,3′,5′-tetrabromo-biphenyl have been studied for this purpose.[242] For the former molecule the two expected conformers can be easily detected from the Br$\cdots$Br distances. The conformers have the same inter-ring angle (45°), and there is no observable energy difference between them. The value 45° was also reported for the inter-ring angle in 3,5,3′,5′-tetrabromobiphenyl but subsequent least-squares refinement of the intensity data indicate that the angle may be smaller, i.e., slightly less than 40°.[243] In order to study the torsional motion the longest Br$\cdots$Br distance peak was analysed. For quantitative results the framework vibration has to be known and, to obtain estimates of this, 3,5,4′-tribromobiphenyl was chosen as a reference substance. The inter-ring angle of this molecule was found to be identical with that of the other two compounds. The Br-3$\cdots$Br-4′ distance in 3,5,4′-tribromobiphenyl is not very different from the Br-3$\cdots$Br-3′ distance in the two other bromo derivatives, but the former does not change with torsion angle. It is therefore reasonable to use the vibration amplitude for the former distance in 3,5,4′-tribromobiphenyl as the framework-vibration amplitude for the long Br$\cdots$Br distances in the other two bromobiphenyl molecules. If this is done, an estimate of the root-mean-square amplitude of the torsional motion ($\sigma_\varphi$) may be obtained. For 3,3′-dibromo- and 3,5,3′,5′ tetrabromo-biphenyl, $\sigma_\varphi$ is in this way found to be 19° and

17°, respectively. Least-squares refinement[243] of the data for the latter compound gave a slightly smaller value, *i.e.*, 15°.

The torsional motion of perfluorobiphenyl has been studied by more advanced electron-diffraction techniques,[29] and both the torsional amplitude and the potential against rotation through the 90° position have been refined. For this molecule the inter-ring angle is 70°, with a torsional amplitude (given as $\sigma_\varphi$) of 10° ± 3°. It seems reasonable that the amplitude is considerably smaller in this case where fluorine–fluorine interaction occurs instead of hydrogen–hydrogen interaction. The barrier in going from the 70° position to the 90° position was found to be between 0.4 and 2.0 kcal/mole, but the barrier through the planar position is too high to be measurable by the electron-diffraction technique.

## C. Molecules with Several Degrees of Freedom for Internal Rotation

The simplest problem involving more than one internal rotor is the case in which the two tops are identical, propane serving as a parent molecule for purposes of illustration. For microwave studies, the interaction between the two rotors and the correlation of the two motions is the key question. The potential function for propane has been written[244, 245] as:

$$V(\varphi_1, \varphi_2) = \tfrac{1}{2} V_3(2 - \cos 3\varphi_1 - \cos 3\varphi_2) + \tfrac{1}{2} V_3' \cos 3\varphi_1 \cos 3\varphi_2 - \tfrac{1}{2} V_3'' \sin 3\varphi_1 \sin 3\varphi_2 \quad \ldots(18)$$

or[246]

$$V(\varphi_1, \varphi_2) = \tfrac{1}{2} V_3(2 - \cos 3\varphi_1 - \cos 3\varphi_2) - \tfrac{1}{2} V_3'[1 - \cos 3(\varphi_1 + \varphi_2)] \quad \ldots(19)$$

where $\varphi_1$ and $\varphi_2$ are the two angles describing the rotation of the two methyl groups. The $V_3$ term, which does not involve interaction between the tops, is only a little larger than that found in ethane. The interaction terms must, however, be included to reproduce the observed microwave spectrum. The results obtained by applying expression (19) are $V_3 = 3.575 \pm 0.1$ kcal/mole and $V_3' = 0.31 \pm 0.04$ kcal/mole, in very good agreement with the results obtained theoretically, *i.e.*, $V_3 = 3.48$ kcal/mole and $V_3' = 0.357$ kcal/mole.[246]

More interesting, from the point of view of conformational studies, is the case in which the two rotors are non-equivalent. In one of the earliest microwave studies on molecules of this type, Hirota[40] analysed the spectrum of $CH_3CH_2CH_2F$. His results are summarized in Figure 11, where the angle variable represents torsion around the $CH_3CH_2$–$CH_2F$ bond. The usual potential function

$$V = \tfrac{1}{2} \sum V_n(1 - \cos n\varphi)$$

was determined from the observed energy levels, the first seven terms being used. The uncertainty in the individual $V$'s was rather large, and the energy difference between the two stable conformers was estimated as $0.47 \pm 0.31$ kcal/mole. The barrier for the other internal motion, the methyl group torsion, was found to be 2.87 kcal/mole for the *gauche* and 2.69 kcal/mole for the *anti* conformer. All interactions between the internal motions were ignored.

Similar, but less complete, results have been reported for $CH_3CH_2CH_2Cl$,[247, 248] $CH_3CH_2CH_2Br$,[248] and $CH_3CH_2CH_2CN$.[249] All of these exist in two conformers, *anti* and *gauche* with respect to torsion around the C–C bond adjacent to the substituent. The two conformers differ little in energy, with the *gauche* form reported as more stable by $0.05 \pm 0.15$ kcal/mole in the chloride and $0.1 \pm 0.2$ in the bromide.

Propionaldehyde, $CH_3CH_2CHO$, also exhibits two stable conformations, with the carbonyl-oxygen *syn* or *gauche* with respect to the methyl group.[250] The *syn* conformer is more stable by $0.9 \pm 0.1$ kcal/mole. The *gauche* conformer is obtained by rotation through about 131°.

A thorough study of the corresponding fluoride, $CH_3CH_2COF$, has recently been completed.[251] Again, two stable conformers exist, a more stable one with the carbonyl-oxygen *syn* to the methyl group and the other a *gauche* form obtained by rotation through $120 \pm 2°$. The observed energy levels and derived potential function are shown in Figure 16. The potential function is analysed in terms of the first four Fourier coefficients with $V_1 = 0.93 \pm 0.30$ kcal/mole, $V_2 = 0.72 \pm 0.25$ kcal/mole, $V_3 = 1.15 \pm 0.18$ kcal/mole, and $V_4 = 0.13 \pm 0.20$ kcal/mole. The two *gauche* forms are, of course, energetically equivalent, but the near equality of the *syn-gauche* and the *gauche-gauche* potential maxima is somewhat unexpected.

In discussing the results of the $CH_3CH_2COF$ experiments, Stiefvater and Wilson[251] suggested that the $V_3$ contribution to the $\varphi$-dependent interaction energy is essentially constant for a group $CX_3$, where X = H, $CH_3$, or F, rotating against a COF framework. The differences between the potential functions and the resulting stability of various conformers are to be attributed to the $V_1$, $V_2$, and $V_4$ terms, which are also assumed to be additive. The potential of

$(CH_3)_2CHCOF$ was then predicted from the standard $V_3$ curve, and the $V_1$, $V_2$, and $V_4$ terms obtained from the investigation of $CH_3CH_2COF$. On this basis, the conformer of $(CH_3)_2CHCOF$ with one methyl group eclipsing the oxygen atom was predicted to be more stable than the symmetric form (H eclipsing O) by 1.33 kcal/mole. This prediction awaits experimental test. While no definite conclusion can be drawn about the nature of the forces responsible for the energy barriers, the authors speculate that the $V_3$ term arises from inter-action of the electrons bonding the attached atoms to the axial atoms while $V_1$ and $V_2$ are largely steric or electrostatic in origin.

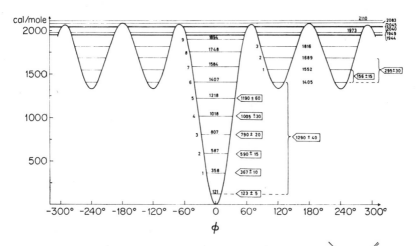

Figure 16. Potential function for rotation around the $-\overset{\diagdown}{C}-\overset{\diagup}{C}\diagdown$ bond in $CH_3CH_2COF$. $\varphi = 0°$ corresponds to the conformation with the carbonyl group *syn* to the methyl group. Calculated and experimental torsional levels are given.

Two microwave studies of butenes have appeared recently. 1-Butene exists in two stable forms with respect to rotation around the central C–C bond.[252] A conformer in which the methyl group is *syn* to the double bond, is separated by a barrier of 1.74 kcal/mole from a slightly more stable form, illustrated in Figure 17, obtained by rotation of 120°. The barrier to rotation of the methyl group is observed to be higher for the *syn* conformer, probably owing to steric interactions between the hydrogen atoms. As the methyl group rotates, a methyl-hydrogen atom comes very close to one of the vinyl-hydrogen atoms, increasing the energy maximum in the potential function. In *cis*-2-butene, mentioned in Section V-B-2, it has been shown[93] that the stable conformation of the methyl group has one C–H bond eclipsing the double bond.

Allyl halides, $CH_2$=$CHCH_2X$,[253] and allyl cyanide, $CH_2$=$CHCH_2CN$,[254] also exist in both *gauche* and *syn* conformations, whereas only *gauche* forms have been reported from microwave studies of allyl alcohol, $CH_2$=$CHCH_2OH$,[255] and 2-propene-1-thiol, $CH_2$=$CHCH_2SH$.[256] The *gauche* preference of these molecules is probably related to the possibility of hydrogen-bond formation. Before this is discussed in detail conformational problems involving hydroxyl groups will be treated.

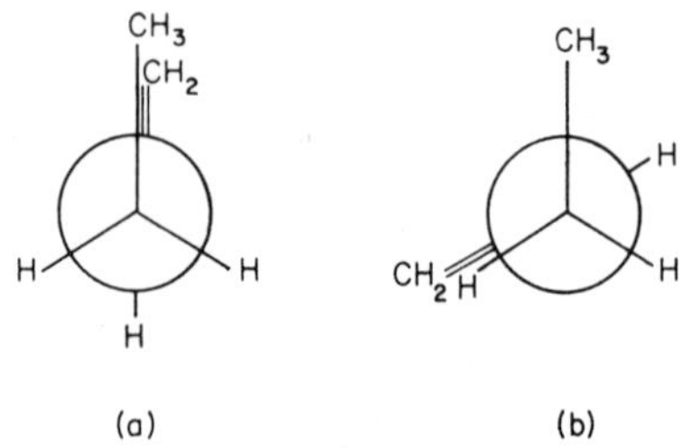

(a)                    (b)

Figure 17. The two conformers of 1-butene. The conformer *b* is the more stable.

In spite of a very large number of investigations, little was known until recently about the conformation of alcohols so far as the orientation of the hydroxyl-hydrogen atom is concerned. The problem is difficult for electron-diffraction techniques because of the small scattering from hydrogen atoms. For $CH_3OH$, microwave spectroscopy can yield the equilibrium conformation only if species with the methyl group partially deuterated are studied, and these have the asymmetric-asymmetric rotor geometry for which theoretical treatments have only recently been developed in the low-barrier case. Preliminary results[257] indicate that $CH_3OH$ has the staggered conformation. Although there are many conflicting reports in the literature, it now appears that $CH_3CH_2OH$ exists in both *anti* and *gauche* forms with respect to torsion around the C–O bond. The *anti* conformer has recently been characterized by microwave spectroscopy in some detail.[258] For propargyl alcohol, $HC$≡$CCH_2OH$, only the *gauche* conformer was detected, and the barrier at the *syn* position is only about 260 cal/mole.[259] Isopropyl alcohol, $(CH_3)_2CHOH$, has been shown to exist in two conformers, one with the C–H and O–H groups *anti* and the other with them *gauche*.[260] A similar result has been obtained for 2-propanethiol, $(CH_3)_2CHSH$.[261]

In addition to the conformational problems connected with torsion of a hydroxyl group, the conformation of the rest of the molecule is often particularly interesting because of the ability of the hydroxyl group to participate in hydrogen-bond formation (see Section III-A). Even before it was generally accepted that two or more conformers may coexist in pure liquids and gases, molecular structures of compounds such as ethylene·glycol, glycerol, and 2-chloroethanol had been studied by a series of investigators. Zahn[262] measured the dipole moments of 2-chloroethanol and ethylene glycol in the gas phase, and proposed several alternative structure models including those with completely free rotation as well as both *syn* and *anti* forms. In spectroscopic studies[263-265] several authors discussed the possibility of coexisting forms, indicating also the possibility of interaction between the hydroxyl group and the chlorine atom of 2-chloroethanol. An electron-diffraction investigation by the sector microphotometer method[13, 266] showed that the *gauche* form prevails in ethylene glycol, 2-chloroethanol, and glycerol. The observed dihedral angles ranged from 71° to 74°. This angle corresponds to formation of a hydrogen bond of length about 2.95 Å between two hydroxyl-oxygen atoms and a hydrogen bond of length about 3.16 Å between the hydroxyl-oxygen and the chlorine atom. These findings were confirmed by visual electron-diffraction studies on 2-chloroethanol,[267] 2-fluoroethanol,[268] and 1,3-dichloro-2-propanol.[269] In the meanwhile, Raman and infrared studies by Mizushima *et al*.[270] had confirmed the prevalence of the *gauche* conformer in gaseous 2-chloroethanol. They found an energy difference between the *anti* and the *gauche* conformer of 0.95 kcal/mole. They also observed that, while the *gauche* and the *anti* conformers coexist in the gas and the liquid phase, the *gauche* seems to be the only conformer present in the solid state. (This may be compared with the results for 1,2-dichloroethane where the *anti* conformer, which is the more stable in the gas phase, persists in the solid state.) The intramolecular hydrogen bond appears therefore to survive both condensation and crystallization. This means that the intramolecular hydrogen bonds compete favourably with the intermolecular engagements, as has been directly shown by X-ray liquid work.[13] This is somewhat surprising since the intramolecular hydrogen bond would be expected to suffer from unfavourable geometry. Recent microwave studies of 2-fluoroethanol[271] and 2-chloroethanol,[272] as well as recent electron-diffraction work[273] on the latter, lead to dihedral angles slightly smaller (about 65°) than those found earlier. The O–H $\cdots$ Cl hydrogen bond length was found to be 3.14 Å. The most recent electron-diffraction work included a high temperature $(T = 473°K)$ study (Figure 6) from which the presence of the *anti* conformer is easily demonstrated. Since the *gauche* to *anti* ratio was determined at two temperatures it is possible to estimate the difference in entropy as well as in energy, if they are assumed to be temperature-independent. The value found for entropy difference is about 3.7 e.u. (the *anti* conformer having the higher entropy) in agreement with spectroscopical results of Mizushima *et al.*,[274] but this value cannot be brought into harmony with the low energy difference obtained by the same authors.[270] The energy difference obtained from the new electron-diffraction study is as large as 2.5 kcal/mole.

The fact that the entropy difference between the *anti* and the *gauche* conformer of 2-chloroethanol is significantly different from zero means that the approximation of putting $\omega$ in Section II equal to unity is invalid for the two 2-haloethanols. This marked difference between the 1,2-dihaloethanes and the 2-haloethanols probably has its origin in the torsional motion around the C–O bond. In the *gauche* conformer of the 2-haloethanols the hydroxyl-hydrogen is, as a participant in hydrogen bonding, probably in a rather well localized position. In the *anti* conformer, however, the hydrogen atom can probably adopt any one of the three staggered positions, as in ethyl alcohol. If the three positions are about equally populated the weight factor ratio between *anti* and *gauche* forms of a 2-haloethanol should be 3:2 rather than 1:2. This simple argument leads to an entropy difference of $\boldsymbol{R} \ln 3 = 2.2$ e.u., to be compared with the measured difference of 3.7 e.u.

An electron-diffraction reinvestigation of ethylene glycol in progress in Oslo may perhaps throw more light on this problem. In ethylene glycol neither of the two hydroxyl-hydrogen atoms in the *anti*-conformer engages in hydrogen bonding. In the *gauche* conformer one of the hydroxyl-hydrogen atoms is engaged in hydrogen bonding and is therefore well localized, and the other is also restricted because of the internal hydrogen bond.

It is interesting that electron diffraction, in spite of its limited ability to localize hydrogen positions in alcohols, may contribute indirectly to the understanding of conformational problems involving hydroxyl groups.

Another case where hydrogen bonding favours the *gauche* conformer is 2-aminoethanol.[275]

Methyl vinyl ether, $H_2C$=$CHOCH_3$, exists predominantly in the *syn* conformation, with a planar heavy-atom skeleton, according to spectroscopic investigations[276, 277] which also gave evidence for a second conformer.[276] A recent electron-diffraction investigation[278] showed that the second conformer is not the *anti* form but is obtained by a rotation of 80–110° around the $\overset{\diagdown}{\underset{\diagup}{C}}$—O— bond from the *syn* form. For the analogous sulphur compound, $CH_2$=$CHSCH_3$, only the *syn* conformer has been found by microwave spectroscopy,[279] but there may well exist a second conformer similar to that of methyl vinyl ether.*

Gas-phase conformational studies have also been made on several esters by microwave spectroscopy. For methyl formate, $HCOOCH_3$, only the form with the carbonyl group *syn* to the methyl group was found.[280] A similar conformation is reported[281] for $FCOOCH_3$ and $NCCOOCH_3$. Ethyl formate, $HCOOCH_2CH_3$, presents a more complicated problem since there are three single bonds about which restricted rotation is possible. Riveros and Wilson[282] have identified two stable conformers, both having the $CH_2$ group *syn* to the carbonyl group considering rotation around the $\overset{\diagdown}{\underset{\diagup}{C}}$—O— bond, but one with the methyl group *anti* and the other with the methyl group *gauche* with respect to rotation around the —O—$\overset{\diagup}{\underset{\diagdown}{C}}$— bond. The two forms are illustrated in Figure 18. The *anti* conformer is more stable by $0.19 \pm 0.06$ kcal/mole. The derived potential function is sketched in Figure 19. As is commonly the case in such studies, the equilibrium conformation of the methyl group with respect to torsion around the C–C bond was not determined.

For methyl nitrite, $CH_3ONO$, two conformers have been reported,[283] with the N=O group severally *anti* and *syn* to the methyl group. The *syn* form is more stable by $0.275 \pm 0.040$ kcal/mole. The

* This has now been confirmed by an electron-diffraction investigation.[431] The energy difference seems to be smaller for methyl vinyl sulphide than for methyl vinyl ether. Preliminary results for anisole show that the conformation with the heavy atoms coplanar $[\varphi(C\text{—}O) = 0]$ is the most stable one. However, there seems to be a small amount of the conformer with $\varphi(C\text{—}O) = 90°$.[432]

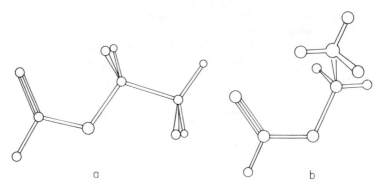

Figure 18. The two identified conformers of ethyl formate,
$HCOOCH_2CH_3$.

barrier to rotation of the methyl group is quite different in the two
forms, being $1.91 \pm 0.19$ kcal/mole in the *syn* conformer but only
0.188 kcal/mole in the *anti* form.

Only one equilibrium conformation of methyl nitrate, $CH_3ONO_2$,
has been observed.[284] All the heavy atoms lie in the same plane, with
a rather high barrier of $9.1 \pm 2.6$ kcal/mole opposing rotation of the

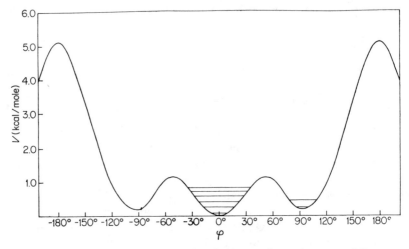

Figure 19. The potential function for internal rotation around the

$-O-C\left<\begin{array}{c}\\\\\end{array}\right.$ bond in ethyl formate, $H\overset{\overset{\displaystyle O}{\|}}{C}-O-CH_2CH_3$. $\varphi = 0°$

corresponds to conformer (a) in Figure 18.

$NO_2$ group out of the plane. The monodeuterated species $CH_2DONO_2$ has been studied, so it is known that the methyl group hydrogens have equilibrium positions staggered with respect to the $O–NO_2$ bond.

A recent microwave study[285] of glycolaldehyde, $CH_2(OH)CHO$, shows that the two oxygen atoms are *syn* with the hydroxyl-hydrogen in the plane formed by the heavy atoms and presumably located between the two oxygen atoms. No evidence for other stable conformers was found.

As an example of a simple chain molecule with more than two degrees of torsional freedom, 1,2,3,4-tetrabromobutane may be mentioned. Both diastereomers have been studied in the gas phase by the electron-diffraction technique.[13] Each pair of neighbouring bromine atoms could in principle be oriented in three different ways, corresponding to one *anti* and two *gauche* orientations, but some of the conformers thus obtainable are highly disfavoured because of interference between non-bonded atoms.

## VI. CYCLIC COMPOUNDS

### A. Introduction

In describing the conformations of cyclic molecules, it is convenient to discuss first the ring-skeleton (Section VI-B) and then the orientation of substituent groups (Section VI-C).

### B. Ring-skeleton Structures

#### 1. *Four-membered rings*

The puckering vibration of four-membered rings may be described by one coordinate. In cyclobutane this vibration has a double minimum making the equilibrium conformation of the carbon skeleton non-planar, as shown by electron diffraction[286, 287] and spectroscopy[288, 289] (see Figure 20). Results of some investigations of four-membered rings are summarized in Tables 9 and 10. In most cases the equilibrium conformation is non-planar. The deviation from planarity may be described by the dihedral angle ($\beta$) between the plane through the atoms C-1, C-2, C-3 and the plane through C-3, C-4, C-1. For cyclobutane itself, $\beta$ is found to be about 35°. It is interesting that the corresponding angle in fused ring systems is considerably smaller. In bicyclo[2.2.0]hexane (**17**) it is 11°,[290] and in

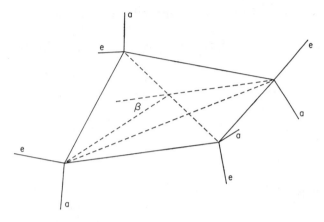

Figure 20. Structure of cyclobutane.

the *syn* and *anti* isomers of tricyclo[4.2.0.0$^{2,5}$]octane **(18)** it is 8–9°.[291]

$$
\begin{array}{ccc}
\text{H} & & \\
\text{H}_2\text{C}-\text{C}-\text{CH}_2 & & \\
| \quad | \quad | & & \\
\text{H}_2\text{C}-\text{C}-\text{CH}_2 & & \\
\text{H} & & \\
\textbf{(17)} & & \\
\end{array}
$$

(17)        (18)        (19)

The four-membered ring in bicyclo[2.1.0]pentane **(19)** seems to be planar.[292]

In substituted cyclobutanes the ring skeleton is frequently found to be planar in the crystalline state (cf. Table 10), but this may be caused by crystal forces. An interesting comparison is given by Margulis *et al.*[293,294] They have shown that *trans*-1,3-cyclobutane-dicarboxylic acid is planar in the crystalline state. However, in the sodium acid-salt

$$\text{Na}^+{}_2\text{C}_4\text{H}_6(\text{COO}^-)_2 \cdot 2\text{C}_4\text{H}_6(\text{COOH})_2$$

the ring in the acid is puckered ($\beta = 25°$), while the dianion contains a planar ring.*

Cyclobutene and the derivatives included in Table 10 are all planar.

The barrier to inversion of unsaturated four-membered rings varies considerably from one molecule to another (Table 9). The potential function may be expressed as

$$V(z) = az^2 + bz^4 \qquad \qquad \dots (20)$$

* Further references to X-ray work on related compounds are given in ref. 294.

Table 9. Dihedral angle ($\beta$) and barrier to inversion of four-membered rings.

| Compound | $\beta$ | Method | Ref. | Barrier | | | Further refs. or remarks |
|---|---|---|---|---|---|---|---|
| | | | | (kcal/mole) | Method | Ref. | |
| Cyclobutane | 35° | ED | 286,287 | | | | 288 |
| | 34.0 ± 0.5° | NIR | 289 | 1.28 ± 0.005 | NIR | 289 | |
| 1,1-Difluorocyclobutane | | | | 0.69 ± 0.02 | MW | 304 | |
| Silacyclobutane | 35.9 ± 2° | FIR | 305 | 1.26 ± 0.01 | FIR | 305 | 306 |
| Azetidine (22) | | | | 1.26 | FIR | 303 | Unsymmetric potential. Energy diff. between the two minima 0.27 kcal/mole |
| Oxetane (20) | (0°) | MW | 298 | 0.044 ± 0.002 | FIR | 300 | 296,297,299,43 |
| Thietane (21) | 28° | MW | 301 | 0.78 ± 0.01 | MW | 301 | 308 |
| | | | | 0.78 | FIR | 307 | |
| Selenetane | 32.5 ± 2° | FIR | 309 | 1.08 | FIR | 309 | |
| Cyclobutanone | (0°) | FIR | 307 | 0.014 | FIR | 307 | |
| | (0°) | MW | 310 | 0.022 ± 0.006 | MW | 310 | |
| Methylenecyclobutane | | | | 0.46 ± 0.11 | MW | 311 | |
| | | | | 0.48 ± 0.03 | MIR | 312 | |
| $\beta$-Propiolactone (23) | Planar | MW | 302 | | | | $V(z) \approx az^2$ |
| Oxetan-3-one (24) | Planar | FIR | 313 | | | | |
| Diketene (25) | Planar | FIR | 313 | | | | $V(z) \approx az^2$ |

ED = electron diffraction. FIR = far-infrared spectroscopy. MIR = mid-infrared spectroscopy. NIR = near-infrared spectroscopy. MW = microwave spectroscopy.

Table 10. Dihedral angle ($\beta$) in some four-membered-ring compounds.

|  | $\beta$ | Method | Ref. |
|---|---|---|---|
| *trans*-1,3-Dibromocyclobutane | 32° | ED | 314 |
| *trans*-1-Bromo-3-chlorocyclobutane | 33° | ED | 314 |
| *cis*-1,3-Dibromocyclobutane | 33° | ED | 314 |
| *cis*-1-Bromo-3-chlorocyclobutane | 33° | ED | 314 |
| Octafluorocyclobutane | 17° | ED | 315 |
| Bromocyclobutane | 29.4° | MW | 316 |
| Chlorocyclobutane | 20.0 ± 1° | MW | 317 |
| Bicyclobutane, $C_4H_7$–$C_4H_7$ | 33° | ED | 318 |
| Cyclobutaneoctaol | Planar | XD | 319 |
| 1,2,3,4-Tetracyanocyclobutane | Planar | XD | 320 |
| *cis*-1,3-Cyclobutanedicarboxylic acid | 31° | XD | 295 |
| *trans*-1,3-Cyclobutanedicarboxylic acid | Planar | XD | 293 |
| *trans*-1,3-Cyclobutanecarboxylic acid in | 25° | XD | 294 |
| $(Na^+{}_2C_4H_6(COO^-)_2)\cdot 2C_4H_6(COOH)_2$ |  |  |  |
| Cyclobutene | Planar | MW | 321 |
| *cis*-3,4-Dichlorocyclobutene | Planar | ED | 322 |
| 3,4-Dimethylenecyclobutene | Planar | ED | 323 |

ED = electron diffraction. XD = X-ray diffraction (solid). MW = microwave spectroscopy.

where $z$ is the ring-puckering vibrational coordinate. A negative $a$ value produces a barrier at the planar conformation.

Extensive microwave[296-298] and infrared[299, 300, 43] studies on oxetane (**20**) have led to the potential shown in Figure 21, where $z = 0$ corresponds to the planar form. The barrier at the planar conformation is so low that the ground-state vibrational level lies above it.

A similar study[301] of thietane (**21**) proved this molecule to be non-planar with an appreciable barrier at the planar form. The experimental data are well reproduced by a potential function of the type (20) with a negative quadratic term. A potential of the form

$$V(z) = az^2 + bz^4 + c \exp(-dz^2) \qquad \ldots(21)$$

has also been investigated for oxetane and thietane. The exponential term was thought[297] to be important for oxetane but was later found to be unnecessary for both compounds.

The conformations of these rings represent a compromise between bond-angle (Baeyer) strain, which favours a planar conformation, and torsional (Pitzer) strain, which favours non-planarity (cf. Section

IV-B). The barrier to inversion in thietane is larger than that in oxetane in spite of the fact that the barrier to internal rotation about a C–S bond is believed to be smaller than about a C–O bond. However, the longer C–S bonds and the smaller "natural" valence angle

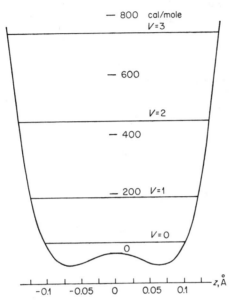

Figure 21. The ring-puckering potential and vibrational levels for oxetane (**20**).

of sulphur relieve the Baeyer strain in the non-planar conformation of thietane. $\beta$-Propiolactone (**23**), where the barriers to rotation about three of the bonds in the ring are reduced compared with cyclobutane, is planar with no detectable barrier.[302] However, the barrier in azetidine (**22**) seems to be larger than one would expect from similar considerations.[303]

$$\begin{array}{ccc} \text{H}_2\text{C}-\text{CH}_2 & \text{H}_2\text{C}-\text{CH}_2 & \text{H}_2\text{C}-\text{CH}_2 \\ | \quad\quad | & | \quad\quad | & | \quad\quad | \\ \text{H}_2\text{C}-\text{O} & \text{H}_2\text{C}-\text{S} & \text{H}_2\text{C}-\text{NH} \\ (\textbf{20}) & (\textbf{21}) & (\textbf{22}) \end{array}$$

(23)  $\text{H}_2\text{C}-\text{C}{\overset{\text{O}}{\diagup}}$ , $\text{H}_2\text{C}-\text{O}$

(24)  $\text{H}_2\text{C}-\text{C}{\overset{\text{O}}{\diagup}}$ , $\text{O}-\text{CH}_2$

(25)  $\text{H}_2\text{C}-\text{C}{\overset{\text{O}}{\diagup}}$ , $\text{C}-\text{O}$ , $\text{H}_2\text{C}{\overset{\diagup}{}}$

## 2. *Five-membered rings*

A description of five-membered rings requires in general two ring puckering coordinates. However, a five-membered ring with a double bond may be approximated by a one-dimensional treatment since one of the ring puckering coordinates is expected to be much higher in frequency than the other. Some results are given in Table 11. Both

Table 11. Five-membered rings with one double bond: dihedral angle and barrier at the planar conformation.

| Compound | Dihedral angle | Barrier (kcal/mole) | Method | Ref. |
|---|---|---|---|---|
| Cyclopentene (Figure 22) | 23.3° ± 1° | 0.66 ± 0.02 | FIR | 324 |
| | 22° | 0.66 ± 0.02 | MW | 44,45 |
| | 22° | 0.69 | NIR | 43 |
| Silacyclopent-3-ene (26) | Planar | | FIR | 325 |
| Silacyclopent-2-ene | Planar | | FIR | 326 |
| 3-Pyrroline (27) | | 0·16[a] | FIR | 303 |
| 2,5-Dihydrofuran (28) | Planar | | FIR | 313,327 |
| | Planar | | NIR | 43 |
| 2,3-Dihydrofuran (29) | 19° | 0.24 | FIR | 328 |
| | | 0.25 | NIR | 43 |
| 2,3-Dihydro-5-methylfuran | | 0.28 | FIR | 313 |
| 2,5-Dihydrothiophene | Planar | | FIR & MIR | 329 |
| | Planar | | NIR | 43 |
| 2,3-Dihydrothiophene | | 0.59 | NIR | 43 |
| 1,3-Dioxol-2-one (30) | Planar | | MW | 39 |
| 1,4-Dimethyl-$\Delta^2$-1,2,3,4,5-tetrazaboroline (31) | Planar | | ED | 330 |

[a] Unsymmetric potential; energy difference between the two minima 0.13 kcal/mole.

ED = electron diffraction. FIR = far-infrared spectroscopy. MIR = mid-infrared spectroscopy. NIR = near-infrared spectroscopy. MW = microwave spectroscopy.

microwave[44,45] and infrared studies[43,324] have shown that cyclopentene is non-planar with a considerable barrier at the planar form (Figure 22). If the torsional strain in the planar form of the ring is reduced, a planar form may be favoured. The difference between 2,5-dihydrofuran (28) and 2,5-dihydrothiophene which are planar, and the corresponding 2,3-dihydro compounds (*e.g.*, 29) which are non-planar, is probably attributable to changes in torsional strain (cf. Table 11).

Silacyclopent-3-ene (**26**) is also planar,[325] in accord with the expected reduction in torsional strain compared with cyclopentene. However, silacyclopent-2-ene [326] is also planar and unusually rigid. It has therefore been suggested that it is the interaction between silicon

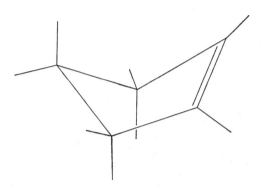

Figure 22. Structure of cyclopentene.

$d$-orbitals and the carbon–carbon $\pi$-bond that is the main cause of planarity in these molecules.

$$\underset{(26)}{\overset{\text{SiH}_2}{\underset{\text{HC}=\text{CH}}{\text{H}_2\text{C}\qquad\text{CH}_2}}}\qquad \underset{(27)}{\overset{\text{NH}}{\underset{\text{HC}=\text{CH}}{\text{H}_2\text{C}\qquad\text{CH}_2}}}\qquad \underset{(28)}{\overset{\text{O}}{\underset{\text{HC}=\text{CH}}{\text{H}_2\text{C}\qquad\text{CH}_2}}}$$

$$\underset{(29)}{\overset{\text{O}}{\underset{\text{H}_2\text{C}-\text{CH}}{\text{H}_2\text{C}\qquad\text{CH}}}}\qquad \underset{(30)}{\overset{\text{CO}}{\underset{\text{HC}=\text{CH}}{\text{O}\qquad\text{O}}}}\qquad \underset{(31)}{\overset{\text{BH}}{\underset{\text{N}=\text{N}}{\text{CH}_3-\text{N}\qquad\text{N}-\text{CH}_3}}}$$

Results for some saturated five-membered rings are given in Table 12. Pitzer and his co-workers [331, 332] showed that the energy minimum for cyclopentane cannot correspond to a planar ring. The highest possible symmetry for a non-planar five-membered ring is $C_2$ or $C_s$ (see Figure 23), and these conformations—as well as intermediate forms—were found to have nearly the same energy. This means that cyclopentane is not found in one well-defined conformation. The puckering moves around the ring (pseudorotation).

Table 12. Five-membered rings with no double bonds in the ring. (References in parentheses.)

| Compound | Structure determ. | Symmetry | Barrier to pseudorotation (kcal/mole) |
|---|---|---|---|
| Cyclopentane | ED (287,333) | — | ~0 (therm) (331,332); ~0 (IR) (345); ~0 (calc) (332,50,54); 0.24 (*ab initio* calc) (76) |
| Silacyclopentane (32) | ED (334,335) | $C_2$ | 3.89 (IR) (346,347); 3.56 (calc) (348) |
| Germacyclopentane | MW (336) | | > 1.5 (MW) (336); 5.9 (IR) (349); 4.5 (calc) (348) |
| Cyclopentanone (33) | MW (337) | $C_2$ | > 1.2 (MW) (337); 2.7 ± 0.8 (FIR) (350); 3.7 (FIR) (313) |
| Tetrahydrofuran (34) | ED (61,62,338) | — | 0.15 (FIR) (351), (MW) (352,353); 0.25 ± 1.0 (calc) (61) |
| Tetrahydrothiophene (35) | ED (339) | $C_2$ | 2.2 (FIR) (354); 2.8 (therm) (355,356); 2.5 ± 0.5 (calc) (339) |
| Tetrahydroselenophene | ED (340) | $C_2$ | 3.65 (calc) (340) |
| Pyrrolidine (36) | | | small (therm) (357) |
| 1,3-Dioxolane (37) | | | 0.15 (FIR) (351,358) |
| 1,2,4-Trioxacyclopentane (38) | ED (63) | $(C_2)$ | > 1.0 (calc) (63) |
| 1,3-Dioxolan-2-one (39) | MW (341) XD (342) | $C_2$ | |
| 1,3,2-Dioxaborolane (40) | MW (343) | $C_2$ | Very low barrier at the planar form |
| 1,2,4,3,5-Trioxadiborolane (41) | MW (344) | $C_{2v}$ | |

Therm = thermodynamic data. IR = infrared spectroscopy. FIR = far-infrared spectroscopy. MW = microwave spectroscopy. ED = electron diffraction. XD = X-ray diffraction. calc = calculated as described in Section IV-B.

The $z$-coordinates for the ring atoms may be given by:[331, 352]

$$z_j = r \cos\left(2\left(\frac{2\pi}{5}j + \varphi\right)\right) \qquad j = 0, 1, 2, 3, 4 \qquad \ldots(22)*$$

$r$ describes the degree of puckering, while variation of $\varphi$ corresponds to rotation of the puckering round the ring. The $C_s$ conformation corresponds to $\varphi = l\pi/10$, where $l$ is an integer ($\varphi = 0$ thus corresponds to a $C_s$ conformation) and the $C_2$ conformation corresponds to $\varphi = (2l + 1)\pi/20$. Adding $\pi/2$ to $\varphi$ changes the signs, but

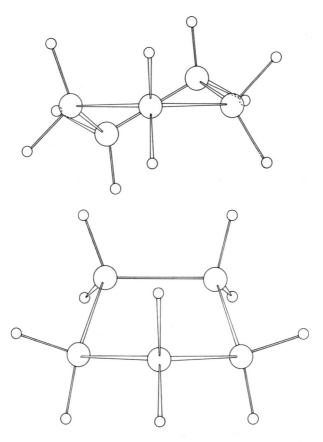

Figure 23. The "half-chair" ($C_2$) (a) and "envelope" ($C_s$) (b) forms of cyclopentane.

* The definitions of $\varphi$ in refs. 331 and 352 differ by a factor of 2; the definition (22) corresponds to ref. 331.

not the absolute values of the $z$-coordinates. An inversion of the molecule is thus possible without passing through any appreciable barrier.

Altona *et al.*[359, 360] have described the pseudorotation in five-membered rings by:

$$\varphi_j = \varphi_m \cos (\Delta/2 + j \cdot 4\pi/5), \qquad j = 0, 1, 2, 3, 4 \quad \ldots (23)$$

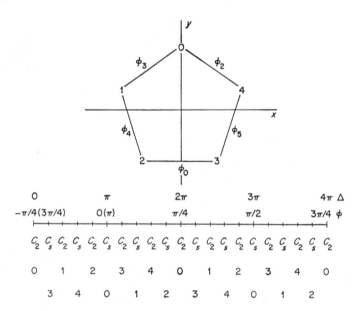

Figure 24. The numbering of the atoms and torsional angles in five-membered rings corresponding to the equations (22) and (23). The lower part of the Figure shows the conformations obtained for various values of $\varphi$ and $\Delta$. The number given below the symmetry symbol shows the atom through which the symmetry axis ($C_2$) or symmetry plane ($C_s$) pass.

The pseudorotation is thus described by the variable $\Delta$; $\varphi_j$ is the torsional angle about the $j$th bond in the ring, and $\varphi_m$ is a constant equal to the maximum possible torsional angle (see Figure 24).

Hendrickson[52] describes pseudorotation as the passage of a ring with a plane of symmetry through an atom to one with an axis of symmetry bisecting the bond adjacent to that atom and *vice versa*.

The pseudorotation in cyclopentane is nearly free, since the $C_2$ and $C_s$ conformations have very nearly the same energy. The potential energy depends in this case practically only on $r$, not on $\varphi$, as has been confirmed by infrared spectroscopy.[345] In other five-membered rings the potential may vary considerably with $\varphi$ (see Table 12).

Five-membered rings containing one or more heteroatoms usually have observably different energies in the $C_2$ and the $C_s$ conformations. Appreciable barriers to pseudorotation are found in both silacyclopentane (32) and germacyclopentane. The reason is mainly the

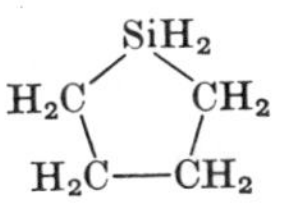

(32) 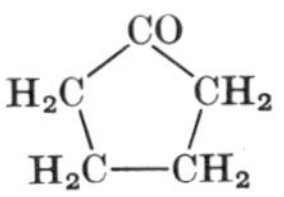        (33) 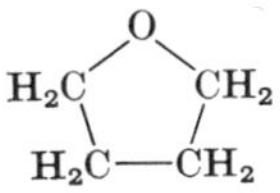        (34)

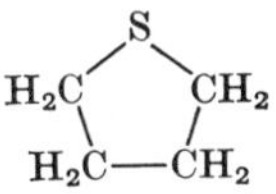

(35) 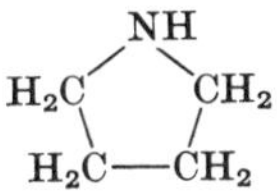        (36) 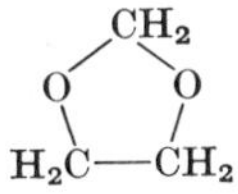        (37)

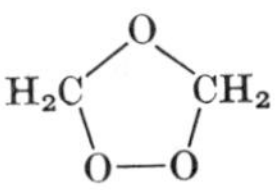

(38) 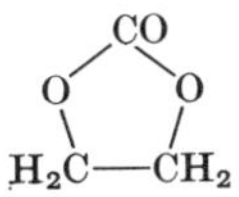        (39) 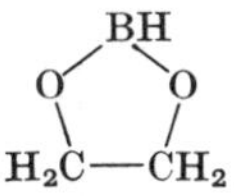        (40) 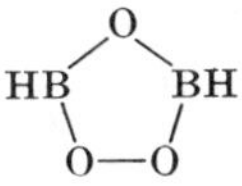        (41)

smaller barriers to rotation about Si–C and Ge–C than about C–C bonds. The difference in torsional strain in the $C_2$- and the $C_s$-conformations of silacyclopentane and germacyclopentane is thus appreciable, and it may be augmented by the greater Baeyer strain in the $C_s$-conformation (see Table 13).

Tetrahydrofuran (34) was predicted by Pitzer and Donath[332] to have a barrier to pseudorotation of about 2.5 kcal/mole. Lafferty et al.[361] found that the far-infrared spectrum of tetrahydrofuran gave evidence for nearly free pseudorotation, as has since been confirmed by further studies (Table 12). The potential found by Gwinn et al.[352, 353] is shown in Figure 25.

The main difficulty in calculating conformational energies of compounds containing oxygen is the assignment of a proper value for the barrier to internal rotation about C–O bonds. The results obtained for tetrahydrofuran by applying the barriers in methanol (1.07 kcal/mole) and in dimethyl ether (2.72 kcal/mole) are given in Table 13. The former value was applied by Pitzer and Donath and gives a too large preference for the $C_2$-form. Inclusion of the Baeyer strain, which was neglected by Pitzer and Donath, improves the agreement with

Table 13. Energy differences (kcal/mole) for $C_s$ and $C_2$ models of some five-membered rings. $E^t$ is the torsional energy, $E^a$ the bond-angle (Baeyer) strain, and $E$ the sum of these two terms and the van der Waals energy. Results for two different barriers about the C–X bonds (taken from the compounds given in parentheses) are reported for the three first compounds. The C–C bonds were assumed to have a barrier of 2.9 kcal/mole.

| | Barrier about C–X bonds | $E^t(C_s) - E^t(C_2)$ | $E^a(C_s) - E^a(C_2)$ | $E(C_s) - E(C_2)$ |
|---|---|---|---|---|
| Tetrahydrofuran [61] | 1.07 ($CH_3OH$) | 2.51 | −1.33 | 1.25 |
| | 2.70 ($CH_3OCH_3$) | −0.26 | −0.57 | −0.73 |
| Tetrahydrothiophene [339] | 1.27 ($CH_3SH$) | 2.90 | 0.09 | 3.03 |
| | 2.18 ($CH_3SCH_3$) | 1.39 | 0.56 | 1.96 |
| Tetrahydroselenophene [340] | 1.01 ($CH_3SeH$) | 3.37 | 0.52 | 3.92 |
| | 1.50 ($CH_3SeCH_3$) | 2.57 | 0.79 | 3.38 |
| Silacyclopentane [348] | 1.70 ($CH_3SiH_3$) | 3.10 | 0.45 | 3.56 |
| Germacyclopentane [348] | 1.2 ($CH_3GeH_3$) | 4.16 | 0.33 | 4.49 |

134

the experimental result (Figure 25). Applying the higher barrier one calculates a slight preference for the $C_s$-form. Pitzer and Donath predicted a barrier to pseudorotation of about 3 kcal/mole in tetra-hydrothiophene (**35**), which is in fairly good agreement with more recent estimates. The barrier in tetrahydroselenophene ($C_4H_8Se$) is probably still greater (see Tables 12 and 13).*

Electron-diffraction investigations have also demonstrated that the barrier is much smaller in tetrahydrofuran [61, 62, 338] than in tetra-hydrothiophene [339] and tetrahydroselenophene.[340] While least-

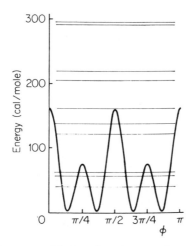

Figure 25. The potential function for pseudorotation in tetra-hydrofuran. (Cf. Figure 24.)

squares refinements of the molecular parameters based on a $C_2$ model converged easily to reasonable results for tetrahydrothiophene and tetrahydroselenophene, a similar procedure for tetrahydrofuran gave very unreasonable results for some of the parameters. The electron-diffraction data for tetrahydrofuran were consistent with a mixture of several conformations including $C_2$ and $C_s$ models.

## 3. Six-membered rings

The conformations of six-membered rings have been mentioned previously in this paper (Section III-A). Extensive discussions have also been given by others.[9, 363, 364]

* [Added in proof] A barrier of 5.4 kcal/mol has now been found in tetrahydro-selenophene.[433]

Like cyclohexane itself, substituted cyclohexanes and heterocyclic analogues prefer in general the chair conformation (see Figure 26). The conformation usually called the boat form does not correspond to a minimum in energy for cyclohexane, although it may do so for other six-membered rings in particular surroundings as, for example, in the complex $PdCl_2$–dimethylpiperazine.[365] Rather, the boat form

Figure 26. Chair (*a*), boat (*b*), and twist or skew boat (*c*) conformations of cyclohexane. The two types of C–H bond are indicated on the chair conformation.

and the skew-boat or twist form represent pseudorotation partners as the "envelope" and "half-chair" conformations of cyclopentane (see Figure 23). According to Hendrickson's descriptions of pseudorotation mentioned previously,[52] every ring with a plane or an axis of symmetry may pseudorotate, except for crown rings, which have simultaneously the alternation of axes and planes that occur in the pseudorotation partners of other rings. Thus the chair form of a six-membered ring cannot undergo pseudorotation.

Calculations of conformational energies by the Westheimer–Hendrickson method (see Section IV-B) have been carried out for six-membered and many larger rings. Hendrickson[51] found the twist form of cyclohexane to be 5.6 kcal/mole higher in energy than the chair form, while the boat form was 0.8 kcal/mole higher still, *i.e.*, 6.4 kcal/mole. The corresponding values found by Allinger *et al.*[54] are 4.9 and 6.6 kcal/mole. As mentioned in Section IV-C-1, *ab initio* calculations give 6–7 kcal/mole for the conformational energy of the twist form. Experimental values range between 4.8 and 5.9 kcal/mole.[9]

The cyclohexane derivatives, 1,4-cyclohexanedione[366–368] and its dioxime,[369] were found to occur in the twist form in the crystalline state (see Figure 26). A recent electron-diffraction investigation has shown that the same is true for the gaseous 1,4-dione.[370] The angle between the two C–O bonds was found to be 144° in the gas phase and about 154° in the crystalline state. The deviation from 180° explains the considerable dipole moment of the compound. From nmr evidence it has been suggested that 1,4-dimethylenecyclohexane also exists in the twist form.[371, 372] Calculations indicate slightly lower enthalpy for the chair conformations of both 1,4-cyclohexanedione[56] and of 1,4-dimethylenecyclohexane,[55] but the flexible twist forms may be favoured by entropy considerations. It is also possible that *cis*-1,4-di-*tert*-butylcyclohexane exists in the twist form. In the chair conformation one of the *tert*-butyl groups has to be axial, and the conformational energy of axial *tert*-butyl is quite large[54, 373] (calculated 5.4 kcal/mole). An electron-diffraction investigation[374] showed that agreement could be obtained for the twist form, but the chair form could not be ruled out.

Six-membered rings in polycyclic molecules may also, of course, exist in the twist form; twistane (**42**) has no other choice.[375]

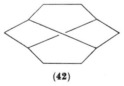

**(42)**

The CCC angles in cyclohexane are about 111.5°.[376]* No significant deviation from this value has been found in the derivatives studied by electron diffraction as listed in Table 14, except in *cis*-1,4-di-*tert*-butylcyclohexane (115.8°). The CCC angles in dodecafluorocyclohexane may also be somewhat larger than in cyclohexane.

The six-membered heterocyclic compounds listed in Table 14 are also most stable in the chair conformation. However, 3,3,6,6-

---

* [Added in proof] A slightly smaller value for the CCC angle (111.05°) has recently been reported.[434]

## Table 14. Six-membered rings studied in the vapour phase.

| Compound | Conformation | Remarks; figures in parentheses are reference numbers |
|---|---|---|
| Cyclohexane (Figure 26) | Chair | $\angle CCC = 111.55 \pm 0.15°$, $r(C–C) = 1.528 \pm 0.005$ Å, ED (376,377) |
| Dodecafluorocyclohexane | Chair | $\angle CCC = 112.6 \pm 0.34°$, $r(C–C) = 1.551$ Å, ED (378) |
| Fluorocyclohexane | Chair | $\angle CCC = 111.5°$, ED (379) |
| Chlorocyclohexane | Chair | $\angle CCC = 111.5°$, ED (380) |
| Cyclohexanethiol | Chair | ED (377) |
| trans-1,2-Dichlorocyclohexane | Chair | ED (381) |
| trans-1,4-Dichlorocyclohexane | Chair | $\angle CCC = 111.5°$, ED (382) |
| trans-1,2-Dibromocyclohexane | Chair | ED (17) |
| trans-1,4-Dibromocyclohexane | Chair | $\angle CCC = 111.5°$, ED (382) |
| 1,4-Cyclohexanedione | Twist | $\angle C_2C_1C_6 = 114.6°$, $\angle C_1C_2C_3 = 112.7°$, ED (370) |
|  | Twist | $\angle C_2C_1C_6 = 117.4°$, $(\angle C_1C_2C_3)_{av} = 111.4°$, XD (367) |
| cis-1,4-Di-tert-butylcyclohexane | ? Twist | $\angle CCC = 115.8°$, ED (374) |
| 1,3,5-Trisilacyclohexane | Chair | $\angle CSiC = \angle SiCSi = 109.5°$, $r(C–Si) = 1.86$ Å, ED (388) |
| 1,4-Dioxane (43) | Chair | $\angle CCO = 109.2 \pm 0.5°$, $\angle COC = 112.45 \pm 0.5°$, $r(C–C) = 1.523 \pm 0.005$ Å, $r(C–O) = 1.423 \pm 0.003$ Å, ED (376) |
| 1,4-Dithiane (44) | Chair | ED (377) |

Table 14 (continued)

| Compound | Conformation | Remarks; figures in parentheses are reference numbers |
|---|---|---|
| Piperazine (45) | Chair | $\angle$CCN = 109.8 ± 0.5°, $\angle$CNC = 112.6 ± 0.5°, $r$(C–C) = 1.527 ± 0.005 Å, $r$(C–N) = 1.471 ± 0.005 Å, ED (376) |
| 1,4-Dimethylpiperazine | Chair | $\angle$CCN = 110.3 ± 1°, $\angle$(CNC)$_{ring}$ = 114.4 ± 1°, ED (376) |
| s-Trioxane (46) | Chair | $\angle$COC = 108.2°, $\angle$OCO = 112.2°, $r$(C–O) = 1.411 Å, MW (389) |
| | Chair | $\angle$COC = 109.2°, $\angle$OCO = 111.0°, $r$(C–O) = 1.411 Å, MW (390) |
| s-Trithiane (47) | Chair | $r$(C–S) = 1.81 Å, ED (377) |
| | Chair | $\angle$CSC = 98.9 ± 0.6°, $\angle$SCS = 114.7 ± 0.7°, $r$(C–S) = 1.814 ± 0.009 Å, XD (391) |
| Cyclohexene (Figure 27) | $C_2$, half-chair | ED (392), MW (393) |
| Tetrachlorocyclohexene | $C_2$, half-chair | ED (394) |
| 1,2-Epoxycyclohexane (49) | $C_2$, half-chair | ED (395,396) |
| 1,3-Cyclohexadiene (50) | $C_2$ | $\varphi$(C$_2$–C$_3$) ~ 18°, ED (397,398,385), MW (399) |
| 1,4-Cyclohexadiene (51) | $C_{2v}$ | Dihedral angle $\beta$ = 20.7°, ED (385), $\beta$ = 15°, NMR (387), $\beta$ = 0°, ED (386) |
| | $D_{2h}$, planar | |

139

ED = electron diffraction. XD = X-ray diffraction. MW = microwave spectroscopy.

tetramethyl-1,2,4,5-tetrathiane (**48**) seems to be more stable in the twist form than in the chair form.[383]

Cyclohexene is known to be most stable in the "half-chair" conformation with $C_2$ symmetry (see Figure 27 and Table 14). Figure 27

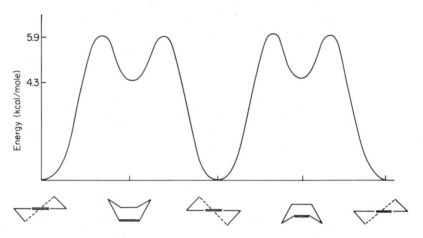

Figure 27. Possible energy profile for the interconversion of the "half-chair" conformations of cyclohexene.

shows the energy profile calculated by Allinger *et al.*[55] for interconversion of the "half-chair" conformations. This curve has maxima for unsymmetric forms and local minima for the boat forms, but the shape of the actual potential is still a matter of controversy.[384]

1,3-Cyclohexadiene (**50**) has been studied by several authors. The two double-bond systems are probably planar, their planes making an angle of about 18°. For 1,4-cyclohexadiene (**51**) (see Table 14) it is difficult to distinguish between a planar model with large out-of-plane oscillations, and a non-planar model. Oberhammer and Bauer[385] report a dihedral angle of 21°, while Dallinga and Toneman[386] favour a planar ring. A dihedral angle of about 15° has been reported from an nmr investigation.[387]

### 4. *Rings with more than six atoms*

Compounds with seven-membered and larger rings studied by electron diffraction are listed in Table 15. Investigation of such compounds in the gaseous state is often extremely complicated. The number of structural parameters becomes large, especially when the root-mean-square amplitudes are included. The possibility that the gas is a mixture of several conformations complicates the problem further. Sometimes the only conclusion that can be drawn is that no single conformation gives satisfactory agreement between experiment and theory. Cyclooctane is such a case.[405] One has to assume highly flexible molecular models where at least one structure parameter must be varied over a wide range in order to obtain acceptable agreement between theoretical and experimental radial distribution curves. A fairly large number of compounds containing medium rings has been studied by X-ray diffraction, and the conformations of these rings have been reviewed recently by Dunitz.[66]

The 1,3,5-cycloheptatriene (**52**) molecule is non-planar, being considerably twisted about two of the double bonds ( ~ 23°). As discussed in Section IV-B, it is not quite clear how much energy is required to

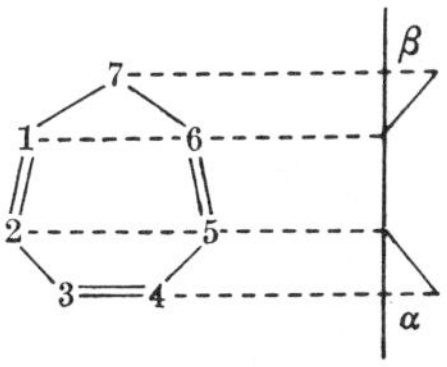

(**52**) 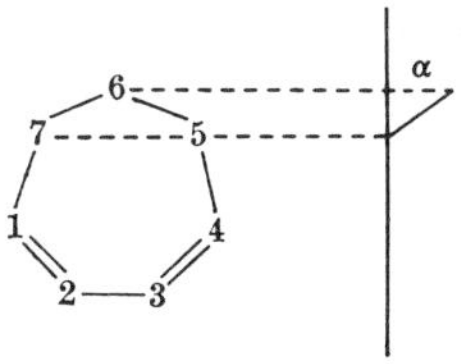          (**53**) 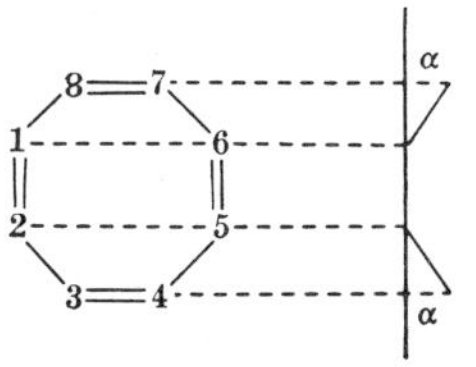          (**54**)

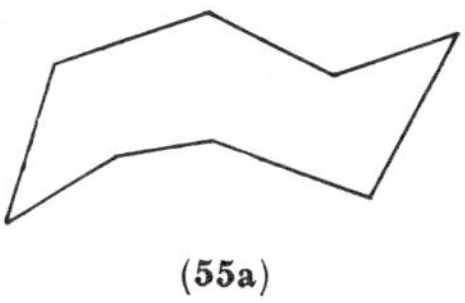

(**55a**) 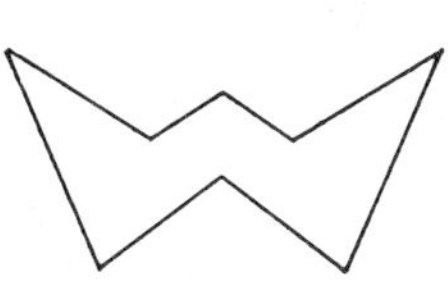          (**55b**)

Table 15. Cyclic molecules with more than six atoms in the ring studied by electron diffraction.

| Compound | Properties | | | | | | | | | | Ref. |
|---|---|---|---|---|---|---|---|---|---|---|---|
| Cycloheptatriene (**52**) | $r$: | 1.356, | 1.446, | 1.356, | 1.446, | 1.356, | 1.505, | 1.505 | | | 400 |
| | $\varphi$: | 23°, | −54°, | 0°, | 54°, | −23°, | −42°, | 42° | | | |
| | $\alpha = 40.5°$, $\beta = 36.5°$ | | | | | | | | | | |
| 1,3-Cycloheptadiene (**53**) | $r$: | 1.35, | 1.48, | 1.35, | 1.54, | 1.55, | 1.55, | 1.54 | | | 401 |
| | $\varphi$: | 0°, | 0°, | 0°, | 55°, | −84°, | +84°, | −55° | | | |
| Cyclooctatetraene (**54**) | $r$: | 1.340, | 1.476, | 1.340, | 1.476, | 1.340, | 1.476, | 1.340, | 1.476 | | 173,402 |
| | $\varphi$: | 0°, | +58-, | 0°, | −58°, | 0°, | +58°, | 0°, | −58° | | |
| 1,3-Cyclooctadiene (Figure 28) | $r$: | 1.347, | 1.475, | 1.347, | 1.509, | 1.542, | 1.542, | 1.509 | | | 403 |
| | $\varphi$: | 0°, | +38°, | 0°, | +18°, | −75°, | +78°, | +32°, | −80° | | |
| 1,5-Cyclooctadiene (Figure 29) | $r$: | 1.339, | 1.508, | 1.548, | 1.508, | 1.339, | 1.508, | 1.548, | 1.508 | | 404 |
| | $\varphi$: | −14°, | +33°, | +52°, | −78°, | −14°, | +33°, | +52°, | −78° | | |
| Cyclooctane | Mixture of several conformers | | | | | | | | | | 405 |
| cis,cis-1,6-Cyclododecadiene (Figure 30) | $r$: | 1.326, | 1.506, | 1.534, | 1.534, | 1.506, | 1.326, | 1.506, | 1.534, | 1.534, | 1.506 | 406 |
| | $\varphi$: | 0°, | 115°, | −58°, | −58°, | 115°, | 0°, | −115°, | 58°, | 58°, | −115° | |
| 1,8-Cyclotetradecadiyne | Mixture of several conformers | | | | | | | | | | 405 |

achieve this twist. Application of the expressions (13) and (14) in Section IV-B gives approximately 9 and 5 kcal/mole, respectively, for each double bond. Even the lower estimate seems too high if the observed twist angle is correct.

1,3-Cycloheptadiene (**53**) seems to be planar, except for the central methylene carbon atom.* The structure may be compared with those of 1,3-cyclohexadiene (**50**) (Table 14) and 1,3-cyclooctadiene (Table 15, Figure 28). The determination of the conformations of these

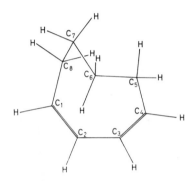

Figure 28. The proposed conformation of 1,3-cyclooctadiene.

seven- and eight-membered rings is on the limit of what can be achieved by the present electron-diffraction technique. It is possible to obtain satisfactory agreement between experimental and theoretical intensities by assuming that the molecule is effectively in one conformation. However, a considerable proportion of other conformations may be present without altering the agreement significantly. The torsional angles around the $=\!C\!-\!C\!=$ bonds are given as 18°, 0°, and 38° in the six-, seven-, and eight-membered rings, respectively. The corresponding angles in cyclooctatetraene are about 58° and in cycloheptatriene about 54°. The corresponding bond lengths are fairly close to the length of the central bond in butadiene (1.465 Å).[173, 174]

1,5-Cyclooctadiene occurs in the gas phase mainly in the conformation with $C_2$ symmetry, illustrated in Figure 29. About 20–30%

---

* This has been confirmed by a new electron-diffraction investigation.[435] However, for some of the molecular parameters the new results deviate considerably from the old ones.

of the alternative form with $C_{2h}$ symmetry may be present.[404] Eclipsing around the C-3/C-4 and C-7/C-8 bonds occurs in both the $C_{2h}$ and the $C_{2v}$ conformation, but the $C_{2v}$ form is fairly flexible and the torsional strain can be reduced by twisting to a form with only $C_2$ symmetry. Calculations[59] indicate that the enthalpy of the $C_2$ form

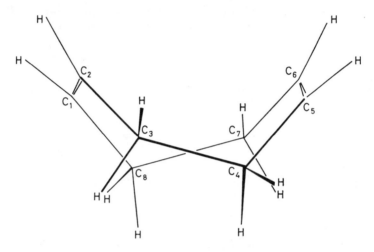

Figure 29. The most stable conformation of 1,5-cyclooctadiene.

is indeed somewhat lower than that of the $C_{2h}$ form, but even if the enthalpies of the two conformations were equal, the flexible $C_2$ form would be favoured by entropy considerations. *cis,cis*-1,6-Cyclo-decadiene occurs at least mainly in the form with $C_{2h}$ symmetry, which is free from torsional strain (Figure 30).

As mentioned in Section IV-B extensive energy calculations have been carried out with considerable success on saturated ring compounds, for example, by Hendrickson,[50, 51] Bixon and Lifson,[57] and by Allinger *et al.*[54] For seven-, eight-, and nine-membered rings the calculations indicate only a small energy difference between the most stable conformation and the form next lowest in energy. For cyclooctane the "boat-chair" conformation (**55a**) is predicted to be the most stable, in agreement with X-ray investigations (cf. ref. (66)). However, Hendrickson[51] predicts six other conformations within 2.0 kcal/mole of the boat-chair. The crown form (**55b**), with a four-fold symmetry axis, was estimated to be 2.8 kcal/mole higher than the boat-chair.

Similar calculations on cyclodecane indicate that the energy difference between the most stable conformation (boat-chair-boat) and the next most stable is somewhat greater than for cyclooctane. Energies 2.1 and 3.1 kcal/mole above the stable form for two conformers with $C_2$ symmetry have been obtained.[407] Preliminary electron-diffraction results[408] indicate that it is possible to obtain satisfactory agreement between experimental and theoretical curves on the basis of just one conformation.

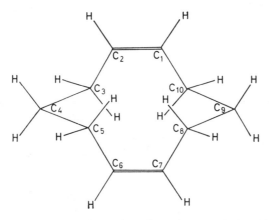

Figure 30. The conformation of *cis,cis*-1,6-cyclodecadiene.

## C. Conformations of Substituted Cyclic Molecules

The puckering of the ring skeleton of cyclic hydrocarbons leads to non-equivalence of the two hydrogen atoms at the same methylene group. The significance of this simple fact was realized by Hassel,[409] who studied cyclohexane and its derivatives. The geometrically different structure models of cyclohexane had been described by Sachse[2] already in 1892. Once the chair form was recognized as the basic ring skeleton of cyclohexane, it became obvious that there were two different kinds of hydrogen atoms, named by Hassel the $\kappa$ and $\epsilon$ hydrogens, now known as the $e$ and $a$ hydrogens[410] (Figure 26). The fact that substitution of one hydrogen atom by another atom or group never results in more than one isomer led to the idea of ring inversion. By the ring inversion an $e$ atom is transferred to an $a$ position and *vice versa*. The barrier to ring inversion is sufficiently small to prevent isolation of either of the two forms. In a mono-substituted cyclohexane the $e$ conformer was found to be the more stable,[411, 17, 379, 380, 412–414] and the lower stability of the *a* conformer was ascribed to the steric interference of the substituent with the 3*a*-hydrogen atoms. In contrast, a preference for the axial position is found for halogen substituents $\alpha$ to oxygen and sulphur in heterocyclic compounds.[364]

Unfortunately experimental gas-phase data for the conformational energy of monosubstituted cyclohexane derivatives are scarce and uncertain. For fluorocyclohexane electron-diffraction studies lead to a value for the conformational energy of 170 cal/mole in favour of the *e*-conformer,[379] while microwave data suggest 400 cal/mole with a large error range of approximately 300 cal/mole.[414] For chlorocyclohexane, electron-diffraction data[380] suggest that the conformational energy is less than 260 cal/mole, while two infrared investigations suggest 340 cal/mole[412] and 520 cal/mole,[413] respectively.

In contrast to the sparsity of gas-phase conformational energy measurements, there exists a vast number of measurements in solution. A compilation, by J. A. Hirsch,[373] of conformational energy data up to 1967 demonstrates that the solvent effect on monosubstituted cyclohexanes must be small. The dipole moment ought to be very much the same for the two conformers, and the only selective effect that might be expected would be a slight extra preference for the *e*-substituent which is more easily available for solvent engagement than the sterically more protected *a* substituent. However, such an effect obtains no support from the measured data, at least if non-polar solvents are used. Systematic attempts to demonstrate a solvent effect were unsuccessful.[415] The "best values" recommended by Hirsch for monosubstituted cyclohexanes may therefore probably be considered as useful data also for gas molecules.

A *cis*-1,2-disubstituted cyclohexane with equal substituents can exist in principle as two enantiomorphous conformers which are, however, rapidly interconverted by ring inversion (Figure 31). The corresponding *trans*-1,2-compound is separable into optically active antipodes, from each of which two geometrically different conformers are obtained by ring inversion (Figure 32). *trans*-1,2-Dibromocyclohexane was the first molecule for which the coexistence of two different conformers in the gas phase was proved experimentally.[17] The electron-diffraction study indicated that the mole ratio of *ee* to *aa* was about 1.5, suggesting that the *ee*-conformer was the more stable by 200–300 cal/mole. Infrared data lead to a value of 610 cal/mole for *trans*-1,2-dichlorocyclohexane.[412] Again for disubstituted cyclohexanes, gas-phase data are sparse and uncertain. Unfor-

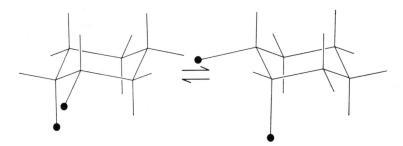

Figure 31. *cis*-1,2-Disubstituted cyclohexane: equilibrium
between two optical antipodes.

tunately, information obtained from pure liquids or solutions is of
little value for gas-phase estimates, as the solvent effect for these
molecules is far from negligible.[416, 417] The dipole moments of the
two coexisting conformers may be rather different: the *ee*-conformer
may have a substantial dipole moment, while the *aa*-conformer must
have a dipole moment close to zero. The solvent effect may be
exemplified by three results for *trans*-1,2-dichlorocyclohexane obtained by Kozima *et al.*[412] from infrared measurements. They find
conformational energies in vapour, in $CS_2$ solution, and in methyl
acetate solution of 610, 170, and $-430$ cal/mole, respectively. This
means that a polar solvent may even reverse the usual stability order.
It is worth noting that *trans*-1,2-dichlorocyclohexane and the corresponding bromo compound crystallize in different conformations.
The former crystallizes as an *ee*- and the latter as an *aa*-conformer.[417]

In the case of the corresponding 1,3-disubstituted cyclohexanes

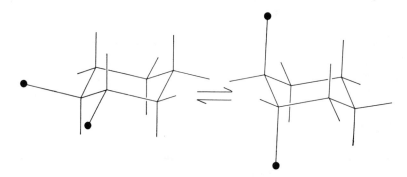

Figure 32. *trans*-1,2-Disubstituted cyclohexane: equilibrium between the *ee* and *aa* conformers.

each optically active antipode of the *trans*-isomer can occur only as a single chiral conformer—ring inversion in this case does not alter the sense of chirality. The *cis*-isomer should offer the possibility of equilibrium between the two conformers *ee* and *aa*. Of these, *aa* is definitely less favourable energetically because of repulsion between the two substituents.

For the 1,4-disubstituted cyclohexanes the *cis*-isomer (*ea*) inverts into an identical molecular species while the *trans*-isomer exists as an equilibrium mixture of *ee*- and *aa*-conformers. Electron-diffraction studies[382] show that for both *trans*-1,4-dibromocyclohexane and *trans*-1,4-dichlorocyclohexane the two conformers (*aa* and *ee*) coexist in a practically 1:1 mixture, corresponding to an energy difference close to zero (Figure 5). An error estimate suggests that the energy difference is less than 170 cal/mole. The *trans*-1,4-dihalocyclohexanes are particularly well suited for energy-difference measurements by electron-diffraction, since the experimental data are rather sensitive to changes in mole ratios for these molecules. Therefore, if high accuracy in energy differences is aimed at, these molecules should be re-examined by modern electron-diffraction technique.

Cyclohexane derivatives with more than two halogen substituents do not introduce any new principles.[418, 419] For molecules with only one kind of substituent the conformer with the smaller number in axial positions will be the more stable. For molecules with more than one kind of substituent it is not obvious which of two possible conformers is the more stable. For example, the 1,2-dichloro-4,5-dibromocyclohexane shown in Figure 33 has two possible conformers *aa,ee* and *ee,aa*; only the former (Figure 33a) exists in the crystalline phase,[409] and the same conformer is the predominating one[420] in the gas phase. From these and similar examples qualitative rules concerning the relative stability of coexisting conformers may be deduced. One may argue, for example, that the bromine atom, as the bulkier substituent, ought to be decisive for the relative conformational stability. Further, it seems that the uncomfortably short 4*e*,5*e*-Br···Br distance will introduce less steric strain than would two bromine atoms in axial positions. The steric strain leads to deviations from the ideal valency angles and torsional angles of cyclohexane itself. The *a*-C–Cl bonds are bent away from the principal axis of the ring by approximately 8°, and the *e*-C–Br bonds are also bent away from each other. Such deviations from ideality are not uncommon in cyclohexane derivatives, in the gas phase or in the crystal.[421]

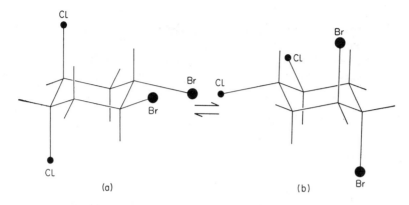

Figure 33. Equilibrium between two conformers of 1,2-dichloro-
4,5-dibromocyclohexane.

The decalins may be considered as 1,2 derivatives of cyclohexane,
*trans*-decalin as an *ee*-derivative and *cis*-decalin as an *ea*-derivative
(Figure 34).[15, 16] For *trans*-decalin ring inversion cannot take place,
for if one of the rings is inverted the other would have to be formed
through two axial bonds, which is impossible. This means that *trans*-
decalin has only one conformer. *cis*-Decalin may invert if the two
rings invert simultaneously, but this inversion only converts the
molecule into its optical antipode. While *trans*-decalin has no more
steric strain than cyclohexane itself, *cis*-decalin is more overcrowded.

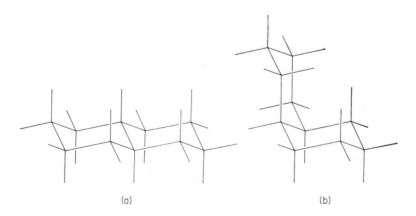

Figure 34. (a) *trans*- and (b) *cis*-Decalin.

In particular, two pairs of hydrogen atoms would come rather close together in an "ideal" structure. This leads to a deformation that distributes itself over the whole carbon skeleton.

Molecules in which 1,3-diaxial substitution cannot be avoided by ring inversion should therefore be highly strained. Examples of this can be found in the 1,2,3,4,5,6-hexachlorocyclohexane series. Of the eight theoretically possible isomers, the first five isolated were studied by electron diffraction[422] and their structures established. The isomers with no, one, or two axial halogens must exist essentially as single conformers in all states of aggregation, since the conformers with four or more axial halogens would certainly be rather unfavourable alternatives. Of the three possible triaxial isomers, two (*aaaeee* and *aeaeae*) have mirror symmetry and invert into identical conformers; the third (*aaeaee*) converts into its optical antipode. To the authors' knowledge the most symmetrical of the triaxial isomers (*aeaeae* with $C_{3v}$ symmetry) has so far not been isolated. This molecule would have three unfavourable $1a$–$3a$ Cl–Cl distances and accordingly should be quite unstable. On the other hand, it is interesting that all eight isomers of 1,2,3,4,5,6-hexahydroxycyclohexane (inositol) are known.[423] Possibly, intramolecular hydrogen bonding stabilizes the inositol molecules in a way analogous to that observed for ethylene glycol and glycerol[13, 266] (see Section V-C). For example, in ethylene glycol, hydrogen bonding seems to stabilize the *gauche* conformer relative to the *anti* conformer. Two neighbouring hydroxyl groups in *ee*- or *ea*-positions are arranged in the same way as two hydroxyl groups in a *gauche* conformer of ethylene glycol. An *aa* arrangement of two neighbouring hydroxyl groups in inositol corresponds to the less favourable *anti* conformer of ethylene glycol. For a triaxial isomer of inositol, as the *eaeaea* isomer, axial hydroxyl groups in 1,3-positions may also get into hydrogen-bonding contact. The steric strain may thus be partially relieved by hydrogen-bond formation.

The structures of 3,4,5,6-tetrachlorocyclohexenes were studied by dipole-moment measurements[424] and electron diffraction.[394] Although the double bond causes four of the carbon atoms to be at least nearly coplanar, it is convenient to retain the designation used for cyclohexane derivatives. Of the six possible isomers, five have been isolated.[425] The missing isomer appears to be the one closest associated with the missing 1,2,3,4,5,6-hexachlorocyclohexane isomer, namely the *aeae*-form. The δ, γ, and ε isomers, which are identified as the *eeee*, *aeee*, and *eaee* isomers, respectively, have all, as expected, only one conformer. The α isomer (*aaee*) inverts into its optical antipode. The only isomer with two geometrically different coexisting conformers, is the $\beta$ isomer characterized by the equilibrium *aeea* $\rightleftarrows$ *eaae*.

The conformational problems of substituted six-membered rings with heteroatoms have been discussed in a recent review by Riddell.[363] If the ring contains a heteroatom (X) with one lone pair and one hydrogen atom, a preference for hydrogen in axial position has been observed for X = P, $S^+$, $Se^+$, and $Te^+$ (see ref. 426 for examples and references). The situation is not so clear when X = N.[363, 426] Energy calculations for piperidine of the type discussed in Section IV-B [427] gave lowest energy for hydrogen in axial position.

While a large number of six-membered ring derivatives has been studied in the gas phase, gas-phase conformation studies of other ring derivatives are less abundant. For five-membered ring derivatives and for derivatives of rings larger than six, the conformational picture is less clear than for cyclohexane derivatives, since these rings usually exhibit more free internal motion than do the cyclohexanes. Rather contradictory results have been reported for halocyclopentanes. Recent spectroscopic studies indicate that the barrier to pseudorotation is considerably larger than in cyclopentane,[428] though estimates by the energy-calculation method described in Section IV-B seem to give a fairly small barrier. A preference for the halogen in the axial position, presumably on the "flap" of an envelope conformation, has been reported from a study of the carbon–halogen stretching frequencies:[429] the energy difference for the chloro compound was given as 0.7 $\pm$ 0.3 kcal/mole in $CS_2$ solution.

Derivatives of cyclobutane are more rigid and can be studied more easily. Four 1,3-dihalocyclobutanes have been studied in the gas phase by electron diffraction,[314] namely, *trans*-1,3-dibromocyclobutane, *trans*-1-chloro-3-bromobutane, *cis*-1,3-dibromobutane, and *cis*-1-chloro-3-bromobutane. Because of the non-planarity of the cyclobutane molecule (see Table 10) the two hydrogen atoms bonded to each carbon atom are non-equivalent. Even if the conditions are somewhat different from the cyclohexanes, it is still convenient to use the notation *e* and *a* also for four-membered rings. At least, this notation seems to leave no ambiguity in practice (Figure 20).

The *trans*-1,3-dibromocyclobutane has only one conformer, the *ea*, that inverts into itself. *trans*-1-Bromo-3-chlorocyclobutane should exhibit two conformers, one having the bromine in *a*-position, the other with the chlorine in *a*-position. The theoretical radial distribution curves for these two conformers are very similar. Only in a limited range of the radial distribution curve, around 3.5 Å, is there a difference. The experimental curve shows clearly that both conformers are present in the gas, and detailed analysis leads to a mixture with 60% of the Br(*a*)Cl(*e*) conformer and 40% of the Br(*e*)Cl(*a*) conformer, corresponding to an energy difference of about 250 cal/mole.

For both *cis*-1,3-dibromocyclobutane and for *cis*-1-bromo-3-chloro-cyclobutane two conformers are possible (*ee* and *aa*). The experimental radial distribution curves for both compounds are in excellent agreement with the corresponding theoretical curve for the *ee*-conformer. A liberal error estimate suggests that the *ee*-conformer must be at least 1.5 kcal/mole more stable than the *aa*-conformer.

## VII. CONCLUDING REMARKS

The activity in the field of gas-phase conformational analysis is both quantitatively and qualitatively more impressive than the present authors imagined at the start of work on this Review. The list of references grew during its preparation beyond expectation. Because of the vast number of papers in the field the Review naturally suffers from a certain arbitrariness in the choice of problems and examples. The authors apologize for possible oversights, as they do for unintentional errors or misjudgements.

Not only is the field of conformational analysis a comprehensive one, it is also a field in rapid growth, which makes prediction particularly difficult. The future of the field depends on the development of the experimental methods and on the possibility of more exact theoretical calculations. The experimental methods that have been in use in conformational analysis in gases were developed for the solution of general structural problems; they are now being developed further in many laboratories, specifically for the solution of conformational problems. For the study of conformation analysis in liquids nmr has appeared very useful, but it has only occasionally been used in the gas phase:[430] development of the nmr gas technique may bring new information of great value.

Much attention is presently being paid to the theoretical side. In spite of the results of semiempirical calculations having often been useful they have occasionally been misleading. The introduction of *ab initio* calculations therefore certainly calls for attention, and the exciting results already obtained may indicate that to a greater extent now than in the past the future of conformational analysis lies in the hands of the theoreticians.

*Acknowledgements:* at the initial stage of work on this Review, Dr. Per Andersen at the University of Oslo made a valuable contribution by collecting literature data. The authors are very grateful for that help.

The Oslo authors (OB and HMS) thank the Norwegian Research Council for Science and the Humanities for financial support. Likewise the Austin author (JEB) thanks the Robert A. Welch Foundation for support.

One of the authors (OB) spent the academic year 1968–1969 as Visiting Professor at the Physics Department of the University of Texas at Austin, during which period this work was started. He expresses his gratitude to colleagues there for hospitality and support, and particularly he thanks the then Chairman of Physics Department, Professor Harold P. Hanson.

# References

1. E. L. Eliel, *Science*, **166**, 718 (1969).
2. H. Sachse, *Z. Phys. Chem.*, **10**, 203 (1892).
3. O. Hassel, *Z. Electrochem.*, **36**, 735 (1930).
4. K. L. Wolf, *Trans. Faraday Soc.*, **26**, 315 (1930).
5. A. Weissberger and R. Sängewald, *Z. Phys. Chem.*, **B9**, 133 (1930).
6. P. Debye, *Phys. Z.*, **31**, 142 (1930).
7. R. Wierl, *Phys. Z.*, **31**, 366 (1930).
8. S. Mizushima, *Structure of Molecules and Internal Rotation*, Academic Press, Inc., New York, 1954.
9. E. L. Eliel, N. L. Allinger, S. J. Angyal, and G. A. Morrison, *Conformational Analysis*, Interscience Publ., New York, 1965.
10. D. H. R. Barton, *Experientia*, **6**, 316 (1950).
11. H. Ll. Bassett, N. Thorne, and C. L. Young, *J. Chem. Soc.*, **1949**, 85.
12. G. Wood and P. E. Woo, *Can. J. Chem.*, **45**, 2477 (1967).
13. O. Bastiansen, *Om noen av de forhold som hindrer den fri dreibarhet om en enkeltbinding*, A. Garnæs' boktrykkeri, Bergen **1948** (in Norwegian with English summary).
14. O. Hassel and B. Ottar, *Arch. Math. Naturvidensk.*, **B XLV,** Nr. 10 (1942).
15. O. Bastiansen and O. Hassel, *Nature*, **157**, 765 (1946).
16. O. Bastiansen and O. Hassel, *Tidsskr. Kjemi, Bergv. Met.*, **6**, 70 (1946).
17. O. Bastiansen and O. Hassel, *Tidsskr. Kjemi, Bergv. Met.*, **6**, 96 (1946).
18. O. Bastiansen and P. N. Skancke, *Advan. Chem. Phys.*, **3**, 323 (1961).
19. A. Almenningen, O. Bastiansen, A. Haaland, and H. M. Seip, *Angew. Chem.*, **77**, 877 (1965); *Angew. Chem. Int. Ed., Engl.*, **4**, 819 (1965).
20. B. Andersen, H. M. Seip, T. G. Strand, and R. Stølevik, *Acta Chem. Scand.*, **23**, 3224 (1969).
21. R. Glauber and V. Schomaker, *Phys. Rev.*, **89**, 667 (1953).
22. H. M. Seip, "Studies on the Failure of the First Born Approximation in Electron Diffraction," in P. Andersen, O. Bastiansen, and S. Furberg,

eds., *Selected Topics in Structure Chemistry*, Universitetsforlaget, Oslo, 1967, p. 25.

23. H. M. Seip, *Acta Chem. Scand.*, **19**, 1955 (1965).
24. A. Almenningen, O. Bastiansen, R. Seip, and H. M. Seip, *Acta Chem. Scand.*, **18**, 2115 (1964).
25. A. Almenningen, S. P. Arnesen, O. Bastiansen, H. M. Seip, and R. Seip, *Chem. Phys. Lett.*, **1**, 569 (1968).
26. R. R. Ryan and K. Hedberg, *J. Chem. Phys.*, **50**, 4986 (1969).
27. Y. Morino and E. Hirota, *J. Chem. Phys.*, **28**, 185 (1958).
28. S. J. Cyvin, I. Elvebredd, B. N. Cyvin, J. Brunvoll, and G. Hagen, *Acta Chem. Scand.*, **21**, 2405 (1967).
29. A. Almenningen, Å. O. Hartmann, and H. M. Seip, *Acta Chem. Scand.*, **22**, 1013 (1968).
30. A. Haaland and J. E. Nilsson, *Acta Chem. Scand.*, **22**, 2653 (1968).
31. A. Clark and H. M. Seip, *Chem. Phys. Lett.*, **6**, 452 (1970).
32. M. Tanimoto, K. Kuchitsu, and Y. Morino, *Bull. Chem. Soc. Japan*, **43**, 2776 (1970).
33. C. H. Townes and A. L. Schawlow, *Microwave Spectroscopy*, McGraw-Hill Book Co., New York, 1955.
34. I. M. Sugden and C. N. Kenney, *Microwave Spectroscopy of Gases*, D. Van Nostrand, New York, 1965.
35. J. E. Wollrab, *Rotational Spectra and Molecular Structure*, Academic Press New York, 1967.
36. J. Kraitchman, *Amer. J. Phys.*, **21**, 17 (1953).
37. C. C. Costain, *J. Chem. Phys.*, **29**, 864 (1958).
38. K. Kuchitsu, *J. Chem. Phys.*, **49**, 4456 (1968).
39. K. L. Dorris, C. O. Britt, and J. E. Boggs, *J. Chem. Phys.*, **44**, 1352 (1966).
40. E. Hirota, *J. Chem. Phys.*, **37**, 283 (1962).
41. K. S. Pitzer and W. D. Gwinn, *J. Chem. Phys.*, **10**, 428 (1942).
42. K. S. Pitzer, *J. Chem. Phys.*, **14**, 239 (1946).
43. T. Ueda and T. Shimanouchi, *J. Chem. Phys.*, **47**, 5018 (1967).
44. C. W. Rathjens, Jr., *J. Chem. Phys.*, **36**, 2401 (1962).
45. L. H. Scharpen, *J. Chem. Phys.*, **48**, 3552 (1968).
46. J. S. Koehler and D. M. Dennison, *Phys. Rev.*, **57**, 1006 (1940).
47. D. G. Burkhard and D. M. Dennison, *Phys. Rev.*, **84**, 408 (1951).
48. E. V. Ivash and D. M. Dennison, *J. Chem. Phys.*, **21**, 1804 (1953).
49. F. H. Westheimer, "Calculation of the Magnitude of Steric Effects," in M. S. Newman, ed., *Steric Effects in Organic Chemistry*, John Wiley and Sons, Inc., New York, 1956, p. 523.
50. J. B. Hendrickson, *J. Amer. Chem. Soc.*, **83**, 4537 (1961).
51. J. B. Hendrickson, *J. Amer. Chem. Soc.*, **89**, 7036 (1967).
52. J. B. Hendrickson, *J. Amer. Chem. Soc.*, **89**, 7047 (1967).
53. K. B. Wiberg, *J. Amer. Chem. Soc.*, **87**, 1070 (1965).
54. N. L. Allinger, J. A. Hirsch, M. A. Miller, I. J. Tyminski, and F. A. Van-Catledge, *J. Amer. Chem. Soc.*, **90**, 1199 (1968).
55. N. L. Allinger, J. A. Hirsch, M. A. Miller, and I. J. Tyminski, *J. Amer. Chem. Soc.*, **90**, 5773 (1968).
56. N. L. Allinger, J. A. Hirsch, M. A. Miller, and I. J. Tyminski, *J. Amer. Chem. Soc.*, **91**, 337 (1969).

57. M. Bixon and S. Lifson, *Tetrahedron*, **23**, 769 (1967).

58. E. J. Jacob, H. B. Thompson, and L. S. Bartell, *J. Chem. Phys.*, **47**, 3736 (1967).

59. G. Favini, F. Zuccarello, and G. Buemi, *J. Mol. Struct.*, **3**, 385 (1969).

60. G. Buemi, G. Favini, and F. Zuccarello, *J. Mol. Struct.*, **5**, 101 (1970).

61. H. M. Seip, *Acta Chem. Scand.*, **23**, 2741 (1969).

62. A. Almenningen, H. M. Seip, and T. Willadsen, *Acta Chem. Scand.*, **23**, 2748 (1969).

63. A. Almenningen, P. Kolsaker, H. M. Seip, and T. Willadsen, *Acta Chem. Scand.*, **23**, 3398 (1969).

64. J. M. Lehn, "Theoretical Conformational Analysis: *ab initio* SCF-LCAO-MO Studies of Conformations and Conformational Energy Barriers–Scope and Limitations," in G. Chirudoglu, ed., "Conformational Analysis. Scope and Limitations, Organic Chemistry." A series of Monographs, Vol. 21, Academic Press, New York–London, 1971.

65. W. H. Fink and L. C. Allen, *J. Chem. Phys.*, **46**, 2261, 2276 (1967).

66. J. D. Dunitz, "Conformations of Medium Rings," in J. D. Dunitz and J. A. Ibers, eds., *Perspectives in Structural Chemistry*, Vol. II, John Wiley & Sons, New York–London–Sydney, 1968.

67. R. M. Pitzer and W. N. Lipscomb, *J. Chem. Phys.*, **39**, 1995 (1963).

68. S. Weiss and G. E. Leroi, *J. Chem. Phys.*, **48**, 962 (1968).

69. C. C. J. Rothaan, *Rev. Mod. Phys.*, **23**, 69 (1951).

70. B. Lévy and M. A. Moireau, unpublished results.

71. R. M. Pitzer, *J. Chem. Phys.*, **47**, 965 (1967).

72. A. Veillard, *Chem. Phys. Lett.*, **3**, 128 (1969).

73. A. Veillard, *Chem. Phys. Lett.*, **4**, 51 (1969).

74. R. H. Hunt, R. A. Leacock, C. W. Peters, and K. T. Hecht, *J. Chem. Phys.*, **42**, 1931 (1965).

75. W. C. Oelfke and W. Gordy, *J. Chem. Phys.*, **51**, 5336 (1969).

76. J. R. Hoyland, *J. Chem. Phys.*, **50**, 2775 (1969).

77. A. Veillard, unpublished result.

78. R. Hoffmann, *J. Chem. Phys.*, **39**, 1397 (1963).

79. L. C. Cusachs, *J. Chem. Phys.*, **43**, S157 (1965).

80. G. Dallinga and P. Ros, *Rec. Trav. Chim.*, **87**, 906 (1968).

81. J. A. Pople, *Trans. Faraday Soc.*, **49**, 1375 (1953).

82. B. Roos and P. N. Skancke, *Acta Chem. Scand.*, **21**, 233 (1967).

83. M. J. S. Dewar and C. de Llano, *J. Amer. Chem. Soc.*, **91**, 789 (1969).

84. J. A. Pople, D. P. Santry, and G. A. Segal, *J. Chem. Phys.*, **43**, S129 (1965).

85. J. A. Pople and G. A. Segal, *J. Chem. Phys.*, **43**, S136 (1965).

86. J. A. Pople and G. A. Segal, *J. Chem. Phys.*, **44**, 3289 (1966).

87. J. A. Pople, D. L. Beveridge, and P. A. Dobash, *J. Chem. Phys.*, **47**, 2026 (1967).

88. N. C. Baird and M. J. S. Dewar, *J. Chem. Phys.*, **50**, 1262 (1969).

89. R. B. Davidson, W. L. Jorgensen, and L. C. Allen, *J. Amer. Chem. Soc.*, **92**, 749 (1970).

90. D. R. Lide, Jr. and D. E. Mann, *J. Chem. Phys.*, **27**, 868 (1957).

91. E. Hirota, *J. Chem. Phys.*, **45**, 1984 (1966).

92. J. R. Hoyland, *J. Chem. Phys.*, **49**, 1908 (1969).

93. S. Kondo, Y. Sakurai, E. Hirota, and Y. Morino, *J. Mol. Spectrosc.*, **34**, 231 (1970).
94. D. R. Lide, Jr. and D. E. Mann, *J. Chem. Phys.*, **29**, 914 (1958).
95. C. C. Lin and J. D. Swalen, *Rev. Mod. Phys.*, **31**, 841 (1959).
96. H. Dreizler, *Z. Naturforsch.*, **23a**, 1077 (1968).
97. H. Dreizler and A. M. Mirri, *Z. Naturforsch.*, **23a**, 1313 (1968).
98. H. D. Rudolph, H. Dreizler, A. Jaeschke, and P. Wendling, *Z. Naturforsch.*, **22a**, 940 (1967).
99. E. Tannenbaum, R. J. Myers, and W. D. Gwinn, *J. Chem. Phys.*, **25**, 42 (1956).
100. W. M. Tolles, E. T. Handelman, and W. D. Gwinn, *J. Chem. Phys.*, **43**, 3019 (1965).
101. H. Dreizler, H. D. Rudolph, and H. Mäder, *Z. Naturforsch.*, **25a**, 25 (1970).
102. H. D. Rudolph, H. Dreizler, and H. Seiler, *Z. Naturforsch.*, **22a**, 1738 (1967).
103. W. H. Flygare, *Ann. Rev. Phys. Chem.*, **18**, 325 (1967).
104. J. Dale, *Tetrahedron*, **22**, 3373 (1966).
105. B. Bak, S. Detoni, L. Hansen-Nygaard, J. T. Neilsen, and J. Rastrup-Andersen, *Spectrochim. Acta*, **16**, 376 (1960).
106. R. S. Wagner and B. P. Dailey, *J. Chem. Phys.*, **26**, 1588 (1957).
107. C. Flanagan and L. Pierce, *J. Chem. Phys.*, **38**, 2963 (1963).
108. E. L. Pace and J. G. Aston, *J. Amer. Chem. Soc.*, **70**, 566 (1948).
109. D. E. Mann and E. K. Plyler, *J. Chem. Phys.*, **21**, 1116 (1953).
110. G. Allen, P. N. Brier, and G. Lane, *Trans. Faraday Soc.*, **63**, 824 (1967).
111. K. G. R. Pachler and P. L. Wessels, *J. Mol. Struct.*, **3**, 207 (1969).
112. J. Ainsworth and J. Karle, *J. Chem. Phys.*, **20**, 425 (1952).
113. O. Bastiansen and J. Brunvoll, unpublished results, 1962.
114. P. Klæboe and J. R. Nielsen, *J. Chem. Phys.*, **33**, 1764 (1960).
115. G. Graner and C. Thomas, *J. Chem. Phys.*, **49**, 4160 (1968).
116. H. J. Bernstein, *J. Chem. Phys.*, **17**, 262 (1949).
117. K. S. Pitzer, *J. Chem. Phys.*, **8**, 711 (1940).
118. R. A. Bonham and L. S. Bartell, *J. Amer. Chem. Soc.*, **81**, 3491 (1959).
119. K. Iimura, N. Kawakami, and M. Takeda, *Bull. Chem. Soc. Jap.*, **42** 2091 (1969).
120. K. Iimura, *Bull. Chem. Soc. Jap.*, **42**, 3135 (1969).
121. B. F. Pedersen and B. Pedersen, *Acta Chem. Scand.*, **18**, 1454 (1964).
122. B. F. Pedersen, *Acta Chem. Scand.*, **23**, 1871 (1969).
123. G. Winnewisser, M. Winnewisser, and W. Gordy, *J. Chem. Phys.*, **49**, 3465 (1968).
124. A. Veillard and J. Demuynck, *Chem. Phys. Lett.*, **4**, 476 (1970).
125. B. Kirtman, *J. Chem. Phys.*, **41**, 775 (1964).
126. P. R. Bunker and H. C. Longuet-Higgins, *Proc. Roy. Soc., London*, **A280**, 340 (1964).
127. W. H. Kirchhoff and D. R. Lide, Jr., *J. Chem. Phys.*, **43**, 2203 (1965).
128. C. V. Berney, L. R. Cousins, and F. A. Miller, *Spectrochim. Acta*, **19**, 2019 (1963).
129. H. C. Longuet-Higgins, *Mol. Phys.*, **6**, 445 (1963).
130. P. R. Bunker, *J. Chem. Phys.*, **47**, 718 (1967).

131. (a) K. Kveseth, H. M. Seip, and R. Stølevik, *Acta. Chem. Scand.*, in the press; (b) O. Bastiansen, K. Kveseth, H. M. Seip, and R. Stølevik, unpublished work.

132. Y. Morino, I. Miyagawa, T. Chiba, and T. Shimozawa, *Bull. Chem. Soc. Japan*, **30**, 222 (1957).

133. K. Kuchitsu, *Bull. Chem. Soc. Japan*, **30**, 391 (1957).

134. A. Almenningen, O. Bastiansen, and F. Harshbarger, *Acta Chem. Scand.*, **11**, 1059 (1957).

135. D. R. Lide, Jr., and D. Christensen, *J. Chem. Phys.*, **35**, 1374 (1961).

136. D. Coffey, C. O. Britt, and J. E. Boggs, *J. Chem. Phys.*, **49**, 591 (1968).

137. R. W. Kilb, C. C. Lin, and E. B. Wilson, Jr., *J. Chem. Phys.*, **26**, 1695 (1957).

138. L. Pauling, *Nature of the Chemical Bond*, 3rd ed., Cornell University Press, Ithaca, New York, 1960.

139. L. Pierce and L. C. Krisher, *J. Chem. Phys.*, **31**, 875 (1959).

140. K. M. Sinnott, *J. Chem. Phys.*, **34**, 851 (1961).

141. L. C. Krisher, *J. Chem. Phys.*, **33**, 1237 (1960).

142. M. J. Maloney and L. C. Krisher, *J. Chem. Phys.*, **45**, 3277 (1966).

143. L. C. Krisher and E. B. Wilson, Jr., *J. Chem. Phys.*, **31**, 882 (1959).

144. Y. Hanyu and J. E. Boggs, *J. Chem. Phys.*, **43**, 3454 (1965).

145. Y. Hanyu, C. O. Britt, and J. E. Boggs, *J. Chem. Phys.*, **45**, 4725 (1966).

146. R. K. Kakar, E. A. Rinehart, and C. R. Quade, and T. Kojima, *J. Chem. Phys.*, **52**, 3803 (1970).

147. F. A. Miller, W. G. Fately, and R. E. Witkowski, *Spectrochim. Acta*, **23A**, 891 (1967).

148. R. Wagner, J. Fine, J. W. Simmons, and J. H. Goldstein, *J. Chem. Phys.*, **26**, 634 (1957).

149. E. A. Cherniak and C. C. Costain, *J. Chem. Phys.*, **45**, 104 (1966).

150. W. G. Dauben and K. S. Pitzer, "Conformational Analysis" in M. S. Newman, ed., *Steric Effects in Organic Chemistry*, John Wiley and Sons, Inc., New York, 1956, p. 57.

151. T. N. Sarachman, *J. Chem. Phys.*, **49**, 3146 (1968).

152. S. Siegel, *J. Chem. Phys.*, **27**, 989 (1957).

153. R. A. Beaudet, *J. Chem. Phys.*, **40**, 2705 (1964).

154. R. A. Beaudet, *J. Chem. Phys.*, **37**, 2398 (1962).

155. R. A. Beaudet, *J. Chem. Phys.*, **50**, 2002 (1969).

156. R. A. Beaudet, *J. Chem. Phys.*, **38**, 2548 (1963).

157. S. L. Hsu and W. H. Flygare, *J. Chem. Phys.*, **52**, 1053 (1970).

158. R. G. Stone, S. L. Srivastava, and W. H. Flygare, *J. Chem. Phys.*, **48**, 1890 (1968).

159. L. Pierce and J. M. O'Reilly, *J. Mol. Spectrosc.*, **3**, 536 (1959).

160. P. Meakin, D. O. Harris, and E. Hirota, *J. Chem. Phys.*, **51**, 3775 (1969).

161. T. Shimanouchi, Y. Abe, and K. Kuchitsu, *J. Mol. Struct.*, **2**, 82 (1968).

162. R. S. Mulliken, *Rev. Mod. Phys.*, **14**, 265 (1942).

163. W. G. Fately, R. K. Harris, F. A. Miller, and R. E. Witkowski, *Spectrochim. Acta*, **21**, 231 (1965).

164. V. Schomaker and L. Pauling, *J. Amer. Chem. Soc.*, **61**, 1769 (1939).

165. W. C. Price and A. D. Walsh, *Proc. Roy. Soc. (London)*, **A, 174**, 220 (1940).

166. A. D. Walsh, *Nature*, **157**, 768 (1942).
167. R. S. Rasmussen, D. D. Tunnicliff, and R. R. Brattain, *J. Chem. Phys.*, **11**, 432 (1943).
168. R. S. Rasmussen and R. R. Brattain, *J. Chem. Phys.*, **15**, 131 (1947).
169. I. Godnev and V. Morozov, *J. Phys. Chem. (U.S.S.R.)*, **21**, 799 (1947); *Chem. Abstr.*, **42**, 2162a (1948).
170. K. Bradacs and L. Kahovee, *Z. Phys. Chem.*, **B**, **48**, 63 (1940).
171. J. G. Aston, G. Szasz, H. W. Woolley, and F. G. Brickwedde, *J. Chem. Phys.*, **14**, 67 (1946).
172. A. Almenningen, O. Bastiansen, and M. Trætteberg, *Acta Chem. Scand.*, **12**, 1221 (1958).
173. W. Haugen and M. Trætteberg, "The Single and Double Bonds between $sp^2$-Hybridized Carbon Atoms, as Studied by the Gas Electron Diffraction Method. I. The Molecular Structures of 1,3-Butadiene and 1,3,5,7-Cyclooctatetraene," in P. Andersen, O. Bastiansen, and S. Furberg, eds., *Selected Topics in Structure Chemistry*, Universitetsforlaget, Oslo, 1967, p. 113.
174. K. Kuchitsu, T. Fukuyama, and Y. Morino, *J. Mol. Struct.*, **1**, 463 (1967–1968).
175. D. J. Marais, N. Sheppard, and B. P. Stoicheff, *Tetrahedron*, **17**, 163 (1962).
176. M. J. S. Dewar and A. J. Harget, *Proc. Roy. Soc. (London)*, **A**, **315**, 443 (1970).
177. A. L. Segre, L. Zetta, and A. Di Corato, *J. Mol. Spectrosc.*, **32**, 296 (1969).
178. O. Bastiansen, K. Hagen, and M. Trætteberg, unpublished work.
179. D. R. Lide, Jr., and M. Jen, *J. Chem. Phys.*, **40**, 252 (1964).
180. S. L. Hsu, M. K. Kemp, J. M. Pochan, R. C. Benson, and W. H. Flygare, *J. Chem. Phys.*, **50**, 1482 (1969).
181. C. F. Aten, L. Hedberg, and K. Hedberg, *J. Amer. Chem. Soc.*, **90**, 2463 (1968).
182. M. Trætteberg, *Acta Chem. Scand.*, **24**, 2295 (1970).
183. D. R. Lide, Jr., *J. Chem. Phys.*, **37**, 2074 (1962).
184. R. A. Beaudet, *J. Chem. Phys.*, **42**, 3758 (1965).
185. C. H. Chang, A. L. Andreassen, and S. H. Bauer, unpublished work.
186. G. Gundersen and K. Hedberg, personal communication.
187. M. Trætteberg, *Acta Chem. Scand.*, **22**, 2294 (1968).
188. M. Trætteberg, *Acta Chem. Scand.*, **22**, 628 (1968).
189. M. Trætteberg, *Acta Chem. Scand.*, **24**, 373 (1970).
190. M. Suziki and K. Kozima, *Bull. Chem. Soc. Japan*, **42**, 2183 (1969).
191. S. L. Hsu and W. H. Flygare, *Chem. Phys. Lett.*, **4**, 317 (1969).
192. N. L. Owen, *J. Mol. Struct.*, **6**, 37 (1970).
193. J. J. Keirns and R. F. Curl, Jr., *J. Chem. Phys.*, **48**, 3773 (1968).
194. D. F. Koster, *J. Amer. Chem. Soc.*, **88**, 5067 (1966).
195. G. L. Carlson, W. G. Fately, and R. E. Witkowski, *J. Amer. Chem. Soc.*, **89**, 6437 (1967).
196. J. E. Katon and W. R. Feairheller, Jr., *J. Chem. Phys.*, **47**, 1248 (1967).
197. G. Allen and H. J. Bernstein, *Can. J. Chem.*, **33**, 1055 (1955).
198. F. Mönnig, H. Dreizler, and H. D. Rudolph, *Z. Naturforsch.*, **20a**, 1323 (1965).

199. F. Mönnig, H. Dreizler, and H. D. Rudolph, *Z. Naturforsch.*, **21a**, 1633 (1966).
200. J. Paldus and D. A. Ramsay, *Can. J. Phys.*, **45**, 1389 (1967).
201. A. de Meijere and W. Lüttke, *Tetrahedron*, **25**, 2047 (1969).
202. H. Günther, H. Klose, and D. Wendisch, *Tetrahedron*, **25**, 1531 (1969).
203. L. S. Bartell and J. P. Guillory, *J. Chem. Phys.*, **43**, 647 (1965).
204. L. S. Bartell, J. P. Guillory, and A. T. Parks, *J. Phys. Chem.*, **69**, 3043 (1965).
205. J. E. Katon, W. R. Feairheller, Jr., and J. T. Miller, Jr., *J. Chem. Phys.*, **49**, 823 (1969).
206. L. V. Vilkov, I. I. Nazarenko, and R. G. Kostygnovskii, *Zh. Strukt. Khim.*, **9**, 1075 (1968) (Engl. transl. p. 960).
207. O. Bastiansen and A. de Meijere, *Acta Chem. Scand.*, **20**, 516 (1966).
208. J. Eraker and C. Rømming, *Acta Chem. Scand.*, **21**, 2721 (1967).
209. P. Rademacher and W. Lüttke, *Angew. Chem.*, **82**, 258 (1970); *Angew. Chem. Int. Ed., Engl.*, **9**, 245 (1970).
210. P. Rademacher, unpublished work.
211. A. A. Bothner-By and D. Jung, *J. Amer. Chem. Soc.*, **90**, 2342 (1968).
212. A. A. Bothner-By and D. F. Koster, *J. Amer. Chem. Soc.*, **90**, 2351 (1968).
213. A. D. Walsh, *Nature*, **159**, 165, 712 (1947).
214. A. D. Walsh, *Trans. Faraday Soc.*, **45**, 179 (1949).
215. C. A. Coulson and W. E. Moffitt, *Phil. Mag.*, **40**, 1 (1949).
216. L. S. Bartell, B. L. Caroll, and J. P. Guillory, *Tetrahedron Lett.*, **1964**, 705.
217. S. Forsén and P. N. Skancke, "A Molecular Orbital Study of the Barrier to Internal Rotation in Benzaldehyde," in P. Andersen, O. Bastiansen, and S. Furberg, eds., *Selected Topics in Structure Chemistry*, Universitetsforlaget, Oslo, 1967, p. 229.
218. G. H. Christie and J. Kenner, *J. Chem. Soc.*, **121**, 614 (1922).
219. E. E. Turner and R. J. W. Le Fèvre, *Chem. & Ind. (London)*, **4**, 831 (1926).
220. F. Bell and J. Kenyon, *Chem. & Ind. (London)*, **4**, 864 (1926).
221. O. Bastiansen, *Acta Chem. Scand.*, **3**, 408 (1949).
222. J. Dahr, *Indian J. Phys.*, **7**, 43 (1932).
223. D. H. Saunder, *Proc. Roy. Soc. (London)*, A, **188**, 31 (1947).
224. J. Toussaint, *Acta Cryst.*, **1**, 43 (1948).
225. J. N. van Niekerk and D. H. Saunder, *Acta Cryst.*, **1**, 44 (1948).
226. J. Trotter, *Acta Cryst.*, **14**, 1135 (1961).
227. A. Almenningen and O. Bastiansen, *Kgl. Norske Vidensk. Selsk. Skr.*, **1958**, No. 4.
228. G. Casalone, C. Mariani, A. Mugnoli, and M. Simonetta, *Acta Cryst.*, B, **25**, 1741 (1969).
229. O. Bastiansen, *Acta Chem. Scand.*, **6**, 205 (1952).
230. M. S. Farag, *Acta Cryst.*, **7**, 117 (1954).
231. A. Almenningen, O. Bastiansen, and P. N. Skancke, *Acta Chem. Scand.*, **12**, 1215 (1958).
232. J. C. J. Bart, *Acta Cryst.*, B, **24**, 1277 (1968).
233. J. Guy, *J. Chim. Phys.*, **46**, 469 (1949).
234. I. Fischer-Hjalmars, *Tetrahedron*, **19**, 1805 (1963).

235. A. Almenningen, O. Bastiansen, and L. Fernholt, *Acta Chem. Scand.*, in the press.
236. A. Skancke, *Acta Chem. Scand.*, **24**, 1389 (1970).
237. A. Almenningen, O. Bastiansen, and P. Svendsås, *Acta Chem. Scand.*, **12**, 1671 (1958).
238. O. Bastiansen and L. Smedvik, *Acta Chem. Scand.*, **8**, 1593 (1954).
239. O. Bastiansen, *Acta Chem. Scand.*, **4**, 926 (1950).
240. E. M. Farbrot and P. N. Skancke, *Acta Chem. Scand.*, **24**, 3640 (1970).
241. H. M. Seip and P. N. Skancke, unpublished work.
242. O. Bastiansen and A. Skancke, *Acta Chem. Scand.*, **21**, 587 (1967).
243. Å. O. Hartmann, H. M. Seip, and A. Skancke, unpublished results.
244. E. Hirota, C. Matsumara, and Y. Morino, *Bull. Chem. Soc. Japan*, **40**, 1124 (1967).
245. A. Trinkaus, H. Dreizler, and H. D. Rudolph, *Z. Naturforsch.*, **23a**, 2123 (1968).
246. J. R. Hoyland, *J. Chem. Phys.*, **49**, 1908 (1969).
247. T. N. Sarachman, *J. Chem. Phys.*, **39**, 469 (1963).
248. C. Komaki, I. Ichishima, K. Kuratani, T. Miyazawa, T. Shimanouchi, and S. Mizushima, *Bull. Chem. Soc. Japan*, **28**, 330 (1955).
249. E. Hirota, *J. Chem. Phys.*, **37**, 2918 (1962).
250. S. S. Butcher and E. B. Wilson, Jr., *J. Chem. Phys.*, **40**, 1671 (1964).
251. O. L. Stiefvater and E. B. Wilson, Jr., *J. Chem. Phys.*, **50**, 5385 (1969).
252. S. Kondo, E. Hirota, and Y. Morino, *J. Mol. Spectrosc.*, **28**, 47 (1968).
253. R. D. McLachlan and R. A. Nyquist, *Spectrochim. Acta*, **24A**, 103 (1968).
254. K. V. L. N. Sastry, V. M. Rao, and S. C. Dass, *Can. J. Phys.*, **46**, 959 (1968).
255. A. N. Murty and R. F. Curl, *J. Chem. Phys.*, **46**, 4176 (1967).
256. K. V. L. N. Sastry, S. C. Dass, W. V. F. Brooks, and A. Bhaumik, *J. Mol. Spectrosc.*, **31**, 54 (1969).
257. H. Test and C. R. Quade, *Symp. Mol. Struct. Spectrosc.*, Columbus, Ohio, paper X 13, 1968.
258. M. Takano, Y. Sasada, and T. Satoh, *J. Mol. Spectrosc.*, **26**, 157 (1968).
259. E. Hirota, *J. Mol. Spectrosc.*, **26**, 335 (1968).
260. S. Kondo and E. Hirota, *J. Mol. Spectrosc.*, **34**, 97 (1970).
261. J. Griffiths and J. E. Boggs, unpublished results.
262. C. T. Zahn, *Phys. Z.*, **33**, 525 (1932).
263. K. W. J. Kohlrausch and G. P. Ypsilanti, *Z. phys. Chem.*, **B**, **29**, 274 (1935).
264. S. Mizushima, T. Kubota, and Y. Morino, *Bull. Chem. Soc. Japan*, **14**, 15 (1939).
265. L. R. Zumwalt and R. M. Badger, *J. Amer. Chem. Soc.*, **62**, 305 (1940).
266. O. Bastiansen, *Acta Chem. Scand.*, **3**, 415 (1949).
267. M. Yamaha, *Bull. Chem. Soc. Japan*, **29**, 865 (1956).
268. M. Igarashi and M. Yamaha, *Bull. Chem. Soc. Japan*, **29**, 871 (1956).
269. M. Yamaha, *Bull. Chem. Soc. Japan*, **29**, 876 (1956).
270. S. Mizushima, T. Shimanouchi, T. Miyazawa, K. Abe, and M. Yasumi, *J. Chem. Phys.*, **19**, 1477 (1951).
271. P. Buckley, P. A. Giguére, and D. Yamamoto, *Can. J. Chem.*, **46**, 2917 (1968).

272. R. G. Azrak and E. B. Wilson, Jr., *J. Chem. Phys.*, **52**, 5299 (1970).
273. A. Almenningen, O. Bastiansen, L. Fernholt, and K. Hedberg, *Acta Chem. Scand.*, in the press.
274. S. Mizushima, T. Shimanouchi, K. Kuratani, and T. Miyazawa, *J. Amer. Chem. Soc.*, **74**, 1378 (1952).
275. R. E. Penn, *Symp. Mol. Struct. Spectrosc.*, Columbus, Ohio, 1968.
276. N. L. Owen and N. Sheppard, *Trans. Faraday Soc.*, **60**, 634 (1964).
277. P. Cahill, L. P. Gold, and N. L. Owen, *J. Chem. Phys.*, **48**, 1620 (1968).
278. N. L. Owen and R. H. M. Seip, *Chem. Phys. Lett.*, **5**, 162 (1970).
279. R. E. Penn and R. F. Curl, Jr., *J. Mol. Spectrosc.*, **24**, 235 (1967).
280. R. F. Curl, Jr., *J. Chem. Phys.*, **30**, 1529 (1959).
281. G. Williams, N. L. Owen, and J. Sheridan, *Chem. Commun.*, **1968**, 57.
282. J. M. Riveros and E. B. Wilson, Jr., *J. Chem. Phys.*, **46**, 4605 (1967).
283. W. D. Gwinn, R. J. Anderson, and D. Stelman, *Second Austin Symp. on Gas Phase Molecular Structure*, 1968.
284. W. B. Dixon and E. B. Wilson, Jr., *J. Chem. Phys.*, **35**, 191 (1961).
285. K. M. Marstokk and H. Møllendal, *J. Mol. Struct.*, **5**, 205 (1970).
286. J. D. Dunitz and V. Schomaker, *J. Chem. Phys.*, **20**, 1703 (1952).
287. A. Almenningen, O. Bastiansen, and P. N. Skancke, *Acta Chem. Scand.*, **15**, 711 (1961).
288. R. C. Lord and B. P. Stoicheff, *Can. J. Phys.*, **40**, 725 (1962).
289. T. Ueda and T. Schimanouchi, *J. Chem. Phys.*, **49**, 470 (1968).
290. B. Andersen, personal communication.
291. B. Andersen and L. Fernholt, *Acta Chem. Scand.*, **24**, 445 (1970).
292. R. K. Bohn and Y.-H. Tai, *J. Amer. Chem. Soc.*, **92**, 6447 (1970).
293. T. N. Margulis and M. Fischer, *J. Amer. Chem. Soc.*, **89**, 223 (1967).
294. E. Adam and T. N. Margulis, *J. Amer. Chem. Soc.*, **90**, 4517 (1968).
295. E. Adam and T. N. Margulis, *J. Phys. Chem.*, **73**, 1480 (1969).
296. S. I. Chan, J. Zinn, and W. D. Gwinn, *J. Chem. Phys.*, **33**, 295 (1960).
297. S. I. Chan, J. Zinn, J. Fernandez, and W. D. Gwinn, *J. Chem. Phys.*, **33**, 1643 (1960).
298. S. I. Chan, J. Zinn, and W. D. Gwinn, *J. Chem. Phys.*, **34**, 1319 (1961).
299. A. Danti, W. J. Lafferty, and R. Lord, *J. Chem. Phys.*, **33**, 294 (1960).
300. S. I. Chan, T. R. Borgers, J. Russell, H. L. Strauss, and W. D. Gwinn, *J. Chem. Phys.*, **44**, 1103 (1966).
301. D. O. Harris, H. W. Harrington, A. C. Luntz, and W. D. Gwinn, *J. Chem. Phys.*, **44**, 3467 (1966).
302. D. W. Boone, C. O. Britt, and J. E. Boggs, *J. Chem. Phys.*, **43**, 1190 (1965).
303. L. A. Carreira and R. C. Lord, *J. Chem. Phys.*, **51**, 2735 (1969).
304. A. C. Luntz, *J. Chem. Phys.*, **50**, 1109 (1969).
305. J. Laane and R. D. Lord, *J. Chem. Phys.*, **48**, 1508 (1968).
306. J. Laane, *Spectrochim. Acta*, **26A**, 517 (1970).
307. T. R. Borgers and H. L. Strauss, *J. Chem. Phys.*, **45**, 947 (1966).
308. J. R. Durig and R. C. Lord, *J. Chem. Phys.*, **45**, 61 (1966).
309. A. B. Harvey, J. R. Durig, and A. C. Morrisey, *J. Chem. Phys.*, **50**, 4949 (1969).
310. L. H. Scharpen and V. W. Laurie, *J. Chem. Phys.*, **49**, 221 (1968).
311. L. H. Scharpen and V. W. Laurie, *J. Chem. Phys.*, **49**, 3041 (1968).

312. T. B. Malloy, Jr., F. Fisher, and R. M. Hedges, *J. Chem. Phys.*, **52**, 5325 (1970).
313. L. A. Carreira and R. C. Lord, *J. Chem. Phys.*, **51**, 3225 (1969).
314. A. Almenningen, O. Bastiansen, and L. Walløe, "The Molecular Structure and Conformation of Four 1,3-Dihalocyclobutanes" in P. Andersen, O. Bastiansen, and S. Furberg, eds., *Selected Topics in Structure Chemistry*, Universitetsforlaget, Oslo, 1967, p. 91.
315. C. H. Chang, R. F. Porter, and S. H. Bauer, *J. Mol. Structure*, **7**, 89 (1971).
316. W. G. Rothschild and B. P. Dailey, *J. Chem. Phys.*, **36**, 2931 (1962).
317. H. Kim and W. D. Gwinn, *J. Chem. Phys.*, **44**, 865 (1966).
318. O. Bastiansen and A. de Meijere, *Angew. Chem.*, **78**, 142 (1966); *Angew. Chem. Int. Ed.*, *Engl.*, **5**, 127 (1966).
319. C. M. Bock, *J. Amer. Chem. Soc.*, **90**, 2748 (1968).
320. B. Greenberg and B. Post, *Acta Cryst.*, **B 24**, 918 (1968).
321. B. Bak, J. J. Led, L. Nygaard, J. Rastrup-Andersen, and G. O. Sørensen, *J. Mol. Struct.*, **3**, 369 (1969).
322. O. Bastiansen and J. L. Derrisen, *Acta Chem. Scand.*, **20**, 1089 (1966).
323. A. Skancke, *Acta Chem. Scand.*, **22**, 3239 (1968).
324. J. Laane and R. C. Lord, *J. Chem. Phys.*, **47**, 4941 (1967).
325. J. Laane, *J. Chem. Phys.*, **50**, 776 (1969).
326. J. Laane, *J. Chem. Phys.*, **52**, 358 (1970).
327. T. Ueda and T. Schimanouchi, *J. Chem. Phys.*, **47**, 4042 (1967).
328. W. H. Greene, *J. Chem. Phys.*, **50**, 1619 (1969).
329. W. H. Greene and A. B. Harvey, *J. Chem. Phys.*, **49**, 177 (1968).
330. C. H. Chang, R. F. Porter, and S. H. Bauer, *Inorg. Chem.*, **8**, 1677 (1969).
331. J. E. Kilpatrick, K. S. Pitzer, and R. Spitzer, *J. Amer. Chem. Soc.*, **69**, 2483 (1947).
332. K. S. Pitzer and W. E. Donath, *J. Amer. Chem. Soc.*, **81**, 3213 (1959).
333. W. J. Adams, H. J. Geise, and L. S. Bartell, *J. Amer. Chem. Soc.*, **92**, 5013 (1970).
334. A. F. Platé, N. A. Belikova, and Yu. P. Egorov, *Dokl. Akad. Nauk SSSR*, **102**, 1131 (1955); cited in ref. 362.
335. K. G. Dzhaparidze, *Soobshch. Akad. Nauk Gruz. SSR.*, **29**, 401 (1962); cited in ref. 362.
336. E. C. Thomas and V. W. Laurie, *J. Chem. Phys.*, **51**, 4327 (1969).
337. H. Kim and W. D. Gwinn, *J. Chem. Phys.*, **51**, 1815 (1969).
338. H. Geise, W. J. Adams, and L. S. Bartell, *Tetrahedron*, **25**, 3045 (1969).
339. Z. Náhlovská, B. Náhlovský, and H. M. Seip, *Acta Chem. Scand.*, **23**, 3534 (1969).
340. Z. Náhlovská, B. Náhlovský, and H. M. Seip, *Acta Chem. Scand.*, **24**, 1903 (1970).
341. I. Wang, C. O. Britt, and J. E. Boggs, *J. Amer. Chem. Soc.*, **87**, 4950 (1965).
342. C. J. Brown, *Acta Cryst.*, **7**, 92 (1954).
343. J. H. Hand and R. H. Schwendeman, *J. Chem. Phys.*, **45**, 3349 (1966).
344. W. V. F. Brooks, C. C. Costain, and R. F. Porter, *J. Chem. Phys.*, **47**, 4186 (1967).
345. J. R. Durig and D. W. Wertz, *J. Chem. Phys.*, **49**, 2118 (1968).

346. J. R. Durig and J. N. Willis, Jr., *J. Mol. Spectroscop.*, **32**, 320 (1969).
347. J. Laane, *J. Chem. Phys.*, **50**, 1946 (1969).
348. H. M. Seip, *J. Chem. Phys.*, **54**, 440 (1971).
349. J. R. Durig and J. N. Willis, Jr., *J. Chem. Phys.*, **52**, 6108 (1970).
350. J. R. Durig, G. L. Coulter, and D. W. Wertz, *J. Mol. Spectrosc.*, **27**, 285 (1968).
351. J. A. Greenhouse and H. L. Strauss, *J. Chem. Phys.*, **50**, 124 (1969).
352. D. O. Harris, G. G. Engerholm, C. A. Tolman, A. C. Luntz, R. A. Keller, H. Kim, and W. D. Gwinn, *J. Chem. Phys.*, **50**, 2438 (1969).
353. G. G. Engerholm, A. C. Luntz, W. D. Gwinn, and D. O. Harris, *J. Chem. Phys.*, **50**, 2446 (1969).
354. D. W. Wertz, *J. Chem. Phys.*, **51**, 2133 (1969).
355. W. N. Hubbard, H. L. Finke, D. W. Scott, J. P. McCullough, C. Katz, M. E. Gross, J. F. Messerly, R. E. Pennington, and G. Waddington, *J. Amer. Chem. Soc.*, **74**, 6025 (1952).
356. G. A. Crowder and D. W. Scott, *J. Mol. Spectrosc.*, **16**, 122 (1965).
357. J. P. McCullough, *J. Chem. Phys.*, **29**, 966 (1958).
358. J. R. Durig and D. W. Wertz, *J. Chem. Phys.*, **49**, 675 (1968).
359. C. Altona, H. J. Geise, and C. Romers, *Tetrahedron*, **24**, 13 (1968).
360. H. R. Buys, C. Altona, and E. Havinga, *Tetrahedron*, **24**, 3019 (1968).
361. W. J. Lafferty, D. W. Robinson, R. V. St. Louis, J. W. Russell, and H. L. Strauss, *J. Chem. Phys.*, **42**, 2915 (1965).
362. N. G. Bokii and Yu. T. Struchkov, *Zh. Strukt. Khim.*, **9**, 722 (1968) (Engl. transl. p. 633).
363. F. G. Riddell, *Quart. Rev. Chem. Soc.*, **21**, 364 (1967).
364. C. Romers, C. Altona, H. R. Buys, and E. Havinga, "Geometry and Conformational Properties of Some Five- and Six-Membered Heterocyclic Compounds Containing Oxygen and Sulfur," in E. L. Eliel and N. L. Allinger, eds., *Topics in Stereochemistry*, Vol. 4, Interscience, New York 1969, p. 39.
365. O. Hassel and B. F. Pedersen, *Proc. Chem. Soc.*, **1959**, 394.
366. P. Groth and O. Hassel, *Proc. Chem. Soc.*, **1963**, 218.
367. P. Groth and O. Hassel, *Acta Chem. Scand.*, **18**, 923 (1964).
368. A. Mossel, C. Romers, and E. Havinga, *Tetrahedron Lett.*, **1963**, 1247.
369. P. Groth, *Acta Chem. Scand.*, **22**, 128 (1968).
370. D. J. Gregory-Allen and K. Hedberg, personal communication.
371. J. B. Lambert, *J. Amer. Chem. Soc.*, **89**, 1836 (1967).
372. F. Lautenschlaeger and G. F. Wright, *Cap. J. Chem.*, **41**, 1972 (1963).
373. J. A. Hirsch, "Table of Conformational Energies—1967," in E. L. Eliel and N. L. Allinger, eds. *Topics in Stereochemistry*, Vol. 1, Interscience, New York 1967, p. 199.
374. A. Haaland and L. Schäfer, *Acta Chem. Scand.*, **21**, 2474 (1967).
375. B. Andersen, personal communication.
376. M. Davis and O. Hassel, *Acta Chem. Scand.*, **17**, 1181 (1963).
377. O. Hassel and H. Viervoll, *Acta Chem. Scand.*, **1**, 149 (1947).
378. K. E. Hjortås and K. O. Strømme, *Acta Chem. Scand.*, **22**, 2965 (1968).
379. P. Andersen, *Acta Chem. Scand.*, **16**, 2337 (1962).
380. V. A. Atkinson, *Acta Chem. Scand.*, **15**, 599 (1961).

381. O. Bastiansen, O. Hassel, and A. Munthe-Kaas, *Acta Chem. Scand.*, **8**, 872 (1954).
382. V. A. Atkinson and O. Hassel, *Acta Chem. Scand.*, **13**, 1737 (1959).
383. C. H. Bushweller, *J. Amer. Chem. Soc.*, **91**, 6019 (1969).
384. J. E. Anderson and J. D. Roberts, *J. Amer. Chem. Soc.*, **92**, 97 (1970).
385. H. Oberhammer and S. H. Bauer, *J. Amer. Chem. Soc.*, **91**, 10 (1969).
386. G. Dallinga and L. H. Toneman, *J. Mol. Struct.*, **1**, 117 (1967–1968).
387. E. W. Garbisch and M. G. Griffith, *J. Amer. Chem. Soc.*, **90**, 3590 (1968), **92**, 1107 (1970).
388. K. G. Dzhaparidze, *Soobshch. Akad. Nauk. Gruz. SSR.*, **23**, 397 (1959); cited in ref. 339.
389. T. Oka, K. Tsuchiya, S. Iwata, and Y. Morino, *Bull. Chem. Soc. Japan*, **37**, 4 (1964).
390. A. H. Clark and T. G. Hewitt, *J. Mol. Structure*, in the press.
391. G. Valle, V. Busetti, M. Mammi, and G. Garazzollo, *Acta Cryst.*, **B, 25**, 1432 (1969).
392. J. F. Chiang and S. H. Bauer, *J. Amer. Chem. Soc.*, **91**, 1898 (1969).
393. L. H. Scharpen, J. E. Wollrab, and D. P. Ames, *J. Chem. Phys.*, **49**, 2368 (1968).
394. O. Bastiansen, *Acta Chem. Scand.*, **6**, 875 (1952).
395. B. Ottar, *Acta Chem. Scand.*, **1**, 283 (1947).
396. V. A. Naumov and V. M. Bezzubov, *Z. Strukt. Khim.*, **8**, 530 (1967) (Engl. transl., p. 466).
397. G. Dallinga and L. H. Toneman, *J. Mol. Struct.*, **1**, 11 (1967–1968).
398. M. Trætteberg, *Acta Chem. Scand.*, **22**, 2305 (1968).
399. S. S. Butcher, *J. Chem. Phys.*, **42**, 1830 (1965).
400. M. Trætteberg, *J. Amer. Chem. Soc.*, **86**, 4265 (1964).
401. J. F. Chiang and S. H. Bauer, *J. Amer. Chem. Soc.*, **88**, 420 (1966).
402. O. Bastiansen, K. Hedberg, and L. Hedberg, *J. Chem. Phys.*, **27**, 1311 (1957).
403. M. Trætteberg, *Acta Chem. Scand.*, **24**, 2285 (1970).
404. K. Hedberg and L. Hedberg, private communication.
405. A. Almenningen, O. Bastiansen, and H. Jensen, *Acta Chem. Scand.*, **20**, 2689 (1966).
406. A. Almenningen, G. G. Jacobsen, and H. M. Seip, *Acta Chem. Scand.*, **23**, 1495 (1969).
407. J. D. Dunitz, H. Eser, M. Bixon, and S. Lifson, *Helv. Chim. Acta*, **50**, 1572 (1967).
408. B. Andersen and A. Marstrander, personal communication.
409. O. Hassel, *Tidsskr. Kjemi, Bergv. Met.*, **5**, 32 (1943) (Eng. transl. by K. Hedberg in E. L. Eliel and N. L. Allinger, eds., *Topics in Stereochemistry*, Vol. 6, Interscience, New York, 1971).
410. D. H. R. Barton, O. Hassel, K. S. Pitzer, and V. Prelog, *Science*, **119**, 49 (1954).
411. O. Hassel and H. Viervoll, *Tidsskr. Kjemi, Bergv. Met.*, **5**, 35 (1943).
412. K. Kozima and K. Sakashita, *Bull. Chem. Soc. Japan*, **31**, 796 (1958).
413. J. Reisse, J. C. Celotti, and G. Chiurdoglu, *Tetrahedron Lett.*, **1965**, 397.
414. L. Pierce and J. F. Beecher, *J. Amer. Chem. Soc.*, **88**, 5406 (1966).

415. P. Klæboe, J. J. Lothe, and K. Lunde, *Acta Chem. Scand.*, **10**, 1465 (1956).
416. P. Klæboe, J. J. Lothe, and K. Lunde, *Acta Chem. Scand.*, **11**, 1677 (1957).
417. P. Klæboe, *Acta Chem. Scand.*, in the press.
418. O. Hassel, *Research*, **3**, 504 (1950).
419. O. Hassel and K. Lunde, *Acta Chem. Scand.*, **4**, 1957 (1950).
420. O. Bastiansen and O. Hassel, *Acta Chem. Scand.*, **5**, 1404 (1951).
421. O. Hassel and E. Wang Lund, *Acta Cryst.*, **2**, 309 (1949).
422. O. Bastiansen, Ø. Ellefsen, and O. Hassel, *Acta Chem. Scand.*, **3**, 918 (1949).
423. S. J. Angyal and D. J. McHugh, *J. Chem. Soc.*, **1957**, 3682.
424. O. Bastiansen and J. Markali, *Acta Chem. Scand.*, **6**, 442 (1952).
425. G. Calingaert, M. E. Griffing, E. R. Kerr, A. J. Kolka, and H. D. Orloff, *J. Amer. Chem. Soc.*, **73**, 5224 (1951).
426. J. B. Lambert, D. S. Bailey, and B. F. Michel, *Tetrahedron Lett.*, **1970**, 691.
427. N. L. Allinger, J. A. Hirsch, and M. A. Miller, *Tetrahedron Lett.*, **1967**, 3729.
428. J. R. Durig, J. M. Karriker, and D. W. Wertz, *J. Mol. Spectrosc.*, **31**, 237 (1969).
429. I. O. C. Ekejiuba and H. E. Hallam, *Spectrochim. Acta*, **26A**, 67 (1970).
430. R. K. Harris and R. A. Spragg, *Chem. Commun.*, **1967**, 362.
431. S. Samdal and H. M. Seip, unpublished results.
432. H. M. Seip and R. Seip, unpublished results.
433. W. H. Greene, A. B. Harvey, and J. A. Greenhouse, *J. Chem. Phys.*, **54**, 850 (1971).
434. H. R. Buys and H. J. Geise, *Tetrahedron Lett.*, **1970**, 2991.
435. K. Hagen and M. Trætteberg, *Acta. Chem. Scand.*, in the press.

[Added in proof] See p. 114. The CNDO/2 method does not give correct results either for biphenyl itself, where a twist angle of 90° is predicted, or for 2,2'-difluorobiphenyl.[241]

# Crystalline π-Molecular Compounds: Chemistry, Spectroscopy, and Crystallography

F. H. HERBSTEIN, Department of Chemistry,
Technion–Israel Institute of Technology,
Haifa, Israel.

## ABBREVIATIONS

*Electron-donors*

| | |
|---|---|
| DAD | Durenediamine |
| DDDT | 13,14-Dithiatricyclo[8.2.1.0$^{4,7}$]tetradeca-4,6,10,12-tetraene |
| DMA | *N,N*-Dimethylaniline |
| HMB | Hexamethylbenzene |
| PD | *p*-Phenylenediamine |
| TAB | 1,3,5-Benzenetriamine |
| TDAE | Tetrakis(dimethylamino)ethylene |
| TDT | 2,4,6-Tris(dimethylamino)-1,3,5-triazine |
| TEA | Triethylammonium |
| TMPD | *N,N,N′,N′*-Tetramethyl-*p*-phenylenediamine |

*Electron-acceptors*

| | |
|---|---|
| BAQ | 2,5-Bis(methylamino)-*p*-benzoquinone |
| BP | 3,4-Benzopyrene |
| BTF | Benzotrifuroxan |
| DDQ | 2,3-Dichloro-5,6-dicyanobenzoquinone |
| DEQ | 2,5-Diethoxy-*p*-benzoquinone |
| DPPH | 2,2-Diphenyl-1-picrylhydrazyl |
| HCBD | Hexacyanobutadiene |
| PMDA | Pyromellitic dianhydride |
| TCNB | 1,2,4,5-Tetracyanobenzene |

| TCNE | Tetracyanoethylene |
| TCNQ | 7,7,8,8-Tetracyanoquinodimethane |
| TCPA | Tetrachlorophthalic anhydride |
| TENF | 2,4,5,7-Tetranitro-9-fluorenone |
| TMU | 1,3,7,9-Tetramethyluric acid |
| TNAP | 11,11,12,12-Tetracyanonaphtho-2,6-quinodimethane |
| TNB | 1,3,5-Trinitrobenzene |
| TNF | 2,4,7-Trinitro-9-fluorenone |
| TNT | 2,4,6-Trinitrotoluene |

*Complex*
| PDC | *p*-Phenylenediamine:chloranil |

### *Prologue*

"I should see the garden far better," said Alice to herself, "if I could get to the top of that hill: and here's a path that leads straight to it— at least, no, it doesn't do that—" (after going a few yards along the path, and turning several sharp corners) "but I suppose it will at last. But how curiously it twists! It's more like a corkscrew than a path! Well, *this* turn goes to the hill, I suppose—no, it doesn't! This goes straight back to the house! Well, then, I'll try it the other way!"

LEWIS CARROLL: *Through the Looking-glass.*

## I. THE SCOPE OF THE REVIEW

Crystallization occurs as the result of secondary interactions (*e.g.*, dispersion forces, dipole–dipole forces, hydrogen bonding) among the molecules of a one-component system. In suitable circumstances two or more different compounds may crystallize from a melt, solution, or vapour phase to form a single new phase instead of a mixture of crystals of the parent phases. In some instances, properties and structure indicate no special interaction between the different kinds of partners, and often the molecules of one component are arranged in the crystal in such a way as to enclose the molecules of the other; typical examples are the quinol:inert-gas and urea:hydrocarbon clathrates. However, there are numerous examples where there is clear evidence of a special action between the different partners. This is often vividly demonstrated by colour in the new phase although the parent partners are both colourless. Such a special interaction can be essentially localized between particular atoms of the two partners, as in hydrogen bonding or in trimethylamine:iodine, can be delocalized in respect to one partner but localized with respect to the other, as in benzene:bromine, or can be delocalized in regard to both partners, as in naphthalene: 1,3,5-trinitrobenzene (TNB).

This review is mostly about the last of these three groups; the electrons principally concerned are π-electrons and we shall therefore call the new phases π-molecular compounds, which is both simple and somewhat non-committal. The partners in π-molecular compounds cannot be chosen at random, but one component must have electron-donor capabilities and the other electron-acceptor capabilities. Thus their interaction is a Lewis base–acid interaction (donor ≡ Lewis base; acceptor ≡ Lewis acid). In this review we shall follow the convention of placing the electron donor first in the name of the molecular compound, as in naphthalene:TNB. We have on occasion strayed outside the strict boundaries of our subject matter as defined above; in some instances, this has been necessary in order to draw attention to unexplored regions, in others to note examples which are chemically but not necessarily structurally similar, or where similarities have a structural rather than an electronic basis.

Our subject was born in St. Petersburg (as it was then) in 1858, when Fritzsche[1] discovered "benzene, naphthalene, and anthracene picrates"; it is still going strong. Our purpose in this review is two-fold. First, we shall attempt to set the subject in a suitable chemical perspective by noting which compounds have been shown to function as partners in π-molecular compounds. The aim here has been to provide a comprehensive survey, but not a complete listing of every π-molecular compound that has ever been prepared. However, attention has been drawn to examples that display special or unusual features. Physical measurements and theoretical studies have been made of only a small fraction of the molecular compounds that have been reported, but our second main purpose has been to attempt to weave into one coherent pattern what is known about the geometrical structures of these compounds, their thermodynamic and spectro-scopic properties, and, briefly, the theoretical interpretation of the interaction between the component molecules. We give more atten-tion to hard experimental facts and their description than to the softer interpretations of experiments or the results of inevitably approximate theoretical calculations.

Emphasis will be placed on examples where the geometrical rela-tion between the donor and acceptor is known or can be determined, and this restricts us almost entirely to a consideration of crystalline compounds. In the past, most attention has been paid to the physical chemistry and spectroscopy of solutions containing donors and acceptors, and geometrical structures have had to be inferred from spectroscopic measurements. The factors involved in bonding the components may be rather different from those that promote the charge transfer which gives rise to the characteristic spectra of $\pi$-molecular compounds (Dewar and Thompson[2]); moreover, there may well be a distribution of configurations in solution (Matsuo and Higuchi[3]). These complications inevitably have their influence both on nomenclature and on definitions. Crystalline molecular compounds have the advantage over solution complexes that their defined geometries permit (at least in principle) detailed theoretical analysis of the interactions involved. Thus we have preferred to cast our net wide in the general review without regard to strict considerations of nomenclature and definition; a broad classification of the molecular compounds considered in this review is given in Section IV-C.

The crystal chemistry of $\pi$-molecular compounds has a number of interesting features, which are only now beginning to be explored in depth. First-order phase transformations and second-order (*i.e.*, order–disorder) transformations occur in many $\pi$-molecular compounds, and there are also examples of disorder due to growth faults. The relation of these features to the chemical nature of the donors and acceptors does not appear to have been considered, nor is much known of any effects they may have on the physical properties of the crystals.

Our subject has been extensively discussed in books[4-10] and reviews;[11-26] in most of these the area covered has been much broader than in the present review, or the emphasis has been rather different. Indeed only the review by Prout and Wright[24] has been principally concerned with the crystallographic results to be discussed here, and this was shorter and less detailed. There are also a number of other less direct but not negligible reasons for an interest in the crystallography of molecular compounds. First, molecular compounds can provide simplified access to molecular structures that might otherwise impose considerable experimental difficulties. Thus, Hanson[27,28] determined the molecular structures of acepleiadylene and azulene in their trinitrobenzene (TNB) compounds, the hydrocarbons themselves being disordered in the solid (acepleiadylene, Hanson;[29] azulene, Robertson *et al.*[30]). Ferguson *et al.*[31] determined the structure of hexahelicene in its equimolar compound with 4-bromo-2,5,7-trinitrofluorenone, the presence of the bromine atom making the analysis simpler. Secondly, many molecular compounds are unstable at relatively low temperatures and decompose thermally by loss of one of the components; Hanson[28] has even reported that the thermal decomposition of azulene:TNB is accelerated by X-irradiation. The defined structures of such compounds make them attractive subjects for the study of some aspects of solid-state thermal decompositions.

## II. THE DEVELOPMENT OF THE SUBJECT

The developments of the first sixty-odd years of interest in π-molecular compounds were summarized by Pfeiffer[4] in a book which still contains much useful and interesting information. At that time there were two rival theories about the structure of molecular compounds; one, put forward by Lowry[32] and by Bennett and Willis,[33] envisaged the formation of covalent bonds between the two partners, whereas the other, due essentially to Pfeiffer, ascribed the cohesion to secondary valence interactions (Restaffinitätskräfte). In 1925 Martinet and Bornard[34] gave a clear description of the way in which different types of substituent influence electron-donor and -acceptor properties of aromatic molecules. The first direct evidence for the Pfeiffer theory was provided in 1943 by Powell, Huse, and Cooke[35] who determined the crystal structure of $p$-iodoaniline:TNB and found that all intermolecular distances were far too long to be ascribed to covalent bonds. Attention thus necessarily shifted to secondary interactions between the components, with major emphasis on finding an explanation for the intense colours developed in crystals and, especially, in solutions.

The currently accepted theoretical interpretation of the stability and spectra of π-molecular compounds is based on the observation that one component of the pair is an electron-donor and the other an electron-acceptor. Early suggestions of this kind were made by Weiss[36] and Brackman,[37] but the most extensive theoretical developments are due to Mulliken.[38–40] The essential idea is that the ground state of the molecular compound can be described by a no-bond wave function of a dative structure involving transfer of an electron from donor to acceptor. In the excited state, on the other hand, the major contribution comes from the dative or charge-transfer structure. The additional, frequently intense, charge-transfer band in the spectra of solutions or suitably oriented crystals is ascribed to excitation of the system from its predominantly no-bond ground state to its predominantly dative excited state. The electron transferred (in whole or part) may be a π-electron or from a lone-pair, and the interaction is thus delocalized or localized. Mulliken's theoretical contributions have been primarily directed towards 1:1 "loose" complexes in solution, but donor:acceptor interactions occur also in the gas,[41,42] liquid (melt), and solid phases. In solution there will be interactions with solute molecules while in the solid longer-range interactions must be taken into account. These additional interactions may be sufficient to stabilize the ionic form as the ground state of the molecular compound.

Crystalline molecular compounds have been defined formally by Hertel and Kleu,[43] and we adopt their definition with some changes. Thus, a molecular compound is a crystalline two-component system in which the centres of gravity of the component molecules are at defined positions in a crystal, whose structure is different from those of the components. In the liquid phase or in solution the molecular compound dissociates into its components, in accordance with the law of mass action, and these are present in the fluid phase in their molecular state. The occurrence of the molecular compound as a (chemically ordered) crystalline phase with structure different from those of the components is the feature that distinguishes molecular compounds from solid solutions. We shall find that amendments are sometimes necessary to this definition: a third component can occasionally be present, disorder of orientation and sometimes of position (to a small extent) can occur, sometimes the structure of the molecular compound appears to be closely related to that of one of the components, and sometimes the components may be ionized in the crystal or in solution.

## III. DESIRABLE AND AVAILABLE INFORMATION

Before we discuss particular substances in any detail it is worth-while considering what information we wish to have and how it can be obtained. We can determine experimentally which substances interact to form crystalline molecular compounds of the type of interest here, their stoichiometry, and the spatial relationship between the components; in addition, the relevant thermodynamic parameters ($\Delta G_f^\circ$, $\Delta H_f^\circ$, $\Delta S_f^\circ$) can be measured, and evidence can be obtained from spectroscopic and other physical measurements about the (charge) states of the components and their interaction in the molecular compound. The task of theory is to explain, or at least to interpret, the experimental facts. This programme differs little in principle from that involved in understanding the combination of atoms to form molecules, except that here we are concerned with secondary (and second-order) chemical interactions.

In general, molecular compounds are synthesized by simple chemical techniques, usually by mixing appropriate solutions, sometimes by simultaneous sublimation of the solid components and subsequent condensation, and occasionally by heating or grinding together intimate mixtures of the solid components. The stoichiometry is overwhelmingly 1:1, but 2:1 or 1:2 ratios are found fairly often and there are a number of established examples of other ratios. There is usually no difficulty in determining the stoichiometry of crystalline molecular compounds by chemical analysis or from measurements of the unit-cell dimensions and crystal density. Melting point–composition diagrams are also often used, but these give only the situation at high temperatures; a more satisfactory, but more laborious, method would be to determine the room-temperature section of the ternary phase diagram for the two components and a suitable solvent. The microscope-fusion method has, however, proved useful for rapid surveys of systems for possible molecular-compound formation, and for detection and study of polymorphic transformations. The situation in solution or in the liquid or gas phase requires more elaborate methods (usually analysis of spectra) and is less certainly defined. The thermodynamic parameters of solid molecular compounds or of those in fluid phases can be determined by various conventional techniques; not much information is available and most of this concerns the compounds in solution. Qualitative comparisons of the relative effectiveness of various "complexing agents" have been made, but a systematic quantitative study of the stabilities of the various molecular compounds remains a task for the future.

The crystal structures of over forty π-molecular compounds have been analyzed, but the crystal chemistry (the types of arrangement and the occurrence of phase and order–disorder transformations) requires more attention. The geometrical structures of molecular compounds in solution are virtually inaccessible by available techniques; electron diffraction could perhaps be used to study the gas-phase structures but no result of such a study has been reported.

There have been numerous theoretical studies of charge-transfer molecular compounds in solution, the primary aims being to account for the interaction between the components and for the appearance of a new band (or bands) in the near-UV or visible region of the spectrum. There have been correspondingly many experimental studies of the solution spectra. Some work has also been done on the spectra (UV, visible, and infrared) of crystalline molecular compounds and correlations have been made, using polarized radiation, of single-crystal spectra with crystal structure and theoretical calculations. More work of this type will certainly be done, because spectra and crystal structures provide basic experimental data for theoretical analysis.

Theoretical treatment of crystalline molecular compounds is more complicated than study of a single pair of component molecules because it is necessary to consider the interactions within the whole three-dimensional array. On the other hand, the spatial relations in the crystals are known, in contrast to the situation in solution where there are also appreciable but unknown interactions between pairs of components and surrounding solvent molecules. In addition, in the crystals the component molecules are almost invariably found in one-dimensional stacks, with stronger interaction within the stacks than between them. Thus, treatment of the crystal on the basis of a one-dimensional approximation is valid, and important progress has been made along these lines by McConnell and his collaborators (Section IV-C).

There are indications of interesting physical and chemical properties among the donor–acceptor molecular compounds. The crystals are markedly anisotropic and show a wide range of semiconductor properties; there are reports of catalytic activity and behavior as polymerization initiators. However, few if any concrete applications have yet emerged.

## IV. SUITABLE PARTNERS

### A. Introduction

Which molecules act as electron-donors and which as electron-acceptors in $\pi$-molecular compounds? How is donor or acceptor quality measured and how is it enhanced or diminished by various substituents? These questions have been considered by Lepley and Thelman[44] and by Hammond,[45,46] particularly in regard to electron-acceptors. Theory suggests that a good donor has a low ionization potential, and a good acceptor a high electron affinity. Particularly effective electron transfer would be expected if both donor and acceptor were planar, although ionization potential and electron affinity are not related to molecular planarity. The broad lines regarding the influence of substituents are clear:[34] electron-donating substituents enhance donor strength, and electron-withdrawing substituents diminish it, and conversely for the electron-acceptors. Type, number, and location of substituent groups are all important factors.

These ideas have so far been tested almost entirely by the qualitative criterion of formation of an isolable crystalline molecular compound or, at least, by its appearance in the binary phase diagram (the disadvantages of the binary phase diagram as an indicator of compound formation have been mentioned above). Neither occurrence of a colour change when solutions of donor and acceptor are mixed, nor formation of a non-crystalline precipitate of doubtful stoichiometry, has been adjudged here to be sufficient proof of existence of a molecular compound.

Often, when the stability of a compound is mentioned in the literature, this refers to volatility of one of the components at room temperature in an open system. Thermodynamic parameters such as free energy of formation have been measured for rather few molecular compounds. Some anomalies in the older literature may be induced to disappear by repeating the work with purer starting materials or checking the products by the more powerful analytical techniques now available.

Finally, one should note that donor and acceptor are relative terms since particular compounds act as donors in some circumstances and as acceptors in others. For example, aromatic hydrocarbons usually behave as reasonably strong electron-donors; however, in the presence of the very powerful donor tetrakis(dimethylamino)ethylene (TDAE), they act as electron-acceptors (Section IV-F).

## B. Acceptors

**1. General.**—One may classify acceptors according to general structural type and according to substituents, which may be unmixed or mixed. The general structure types are: (i) Inorganic and aliphatic compounds: the few examples in this group have been placed outside the boundaries of the present treatment; however, it is desirable to draw attention to their existence. (ii) Olefinic compounds: tetracyanoethylene (TCNE) and hexacyanobutadiene are the most important examples. (iii) Substituted aromatic compounds: substituted benzenes have been used most but some polynitronaphthalenes form molecular compounds. (iv) Substituted aromatic ketones and quinones: most attention has been paid to substituted benzoquinones, but fluorenones, naphthaquinones, and anthraquinones have also been studied. In a few instances we shall find it convenient to deviate from the strict canons of this classification.

Appropriate substitution is generally essential if a molecule is to perform adequately as an electron-acceptor. For example, maleic anhydride, phthalic anhydride, and hexachlorobenzene do not form stable molecular compounds with most aromatic hydrocarbons, but tetrachlorophthalic anhydride (TCPA) is a powerful electron-acceptor. The details of such synergic action have not yet been elucidated by theoretical or experimental studies. Experience, however, suggests that mixing of substituents produces the most powerful electron-acceptors; for example, 2,3-dicyano-5,6-dichloro-benzoquinone (DDQ) is a more powerful electron-acceptor than either chloranil or cyananil, both of which are much more powerful than *p*-benzoquinone itself. Unsymmetrical substitution also enhances the effectiveness of electron-acceptors, presumably because dipole and polarization forces can now contribute, in addition to charge-transfer interactions.

Many typical electron acceptors are brought together in formulae (**1**)–(**34**). The classification here is in terms of fundamental structures to which substituents have been appended in various ways. The fundamental structures are ethylene, *p*- and *o*-benzoquinone, and cyclopentadienone, benzene, and naphthalene while the most important electron-withdrawing substituents are nitro-, cyano-, halo-, anhydride, furoxan, and furazan groups.

**2. Electron affinities.**—The electron affinities of organic molecules have recently been critically reviewed by Briegleb.[47] The electron affinity ($EA$) of an electron-acceptor molecule A is equal to the difference in energy between the anion $A^-$ and the uncharged molecule A plus an electron infinitely separated from it. Thus

$$EA = E_{(A+e)} - E_{A^-} \qquad \ldots (1)$$

The ionization energy of an electron-donor will be needed below where the various donors are discussed. The ionization energy ($I$) of an electron-donor molecule D is equal to the difference in energy between the uncharged molecule D and the cation $D^+$ plus an electron infinitely separated from it:

$$I = E_{(D^+ + e)} - E_D \qquad \ldots (2)$$

*Substituted ethylenes*

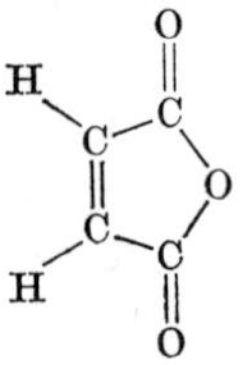


**(1)**

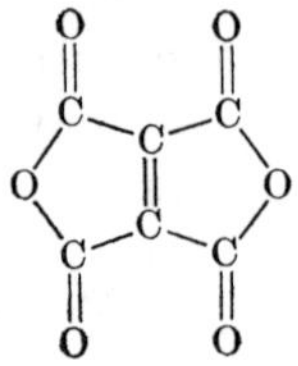


**(2)**

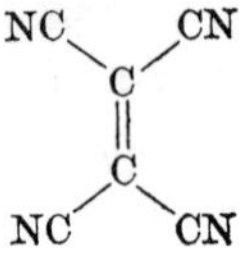


**(3)**

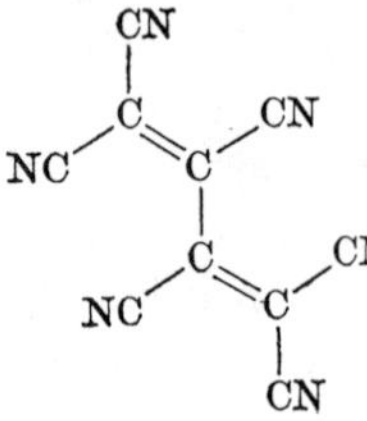


**(4)**

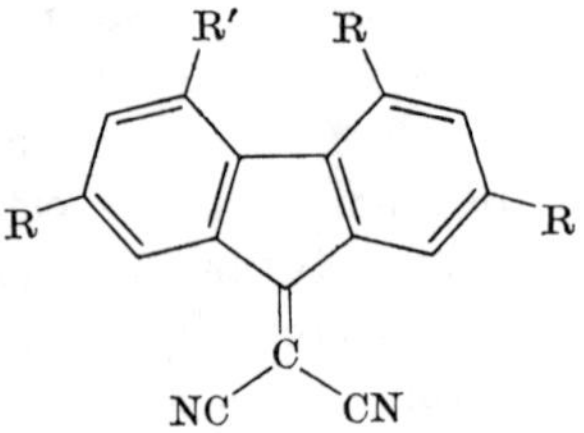


**(5)**

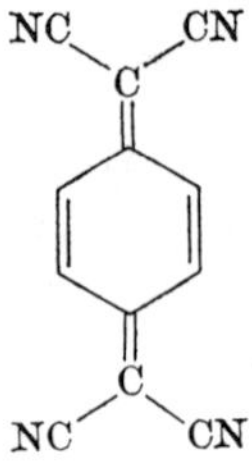


**(6)**

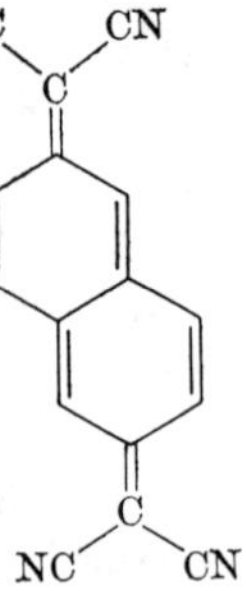


**(7)**

**a**: R = R′ = H
**b**: R = $NO_2$, R′ = H
**c**: R = R′ = $NO_2$

*Substituted* p-*benzoquinones*

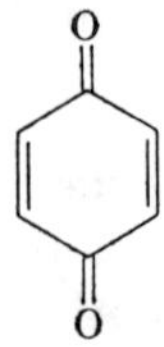


**(8)**

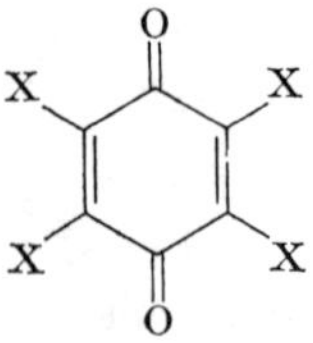


**(9)**

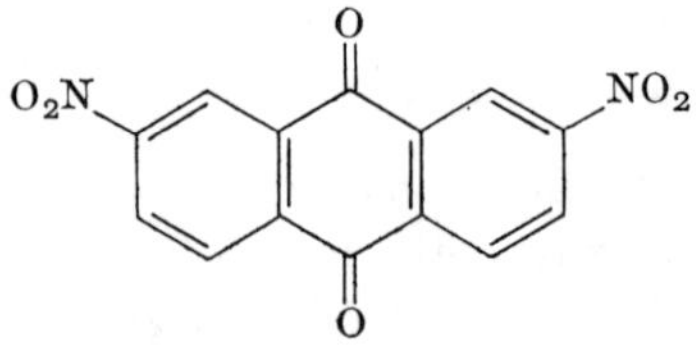


**(10)**

X = F, Cl, Br, I, CN, or $CH_3$

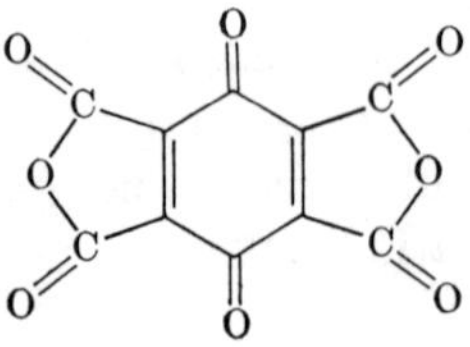


**(11)**

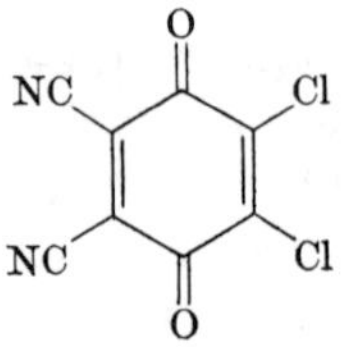


**(12)**

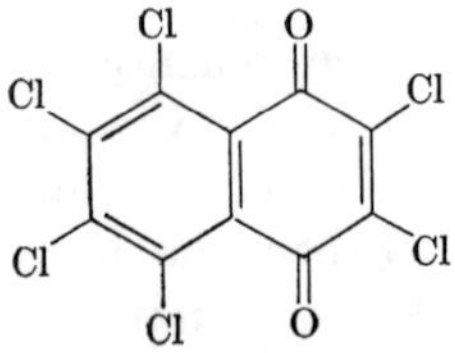


**(13)**

## Substituted o-benzoquinones

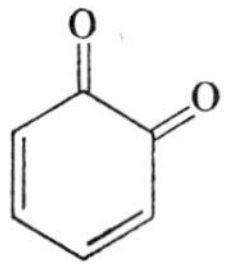

**(14)**

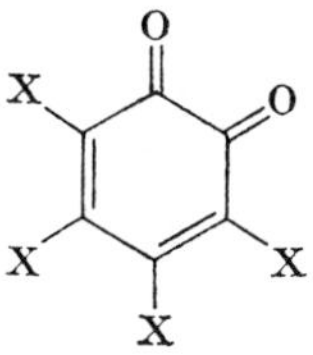

**(15)**

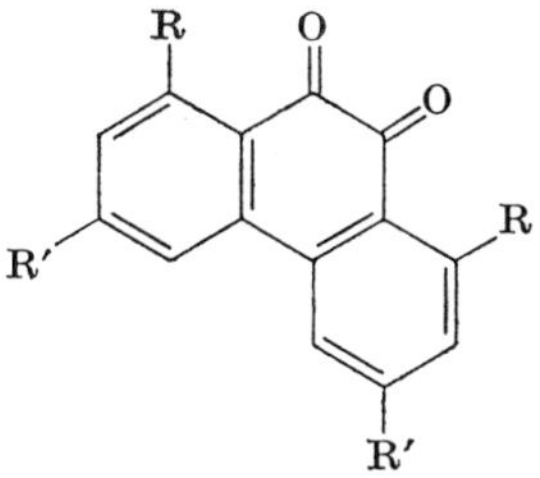

**(16)**

a: R = NO$_2$, R' = H
b: R = H, R' = NO$_2$

## Substituted benzenes

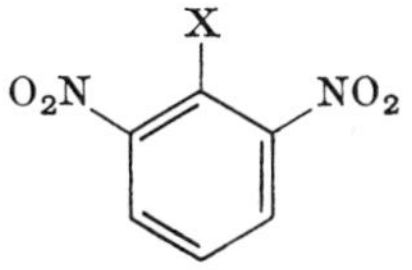

**(17)**

X = H, OH, or CONH$_2$

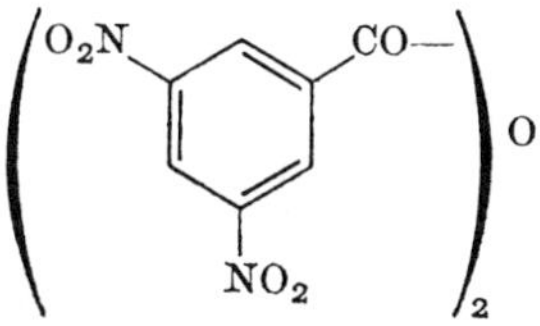

**(18)**

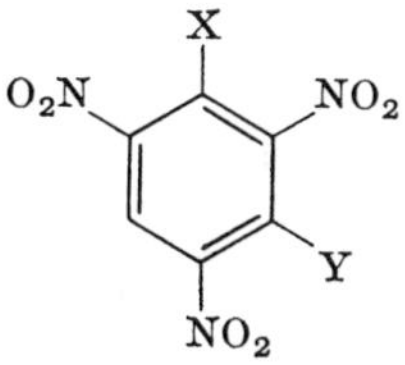

**(19)**

X = H, OH, CH$_3$, Cl,
Br, I, NH$_2$, or N≡N;
Y = H; or X = Y = OH

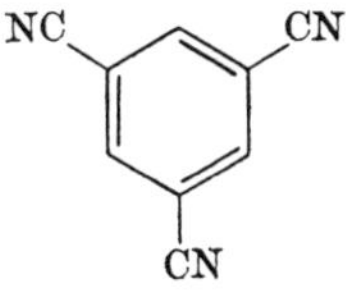

**(20)**

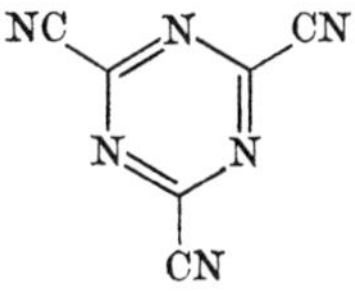

**(21)**

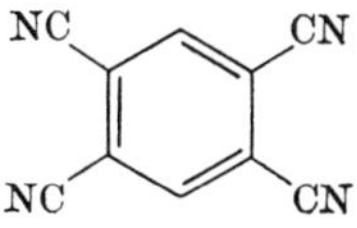

**(22)**

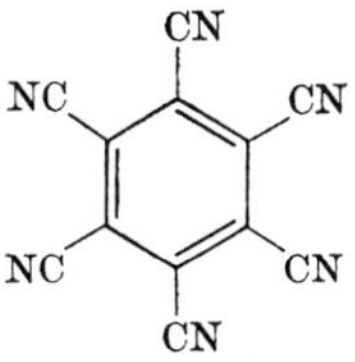

**(23)**

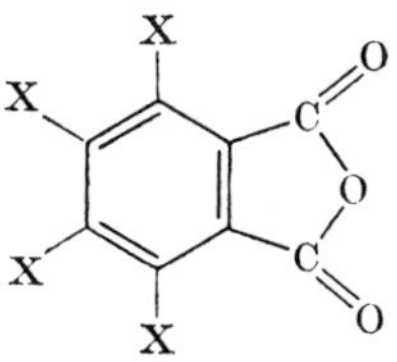

**(24)**

X = H, F, Cl, Br, or I

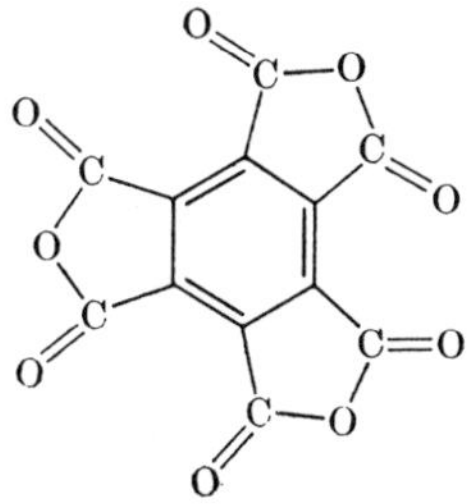

**(25)**

179

## Substituted benzenes (continued)

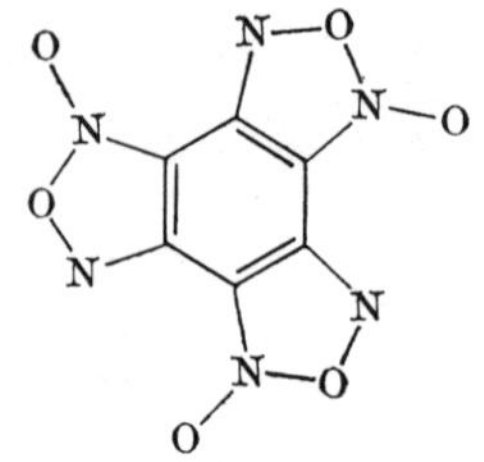

(26)

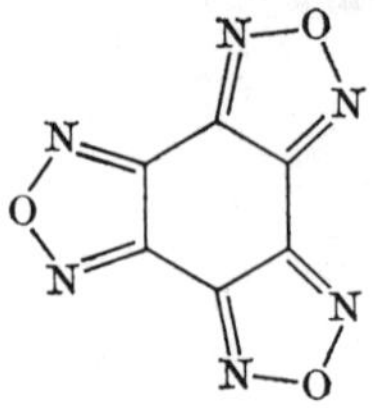

(27)

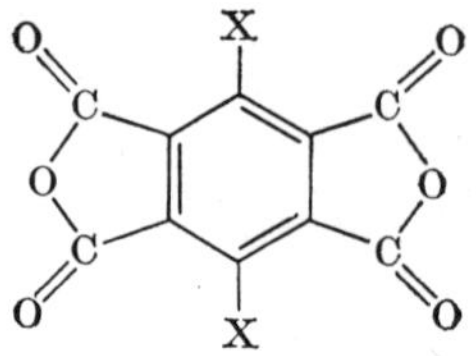

(28)

X = H or Br

## Substituted naphthalenes

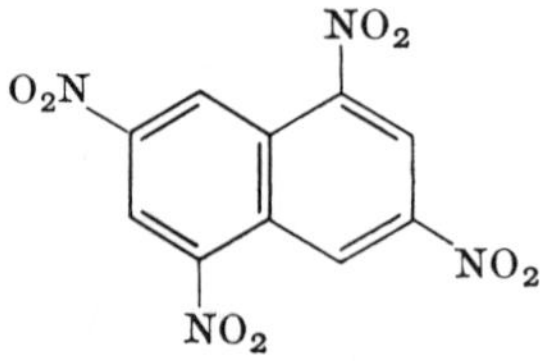

(29)

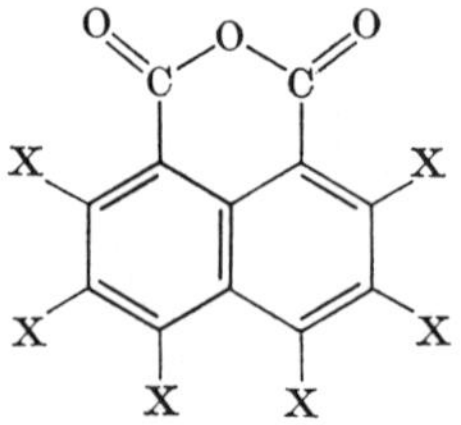

(30)

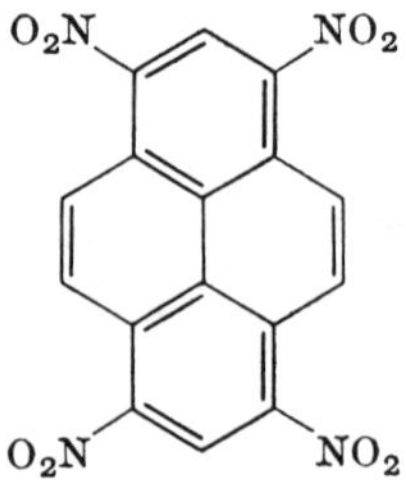

(31)

## Cyclopentadienone systems

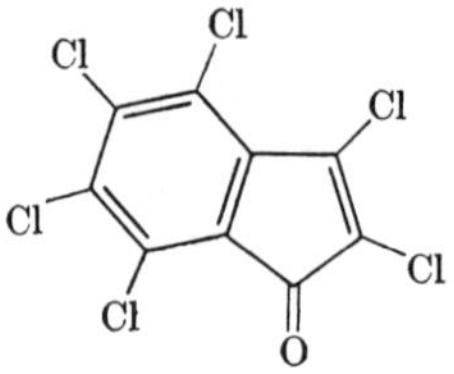

(32)

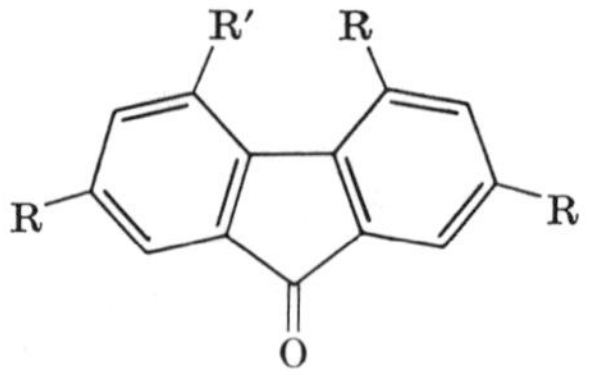

(33)

a: R = NO$_2$, R' = H
b: R = R' = NO$_2$

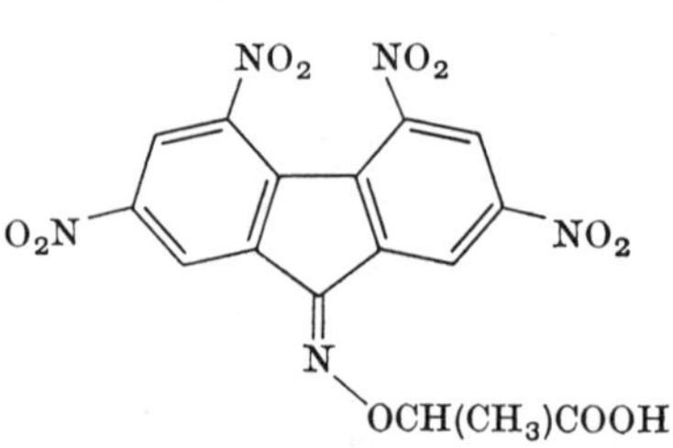

(34)

180

Table 1. Electron affinities of various electron acceptors.[a]

| Electron acceptor | Electron affinity (ev) | Formula no. |
|---|---|---|
| Tetracyanoethylene (TCNE) | 1.80 | **3** |
| 7,7,8,8-Tetracyanoquinodimethane (TCNQ) | 1.7 | **6** |
| 2,3-Dichloro-5,6-dicyanobenzoquinone (DDQ) | 1.95 | **12** |
| Tetracyano-*p*-benzoquinone | 1.7 | **9** (X = CN) |
| Tetrabromo-*p*-benzoquinone (bromanil) | 1.4 | **9** (X = Br) |
| Tetrachloro-*p*-benzoquinone (chloranil) | 1.37 | **9** (X = Cl) |
| Tetraiodo-*p*-benzoquinone (iodanil) | 1.36 | **9** (X = I) |
| 2,6-Dinitro-*p*-benzoquinone | 1.75 | |
| 2,6-Dibromo-*p*-benzoquinone | 1.2 | |
| 2,3-Dicyano-*p*-benzoquinone | 1.7 | |
| 2,3-Dichloro-*p*-benzoquinone | 1.1 | |
| Tetrabromo-*o*-benzoquinone (*o*-bromanil) | 1.6 | **15** (X = Br) |
| Tetrachloro-*o*-benzoquinone (*o*-chloranil) | 1.55 | **15** (X = Cl) |
| Phenanthroquinone | 0.7 | **16** (R = R′ = H) |
| 1,4-Naphthoquinone | 0.7 | |
| *p*-Benzoquinone | 0.6 | **8** |
| Duroquinone | 0.6 | **9** (X = $CH_3$) |
| 9,10-Anthraquinone | 0.5 | |
| 1,3,5-Trinitrobenzene (TNB) | 0.7 | **19** (X = Y = H) |
| *p*-Dinitrobenzene | 0.7 | |
| Nitrobenzene | 0.5 | |
| *m*-Dinitrobenzene | 0.3 | **17** (X = H) |
| *o*-Dinitrobenzene | 0 | |
| 2,4,6-Trinitrotoluene (TNT) | 0.6 | **19** (X = $CH_3$, Y = H) |
| 2,4,6-Trinitroxylene | 0.4 | **19** (X = Y = $CH_3$) |
| 1,2,4,5-Tetracyanobenzene (TCNB) | 0.4[b] | **22** |
| 1,3,5-Tricyanobenzene | 0.1 | **20** |
| 1,4-Dicyanobenzene | 0 | |
| Mellitic trianhydride | 1.17 | **25** |
| Dibromopyromellitic dianhydride | 1.16 | **28** (X = Br) |
| Pyromellitic dianhydride (PMDA) | 0.85 | **28** (X = H) |
| Tetrachlorophthalic anhydride (TCPA) | 0.58 | **24** (X = Cl) |
| Maleic anhydride | 0.57 | **1** |
| Phthalic anhydride | 0.15 | **24** (X = H) |
| 9-Fluorenylidenemalononitrile | 0.70[c] | **5a** |
| 2,4,7-Trinitro-9-fluorenylidenemalononitrile | 1.13 | **5b** |
| 2,4,5,7-Tetranitro-9-fluorenylidene-malononitrile | 1.52 | **5c** |
| 2,4-Dinitro-9-fluorenone | 0.93[c] | |
| 2,4,7-Trinitro-9-fluorenone (TNF) | 0.95 | **33a** |
| 2,4,5,7-Tetranitro-9-fluorenone (TENF) | 1.16 | **33b** |

[a] Experimental values are from Briegleb[47] except for [b] Rosenberg *et al.*[48] and [c] Mukherjee.[49]

Table 2. Ionization energies ($I$) of some unsubstituted and substituted aromatic hydrocarbons.

| Molecule | $I$ (ev) | Remarks |
|---|---|---|
| Benzene | 9.2 | E |
| Naphthalene | 8.2 | E |
| Anthracene | 7.5 | E |
| Tetracene | 7.0 | E |
| Pentacene | 6.6 | C |
| Phenanthrene | 8.1 | E |
| Triphenylene | 8.1 | E |
| Chrysene | 7.8 | E |
| trans-Stilbene | 8.0 | E |
| Dibenz[a,h]anthracene | 7.6 | C |
| Benzo[c]phenanthrene | 7.9 | C |
| Perylene | 7.1 | E |
| Coronene | 7.6 | E |
| Biphenyl | 8.4 | E |
| Tetrabenzonaphthalene | 7.6 | C |
| Pyrene | 7.6 | E |
| Hexamethylbenzene | 8.0 | D |
| TMPD | 6.7 | D |

$E$ refers to experimentally measured values, references being given by Briegleb.[47]

C refers to values calculated from Briegleb's equation (21):

$$I = 5.11 + 0.701\tilde{\nu}_0 \ (\text{ev})$$

where $\tilde{\nu}_0 \ (= E_j - E_1)$ is the energy difference between the highest occupied and the lowest unoccupied orbital according to Hückel molecular orbital theory (the equation has been shown to be valid for the range $\tilde{\nu}_0 \approx 2.5–4$ ev).

Values labelled D are taken from Briegleb's paper,[47] p. 626.

The difference in energy between the lowest vibrational level of the ground state of the neutral molecule and the corresponding level of the cation or anion is termed the adiabatic ionization energy, or the adiabatic electron affinity, respectively. The energy difference between the energy level of this ground state and that part of the potential curve of the ionized state to which, on applying the Franck–Condon principle, transition is most likely to occur is denoted as the vertical ionization energy or the vertical electron affinity, as the case may be. Mulliken[21] considers that the vertical ionization energies and electron affinities usually apply to charge-transfer interactions. It should also be noted that it is not yet certain (ref. 47, p. 619) whether the various methods for measuring ionization energies give the vertical or the adiabatic ionization energy. Further, the values cited are seldom accurate to better than about 0.2 ev. Values of the electron affinity for many of the electron acceptors considered here are collected in Table 1; analogous values of ionization energy are in Table 2. It is of interest to compare typical values of $I$ and $EA$ for organic donors and acceptors with those for alkali metal and halogen atoms. For a typical organic donor $I \approx 7$ ev whereas for Li $I_1 = 5.4$ ev and for Cs $I_1 = 3.9$ ev; for the best organic electron-acceptor $EA = 2$ ev, whereas $EA$ for fluorine $= 3.4$ ev and for iodine 3.23 ev. Thus, it is clearly more difficult to form cations from organic donors than from alkali-metal atoms, and also less advantageous to form anions from organic acceptors than from halogen atoms.

## C. Different Kinds of Donor–Acceptor Interaction

The ground states of most crystalline molecular compounds are non-ionic, but evidence has accumulated in recent years that crystals with ionic ground states also occur and that these often have particularly interesting physical properties. In one kind of molecular compound—the "isomeric complexes" (Komplex-Isomeren)—a particular pair of partners can interact in two chemically distinct ways, depending on temperature and pressure, so that two polymorphs of the same (gross) chemical formula are obtained. One of these is a donor–acceptor molecular compound with a non-ionic ground state while the other is salt-like. The experimental evidence is first summarized and then used to provide a general framework for classification of the molecular compounds of different types.

**1. Distinction between molecular compounds with non-ionic and ionic ground states.**—The suggestion by Weiss[36] in 1942 (see also Weiss[50]) that all $\pi$-molecular compounds are actually charge-transfer salts, involving complete transfer of an electron from donor to acceptor, was prescient but extreme. Experience has shown that the "weak complexes" do not have complete electron transfer in their ground states and are better described by the appropriate parts of Mulliken's theory;[38–40] nevertheless, there are an appreciable number of molecular compounds where the components are ionized in the ground state. The distinction between the two groups can be demonstrated in a number of ways.

*a. Spectroscopy in the near-UV, visible, and IR regions.* The spectra of $\pi$-molecular compounds are given to a good approximation by the sum of the spectra of the individual components, with an additional absorption band, the charge-transfer (CT) band, which occurs in the near-UV, visible or near-IR region, between about 5 and 30 kK. (Sometimes more than one CT band is observed.) The characteristic intense colours of the π-molecular compounds are due to the CT bands; their position is such that they do not affect the IR spectrum between 4 and 0.7 kK (2–15 $\mu$). The components may be present in the crystalline molecular compound as uncharged or charged species, and their contributions to the spectrum will differ accordingly; conversely, the nature of the ground state, whether ionic or non-ionic, can be determined by overall comparison of the spectrum of the molecular compound with spectra of the individual components in ionized and neutral conditions. The example of TMPD:chloranil is discussed in some detail in Section VI-J.

There are small differences in frequencies and intensities between the IR spectra of molecular compounds and the sum of the contributions from the components; Kross and Fassel[51] have found some correlation between frequency shifts and the chemical nature of the donor for a group of picric acid molecular compounds.

*b. Magnetic susceptibilities.* Kainer and Überle[52] have measured the (bulk) magnetic susceptibilities of some TMPD:haloanil molecular compounds at various temperatures between 295°K and 77°K. Paramagnetism was detected in a number of samples.

*c. ESR measurements.* ESR measurements on many molecular compounds have demonstrated the presence of free radicals, but usually only in very small amounts. Consequently, the results were at first regarded with considerable scepticism, and the free radicals were ascribed to impurities produced by decomposition of the rather unstable components. ESR signals are certainly produced by impurities in some instances, but there are enough established examples of reproducible behaviour to demand an explanation connected with the properties of the molecular compounds themselves. The most convincing explanations are based on single-crystal studies, especially as a function of temperature. The possible situations are sketched here; they are discussed below in more detail together with the results of other measurements for particular molecular compounds.

Paramagnetic species derived from organic molecules are almost always either in a doublet state (one unpaired electron, $S = \frac{1}{2}$) or a triplet state (two unpaired electrons, $S = 1$). The species discussed here have one unpaired electron per radical; if there is no interaction between unpaired electrons on neighbouring radicals in the crystal, then the crystal as a whole will be in a doublet state. The intensity of the ESR signal will then decrease with increasing temperature, because of progressive equalization of populations in the two energy levels of the electron in an external magnetic field. This is normal Curie-type temperature-dependence.

However, it has been found that the ESR signals from many molecular compounds increase in intensity with increasing temperature, and these crystals thus do not consist of a simple collection of radicals in the doublet state. The usual explanation for this type of behaviour is that there is a temperature-dependent equilibrium between two species, a diamagnetic ground state and a thermally accessible, paramagnetic, excited state. One possibility is establishment of equilibrium between the radicals and their diamagnetic parents; another is that adjacent *pairs* of radicals could interact to give a diamagnetic singlet ground state and a paramagnetic triplet excited state. However, for many crystals the coupling between adjacent radicals is so strong that a collective description of the spin excitations is essential, and this has been worked out by McConnell and his co-workers in a number of papers. The principal conclusions are summarized by McConnell, Hoffman, and Metzger[54] for crystalline donor–acceptor molecular compounds, where the isolated components are diamagnetic. The D and A molecules were assumed to be arranged in alternating fashion in linear stacks, in accordance with the experimental evidence summarized in Section V. The elementary excitations in such stacks depend on the nature of the ground state of the crystal. Interactions between different stacks are ignored. The lowest-energy elementary electronic excitations above the ground state in the non-ionic crystal are expected to be triplet excitons, corresponding to a Coulomb-bound pair of parallel spins propagated through the crystal lattice, *i.e.*:

$$----\mathrm{D\ A\ D\ A\ D^+ \uparrow} + \mathrm{A^- \uparrow} + \mathrm{D\ A\ D\ A}----$$

In general, such crystals are expected to be diamagnetic or very feebly paramagnetic since the magnetic excitations are not thermally accessible. Optical excitation would be necessary to populate such states sufficiently for detection by ESR.

The donor–acceptor crystal with an ionic ground state can be regarded as composed of an array of non-interacting linear Heisenberg antiferromagnets. The ESR spectra of the crystals are quite different from those of the individual, singly charged donor or acceptor ions isolated in a solid matrix, but they may be understood on the basis of collective, or crystal, states.[55–57] For an even number of radicals in a stack, the ground state is a singlet with alternation of spins on adjacent radicals. The excited states of the linear Heisenberg antiferromagnet are not well understood, but it is known that the lower excited states are triplets. The two parallel spins forming the triplet state behave differently in the two extreme situations encountered. If the strength of the coupling between adjacent radicals alternates strongly along a stack, then the two spins are strongly correlated and move together as a pair. Hughes and Soos[58] call these correlated spin pairs "Frenkel spin excitons" and note that they can be unambiguously identified by the fine-structure splitting found at low exciton densities and resulting from the electron–electron dipolar interaction between the spins moving together as a pair. Frenkel spin excitons have been studied in ion-radical salts such as Würster's blue perchlorate, $TMPD^+ ClO_4^-$ and various TCNQ salts (TMPD = $N,N,N',N'$-tetramethyl-$p$-phenylenediamine, TCNQ = 7,7,8,8-tetracyanoquinodimethane).

The second situation occurs when the coupling between adjacent radicals in a stack is essentially equal throughout the stack. The two spins now move independently, and Hughes and Soos[58] refer to this as a "Wannier spin exciton." No fine-structure splitting is expected for Wannier spin excitons. The adjacent radicals are coupled so strongly that even at room temperature most of the spins are in the diamagnetic singlet state and there is a thermal-equilibrium density of Wannier spin excitons. Regular linear Heisenberg antiferromagnets are found in ion-radical molecular compounds such as $p$-phenylenediamine:chloranil (PDC) (Hughes and Soos[58]) and TMPD:chloranil,[59,60] which will be discussed in more detail below.

The effect of hydrostatic pressure on the ESR intensity should provide another means of demonstrating the ionic character of the ground state.[54] Compression of the crystal will bring D and A units closer together, thus decreasing the $DA \rightarrow D^+A^-$ excitation energy but increasing the $D^+A^- \rightarrow DA$ excitation energy. Thus ESR intensity should decrease with increasing pressure in crystals with ionic ground states, and conversely for crystals with non-ionic ground states. A concrete illustration is provided by PDC[58] (see Section VI-J).

**2. Isomeric complexes.**—Pfeiffer[4] suggested many years ago that the interaction between the two components in molecular compounds could be considered as an interaction between particular force fields localized in different regions of the component molecules. If more than one type of force field were present in each of the components, then isomeric molecular complexes could be obtained depending on which pair of force fields predominated. The idea was elaborated by Hertel and his co-workers in a number of papers in the period 1924–34. The phenomenon was called "Komplexisomerie." The most extreme example reported is that of $N,N$-dimethyl-2-naphthylamine and 2,4,6-trinitroanisole.[61] Crystallization from solution under different conditions gives either red crystals (m.p. 65°) or yellow crystals (m.p. 190°). The red crystals correspond to the $\pi$-molecular compound, which dissociates into its components on dissolution, while the yellow crystals are $N,N,N$-trimethyl-2-naphthylammonium picrate, a typical organic salt giving an ionic solution in which the original molecular components are no longer present.

$N(CH_3)_3{}^+ \quad O_2N \quad \text{O}^- \quad NO_2$  
yellow  
m.p. 190°  
$NO_2$

$N(CH_3)_2 \quad O_2N \quad OCH_3 \quad NO_2$  
red  
m.p. 65°  
$NO_2$

The differences between the two isomers in the other examples discussed by Hertel *et al.* are not as marked as in the above example but depend on analogous interactions between aromatic amines and polynitrophenols. Hertel[62] has suggested that progressive weakening of the acid–base (hydroxyl–amine) interaction and strengthening of the donor–acceptor interaction should produce binary compounds that range in type from "true phenoxides" at one end of the series to "true molecular compounds" at the other. Intermediate situations should also be found, where the same partners form either phenoxides or molecular compounds, depending on ambient conditions. This is illustrated by the following summary of Hertel's experimental results.[62]

The strong bases aniline, *m*- and *p*-chloroaniline, *m*- and *p*-bromoaniline, the toluidines, 2-methyl-4-bromoaniline and the naphthylamines give "true phenoxides" with picric acid (**19**; X = OH, Y = H) and 2,6-dinitrophenol (**17**, X = OH). 4-Bromo-1-naphthylamine and 4-chloro-1-naphthylamine give only "true phenoxides" with picric acid, but give both isomers with 2,6-dinitrophenol; similar results were obtained for *o*-chloroaniline, 2,4-dichloroaniline 2,4-dibromoaniline, and 2-bromo-4-methylaniline. The weak bases *o*-bromo- and *o*-iodoaniline and 1-bromo-, 1-chloro-, and 1,6-dibromo-2-naphthylamine give both isomers with picric acid, but only molecular compounds with 2,6-dinitrophenol. Extremely weak bases such as skatole and indole form only molecular compounds with both picric acid and 2,6-dinitrophenol. The colours, melting points, and types of the binary compounds formed by some aromatic amine–polynitrophenol combinations are given in Table 3.

Table 3. Characteristics of binary compounds formed by various aromatic amine–polynitrophenol combinations (Hertel[62]).
I = "true phenoxide"; II = molecular compound.

| Acceptor/acid $K_a$ | Picric acid $1.6 \times 10^{-3}$ | 2,6-Dinitro-phenol $2.7 \times 10^{-4}$ | 2,4-Dinitro-phenol $1.0 \times 10^{-4}$ | 2,5-Dinitro-phenol $7 \times 10^{-6}$ |
|---|---|---|---|---|
| Donor/base $K_b$ | | | | |
| Aniline $3.2 \times 10^{-10}$ | I Bright yellow | I Golden yellow | I Lemon-yellow | II Orange |
| *p*-Bromoaniline $8.8 \times 10^{-11}$ | I Yellow M.p. 180° | I Yellow M.p. 92° | II Red M.p. 70° | II Violet M.p. 72° |
| *m*-Bromoaniline $1 \times 10^{-12}$ | | I Yellow M.p. 84° | | |
| *o*-Bromoaniline | | II Red M.p. 34° | | |
| *o*-Chloroaniline $9.2 \times 10^{-13}$ | I Yellow M.p. 134° | II Red M.p. 41° | | |
| 2,4-Dibromoaniline | | II Red M.p. 58° | | |

The aromatic amine:polynitrophenol combinations that give isomers are listed in Table 4. In normal crystal-chemical nomenclature these isomers are polymorphs, which can be either enantiotropically or monotropically related, *i.e.*, in some instances each polymorph is stable within a certain range of temperature (enantio-

Table 4. Isomeric complexes (polymorphs) for various aromatic amine–polynitrophenol combinations (after Hertel[62]).

| Components | Colour | Transn. pt. | M.p. | Type |
|---|---|---|---|---|
| *Picric acid as acceptor* | | | | |
| 1-Chloro-4-naphthylamine | Yellow | 162°? | — | I }? |
| | | | 180° | II? |
| 1-Bromo-2-naphthylamine[a] | Yellow | 117° | — | I } enantiotropic |
| | Dark red | | 178° | II |
| 1-Chloro-2-naphthylamine | Yellow | 130° | — | I } enantiotropic |
| | Dark red | | 174° | II |
| 1,6-Dibromo-2-naphthyl-amine | Yellow | 0°? | — | I }? |
| | Dark red | | 131° | II |
| *o*-Bromoaniline | Yellow | 95° | — | I } enantiotropic |
| | Orange | | 128° | II |
| *o*-Iodoaniline | Yellow | 96° | — | I } enantiotropic |
| | Orange | | 112° | II |
| *o*-Toluidine[c] | Yellow | | 219° (explodes) | I? monotropic(?) |
| | Red | | 110° | II? |
| *2,6-Dinitrophenol as acceptor* | | | | |
| 4-Bromo-1-naphthylamine[b] | Yellow | — | 91.5° | I } monotropic |
| | Red | | 84.5° | II |
| 1-Chloro-4-naphthylamine | Yellow | 76° | — | I } enantiotropic |
| | Red | | 81° | II |

[a] See Figure 1.  [b] See Figure 2.  [c] From Râscanu.[63]

tropy), whereas in others one polymorph is always metastable with respect to the other (monotropy). In the examples examined so far it has always been found that the "true phenoxide" is stable at lower temperatures than the molecular compound (where the polymorphs are enantiotropically related); only one example of a monotropic system has been studied (4-bromo-1-naphthylamine: 2,6-dinitrophenol) and here the "true phenoxide" is the stable form and the molecular compound metastable. These facts suggest that the interaction energy between the components has the dominant role in stabilizing the "true phenoxide" structure, but that, at higher temperatures, entropy considerations become more important and stabilize the molecular-compound polymorph. This entropy increment may be derived partly from disorder in the molecular compounds, and partly from additional motion consequent on a reduction of the forces binding the components. No work has yet been done on the polymorphs at temperatures below 0°c, but one is tempted to suggest that molecular-compound polymorphs stable at room temperature and above may transform into "true phenoxide" polymorphs on cooling.

The phase diagram has been determined for one enantiotropic system (1-bromo-2-naphthylamine:picric acid,[61] Figure 1)* and one monotropic system (4-bromo-1-naphthylamine:2,6-dinitrophenol,[61] Figure 2), and some thermodynamic and preliminary crystal-

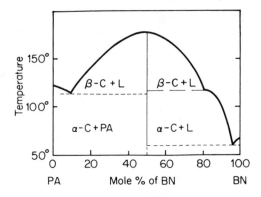

Figure 1. Phase diagram for the system 1-bromo-2-naphthylamine (BN):picric acid (PA). α-C is the yellow 1:1 molecular compound and β-C is the red 1:1 molecular compound. - - - Thaw temperature. – – Transformation temperature. [After Hertel.[62]]

lographic studies have been made on the latter. Its phase diagram consists of two separate parts, showing the independent behaviour of the stable and the metastable system. From the measured heats of solution and solubilities of both polymorphs[64] the transition red → yellow is accompanied by changes $\Delta H = -3.3$ kcal mole$^{-1}$, $\Delta S \approx -9.3$ cal mole$^{-1}$ deg$^{-1}$, $\Delta G \approx -0.5$ kcal mole$^{-1}$ at 300°K. These values cannot be regarded as accurate, but they indicate that the yellow ("true phenoxide") polymorph has both a lower enthalpy

* The crystal structure of this molecular compound has been determined (Addenda to Section VI-E).

and a lower entropy than the red (molecular compound) polymorph. Cell dimensions and space groups have been determined for both polymorphs:[65]

"True phenoxide": yellow plates, $a = 14.0$, $b = 8.0$, $c = 14.5$ Å, $\beta = 102°6'$, $\rho_{meas} = 1.654$ g cm$^{-3}$; $Z = 4$; $P2_1/c$.
Molecular compound: red needles, $a = 9.5$, $b = 13.5$, $c = 13.8$ Å, $\beta = 105°$, $\rho_{meas} = 1.56$ g cm$^{-3}$; $Z = 4$; $P2_1/c$.

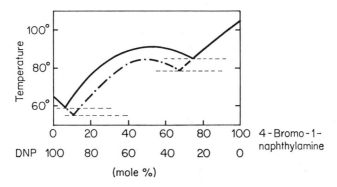

Figure 2. Phase diagram for the system 4-bromo-1-naphthyl-amine:2,6-dinitrophenol (DNP). The full curve shows the behaviour of the stable system; here mixtures of the "true phen-oxide" polymorph and one of the components were used to determine the thaw points and liquidus. The dot–dash curve shows the behaviour of the metastable system; here mixtures of the molecular compound polymorph and one of the components were used to determine thaw points and liquidus. [After Hertel.[61,62]]

The volumes per formula unit are yellow 396 Å$^3$, red 426 Å$^3$, the difference being unexpectedly large. No structural work has been done, but Briegleb and Delle have provided strong evidence, by reflection spectroscopy in the visible and UV region[66] and by IR spectroscopy in the 2–5 $\mu$ region,[67] that the "true phenoxides" are salts while the molecular compounds show CT bands somewhat similar to those found in typical $\pi$-molecular compounds. If these conclusions are substantiated by crystal-structure analyses, the polymorphic transformations must involve rather subtle cooperative proton-transfer reactions between the two phases, where N–H bonds are broken in one phase (the salt) and O–H bonds are formed in the second phase (the molecular compound). Hydrogen bonding of the ions, as envisaged by Briegleb and Delle,[67] however, facilitates such proton-transfer reactions.

Some information on this point is available for piperidine picrate, which Briegleb and Delle[66,67] classify as a salt on the basis of its spectrum. The crystals are triclinic with [001] $\simeq$ 6.92 Å, and polarized infrared absorption studies (Kross, Nakamoto, and Fassell[68]) suggest that an N–H bond of the piperidinium ion is coplanar with the picrate ring, in agreement with a hydrogen-bonded salt-like structure.

Another system that may involve features similar to those found in the isomeric complexes is guaiacol:picric acid.[69] Here red and yellow polymorphs have been reported, the red being less stable than the yellow, although it is not known whether there is an enantio-tropic or monotropic relation between them. The cell dimensions of the two forms are:

Yellow: $a = 25.35$, $b = 8.57$, $c = 13.7$ Å, $\beta = 96°$, $Z = 8$, $P2_1/a$, needle axis [001].

Red:   $a = 13.1$,  $b = 6.53$,  $c = 8.60$ Å, $\beta = 92°$, $Z = 2$, $P2_1$, needle axis [010].

The volumes per formula unit are 370 Å$^3$ (yellow) and 365 Å$^3$ (red), which do not differ significantly.

It has been reported[70] that aniline picrate occurs in two poly-morphic forms, the metastable form being 28% the more soluble in water at 18°. As the two forms are different shades of yellow, this is probably an ordinary example of polymorphism not involving isomerism; aniline is too strong a base to form a molecular compound with picric acid (Table 2). The cell dimensions of aniline picrate have been reported (Hertel and Schneider[71]), but the crystal structure is not known, nor which polymorph was studied.

### 3. Family tree of π-molecular compounds and related species.—

In this Section we consider the different types of interaction between two molecules, one acting formally as a donor and one as an acceptor. The entity transferred may be an electron; a small degree of transfer corresponds to a weak or Mulliken-type molecular compound, with essentially non-ionized components in the ground state, while a large amount of transfer gives ionized components in the ground state. We call these Weiss-type molecular compounds. In another group of molecular compounds (the "isomeric complexes") the entity transferable is either an electron or a proton. The partners participating in these molecular compounds give, depending on circumstances, either Mulliken-type molecular compounds or the so-called "true phenoxides", which appear to have salt-like structures with hydrogen-bonding between the ions.

An attempt has been made to represent these possibilities in the family tree of Figure 3. The question now arises whether the various

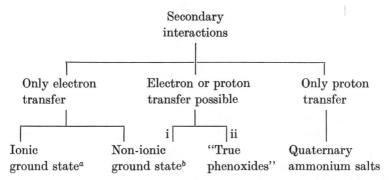

Figure 3. Family tree showing relationships between the various groups in the broad category "$\pi$-molecular compounds."

[a] Weiss type molecular compounds. [b] Mulliken-type molecular compounds.
i, Electron transfer. ii, Proton transfer.

groups in this Figure are separated by sharp boundaries or whether they merge gradually. Qualitative considerations based on spectral studies (Davis and Symons[53]) and semiquantitative calculations (McConnell, Hoffman, and Metzger[54]) suggest that crystalline 1:1 donor–acceptor molecular compounds are quite sharply divided into two classes, with non-ionic and ionic ground states severally. This clear-cut distinction is only valid at such low temperatures that the concentration of excitations is small. The group into which a particular molecular compound falls is determined by the relative magnitudes of the ionization potentials and the electron affinities of the components, on the one hand, and the Madelung interaction energy of the crystal structure on the other. Pott and Kommandeur,[59,60] however, have suggested that in certain circumstances a "molionic" crystal, containing both charged and uncharged varieties of the donor and acceptor molecules, will be more stable than either the purely molecular or purely ionic crystal. They suggest that polarization effects were not adequately considered by McConnell *et al.*[54] and have identified TMPD:chloranil as an example of a molionic crystal. It is difficult to assess the reliability of the rather generalized calculations that have been reported, but it is clear that the weight of current experimental evidence is entirely on the side of McConnell *et al.*[54] Many examples of molecular compounds with non-ionic ground states are known, whereas no molionic crystal has yet been found (as we show in Section VI-J, TMPD:chloranil is not a molionic crystal).

There has not been any theoretical discussion of "isomeric complexes." Here too, the experimental evidence favours a sharp distinction between the two types of "isomeric complex." Despite this, there can be a delicate balance between the stabilities of the two types, as shown by the occurrence of phase transformations accompanied by a change of structure type. McConnell[55] suggested that a similar situation could occur for π-molecular compounds with ionic and non-ionic ground states if the energy required to create ions were very similar to their Madelung interaction energy. No concrete example has yet been found, but one would expect the ionic ground state for a molecular compound to be favoured by low temperature and/or high pressure (cf. the behavior of isomeric complexes discussed below).

## D. Classification of the Known Acceptors

In this Section we consider in more detail the various electron-acceptors, classified according to chemical type (cf. Table 1), and the molecular compounds they form with different donors. We restrict ourselves here to experimental results concerning the formation of molecular compounds; structures and inferences from physical measurements about interactions between components are discussed later. Some workers have attempted to draw conclusions regarding the effectiveness of various electron-acceptors on the basis of experimental results on the formation of molecular compounds, and we discuss these generalizations briefly.

**1. Inorganic and aliphatic acceptors.**—*a. Inorganic compounds.* Dinitrogen tetroxide, which is electron-deficient, acts as an electron-acceptor towards both *n*- and π-donors (Addison and Sheldon[73]). The *n*-donors generally give 2:1 molecular compounds and will not be considered further. The π-donors give 1:1 molecular compounds. Addison and Sheldon[73] determined binary phase diagrams for the systems benzene–$N_2O_4$, mesitylene–$N_2O_4$, and tetralin–$N_2O_4$, and found congruently melting 1:1 molecular compounds only. The respective melting points were $-8°$, $-19°$, and $-39°$. These molecular compounds are coloured at low temperature but lose their colours when heated towards their melting points. *p*-Xylene does not form a molecular compound; more complex aromatic compounds appear not to have been studied. The crystal structure of benzene–$N_2O_4$ has been determined[74] but will not be discussed here. Other important inorganic acceptors are the halogens[75] and, perhaps, sulphur dioxide, (*e.g.*, ref. 196) but these also will not be discussed here.

*b. Nitroalkanes.* Tetranitromethane and hexanitroethane are the only two representatives of this group to have been investigated. Colour formation on addition of tetranitromethane to solutions of aromatic hydrocarbons was first reported by Werner.[76] No solid molecular compound has yet been found; benzene, for example, gives a simple eutectic freezing-point diagram.[77] Any studies made with more complex aromatic compounds have not been reported.

Hexanitroethane, on the other hand, has been reported[78] to give an easily dissociable, red molecular compound with naphthalene, and analogous compounds with phenol and resorcinol. The properties, of hexanitroethane have been reviewed recently[79] and formation of unstable, coloured molecular compounds was noted, but without details.

**2. Substituted olefinic compounds.**—Two powerful electron-acceptors come directly into this category, namely, tetracyanoethylene (TCNE) (**3**) (Cairns *et al.*[80]) and 1,3-butadienehexacarbonitrile (hexacyanobutadiene) (HCBD) (**4**) (Webster;[81] McKusick and Webster[82]). There are also some other electron-acceptors that can be considered formally as ethylene derivatives, *e.g.*, 7,7,8,8-tetracyanoquinodimethane (TCNQ) (**6**) (Acker and Hertler;[83] Melby *et al.*[84]), and 9-fluorenylidenemalononitrile (**5a**).

Many $\pi$-molecular compounds have been prepared by using TCNE and TCNQ as electron-acceptors; they generally have an equimolar composition. The various crystallographic and other studies made on them will be discussed later. HCBD invariably forms 2:1 compounds, *e.g.*, (pyrene)$_2$:HCBD (Webster[81]), and these have ionic ground states, but no extensive study of this has been reported. 9-Fluorenylidenemalononitrile has not been studied at all in this connexion. 11,11,12,12-Tetracyanonaphthalene-2,6-quinodimethane (**7**) (TNAP) has been synthesized (Dickman *et al.*[85]) and shown to be a stronger $\pi$-acid than TCNQ and TCNE; ion-radical salts analogous to those of TCNQ have been prepared from it, but no $\pi$-molecular compound has been reported. 11,11,12,12-Tetracyano-1,4-napthalenequinodimethane has been synthesized (Chatterjee[72]); it has an electron affinity of 1.49 eV. Charge-transfer bands were observed in its solution with various aromatic hydrocarbons but neither a crystalline molecular compound nor an ion-radical salt was reported.

**3. Substituted aromatic compounds.**—*a. Polynitrobenzenes, without and with substituents.* Many, if not most, polynitroaromatic molecules* form π-molecular compounds with electron-donors. For example, Dermer and Smith[87] found that 56 out of 91 polynitroaromatic compounds available to them (in 1939) formed well-defined π-molecular compounds with naphthalene. A considerable but unsystematized body of information exists concerning the relation between molecular structure and ability to form π-molecular compounds. Sinomiya[88–96] made extensive efforts to draw general conclusions from the phase diagrams of the systems he studied. The more fundamental approach is *via* the thermodynamic properties of the various donor–acceptor systems, and the limited information available, still too meagre to permit extensive generalization, is summarized in Section IVG.

(i) Unsubstituted polynitrobenzenes. Neither nitrobenzene nor *o*-dinitrobenzene gives crystalline molecular compounds with electron-donors. *m*-Dinitrobenzene forms equimolar compounds with 1- and 2-naphthol,[91,97] with some aromatic hydrocarbons and aromatic amines (Buchler and Heap[98]), and with fluoranthene, and both 1:1 and 1:2 compounds with pyrene.[94] *p*-Dinitrobenzene forms an equimolar compound with pyrene[94] and with 1- and 2-naphthol.[93,97] However, the report[99] of a 3:1 phenanthrene:*p*-dinitrobenzene molecular compound would appear to require further investigation. Both *m*- and *p*-dinitrobenzene give equimolar compounds with 1,2,3,4-tetrahydro-1-oxocarbazole (**35**).[100]

(35)

1,3,5-Trinitrobenzene (TNB) is one of the most effective complex-forming reagents, and many molecular compounds (generally equimolar) have been reported with aromatic hydrocarbons[101]

* Much information on the chemistry of polynitroaromatic compounds has been summarized by Urbanski.[86]

and arylmines.[102] Various *cis*- and *trans*-stilbene derivatives form molecular compounds with TNB,[103] (see also p. 222) as do polyenes,[104] bithiophenes, and cyclopentadithiophenes.[107] References to the preparation of TNB molecular compounds are scattered through the literature; their occurrence, together with that of picrates, is usually noted in encyclopædias such as Heilbron's "Dictionary of Organic Compounds." Many binary phase diagrams have been determined, *e.g.*, by Kremann and Muller[108] and Kofler,[109] as well as many crystal structures. The latter are to be discussed later in this review.

The only tetranitrobenzene that has been reported to form molecular compounds is 1,2,4,6-tetranitrobenzene, which shows somewhat irregular behaviour (Sinomiya [92,93]). Equimolar compounds are formed with pyrene, fluoranthene, acenapthene, phenanthrene, 1- and 2-naphthol, a 1:2-compound is formed with fluorene, a 2:3 compound with phenanthrene, and a 3:2 compound with naphthalene. Evidently the presence of three nitro groups *meta* to one another offsets the effects of the three nitro groups *ortho* to one another.

Polynitrobenzenes have been shown to form molecular compounds with one another; for example, nitrobenzene and *m*-dinitrobenzene form an equimolar compound with an incongruent melting point, and $(C_6H_5NO_2)_2$:TNB melts congruently at 66°.[110,111] Another example is the incongruently melting, equimolar compound formed between TNB and TNT.[112] No structural information has been reported for any of these molecular compounds, but they could be packing compounds, perhaps like that formed between 1,2,4,5-tetrabromobenzene and hexabromobenzene.[113] However, it is probable that the 1:1 and 2:1 molecular compounds between 2-nitronaphthalene and TNB [95] (the equimolar compound was also reported by Sudborough[101]) and the 1:1 compound between 2-nitronaphthalene and TNB [94] are $\pi$-molecular compounds.

(ii) Substituted polynitrobenzenes. The most important of the present group is picric acid, involved in the first $\pi$-molecular compounds to be discovered.[1]

Monosubstituted nitrobenzenes form relatively few molecular compounds; a fairly comprehensive series of substituents (COOH, CHO, OH, Cl, Br, Me, $NH_2$, in *o*-, *m*-, *p*-positions) was studied by Sinomiya,[91] but molecular compounds were found only for 1-naphthol (and in some instances 2-naphthol) with *m*- and *p*-nitrobenzoic acid and *m*- and *p*-chloronitrobenzene. An equimolar compound between anthracene and *m*-nitrophenol (**36**) has also been reported.[108] Disubstituted nitrobenzenes perform rather better, especially when the substituents are *meta* to the nitro group. For example 3,5-dicyanonitrobenzene (**37**), 5-nitroisophthaloyl chloride (**38**) and 1,3-bis(methylsulphonyl)-5-nitrobenzene (**39**) all form compounds with some aromatic hydrocarbons and arylamines.[114]

NO$_2$　　　　　　NO$_2$　　　　　　NO$_2$　　　　　　NO$_2$

OH　NC　　　　CN　ClOC　　　COCl　CH$_3$SO$_2$　　　SO$_2$CH$_3$

(**36**)　　　　　(**37**)　　　　　　(**38**)　　　　　　(**39**)

The effect of single substituents (OH, CH$_3$, NH$_2$, Cl, Br, COOH) on various dinitrobenzenes has been studied by Sinomiya;[89,90,93,94] rather similar studies, but restricted to substitution in *m*-dinitrobenzene, have been carried out by Dermer,[115] Buehler and Heap,[98] and Buehler, Hisey, and Wood.[116] Many equimolar molecular compounds were reported; the general conclusions are summarized at the end of this Section. Sinomiya described 3:2 black molecular compounds between 1- and 2-naphthylamine and 2,3-dinitrophenol,[93] a similar ratio for 2-naphthol: "dinitroaniline" and a 3:1 ratio for naphthalene: "dinitroaniline".[90]

Reichstein[117] and Buu-Hoï, Jacquignon, and Roussel[118] have shown that 3,5-dinitrobenzamide and various esters of 3,5-dinitrobenzoic acid give equimolar molecular compounds with a number of electron-donors, including pyrene and 1- and 2-naphthylamine. Reichstein[117] also prepared 1:2 compounds with benzidine (benzidine:ester = 1:2) while Buu-Hoï *et al.*[118] obtained 2:1 donor–acceptor compounds with 3,5-dinitrobenzoic anhydride (**18**), the particular donors studied being pyrene, 3-methoxypyrene (**40b**), 3-pyrenol (**40a**), and benzo[*a*]carbazole (**41**). It is tempting to assume that each donor molecule interacts with a single benzene nucleus of the acceptor, but this has not been proved [the converse situation has been demonstrated in crystals of copper oxinate: (TCNB)$_2$ (see Section VI-I)]. Molecular compounds of 3,5-dinitrobenzoic acid have been studied,[90] and the crystal structure of phenothiazine: 3,5-dinitrobenzoic acid has been described briefly.[119]

Picric acid forms crystalline "picrates" with very many aromatic hydrocarbons, phenols, and amines (aliphatic and aromatic). Widespread use is made of these substances for identification of the

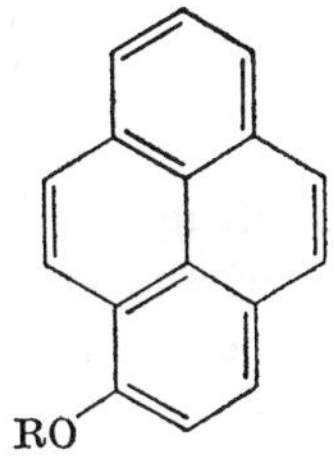

(40)
a: R = H
b: R = CH₃

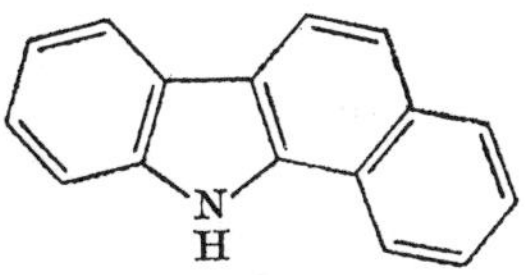

(41)

variable component. Some are true π-molecular compounds and are written here as, for example, naphthalene:picric acid, while others are formulated as salt-like and noted here as picrates (cf. Section IV-C).

Melting points and colours of picric acid molecular compounds of aromatic hydrocarbons, and polymethyl- and polyhydroxy-benzenes and -naphthalenes, were compared by Baril and Hauber.[120] Methyl substitution raises the melting point (increases stability), whereas hydroxyl substitution generally lowers it. Similar studies have been made of the picrates of phenyl and naphthyl ethers.[121,122]

Work on picrates up to about 1940 has been summarized in three publications (Râscanu;[63,123] Baril[124]), unfortunately all in rather obscure journals. The effects of additional substitution in picric acid by alkyl (methyl, dimethyl, ethyl, *tert*-butyl), halogen (Cl, Br, I), and methoxy groups on the formation of picric acid molecular compounds with aromatic hydrocarbons and salts with various amines has been studied in some detail[125,126] by determining binary phase diagrams and, when possible, by isolating crystalline products. For example, all eight aromatic hydrocarbons used (acenaphthene, anthracene, fluorene, hexamethylbenzene, 2-methyl-naphthalene, naphthalene, phenanthrene, and pyrene) gave isolable methylpicric acid molecular compounds; corresponding molecular compounds of ethylpicric acid were not isolated with anthracene, fluorene, or hexamethylbenzene, and none with dimethylpicric acid, but all these appeared in the phase diagrams. This somewhat complicated pattern of results indicates the pitfalls involved in drawing conclusions only from success in isolating crystalline molecular compounds.

Studies have been made of binary phase diagrams (for example, by Efremov,[127] Pushin and Kozuhar,[128] Kofler,[109] and Mindovich[129]) of the thermodynamics of crystalline picric acid molecular compounds (Bell and Fendley,[130] Dimroth and Bamberger,[131] and Gorbachev and Mindovich[132]), and of their physical properties (Mindovich and Gorbachev[133]). Surprisingly, only one crystal-structure determination has been reported for a picric acid molecular compound (Addenda to Section VI-E), and there have been only one or two single-crystal spectroscopic studies. Thermodynamic measurements on the molecular compounds of aromatic hydrocarbon with TNB, TNT, and picric acid will be discussed in Section IV-G).

Among the singly substituted 1,3,5-trinitrobenzenes that form molecular compounds with aromatic hydrocarbons are TNT, picryl azide (**19**, X = N≡N, Y = H), picramide (**19**, X = NH$_2$, Y = H;),[134] and tetryl (*N*-methyl-2,4,6,*N*-tetranitroaniline) (**42**).[134] Dermer[115] prepared various molecular compounds with styphnic acid (**19**, X = Y = OH) and 3-methyl-2,4,6-trinitrophenol (**43**); styphnic acid is the most widely used disubstituted polynitrobenzene electron-acceptor (cf. ref. 93). On the other hand, 2,4,6-trinitro-*m*-xylene (**44**) did not form molecular compounds with naphthalene, anthracene, phenanthrene, fluorene, or acenaphthene.[135]

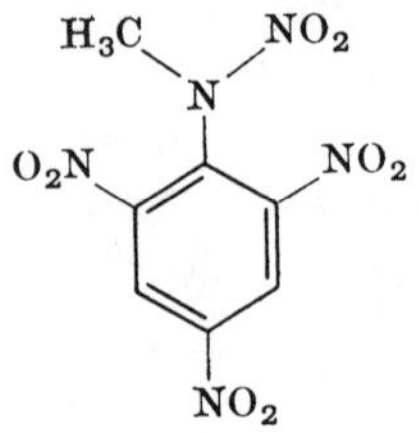

(**42**) 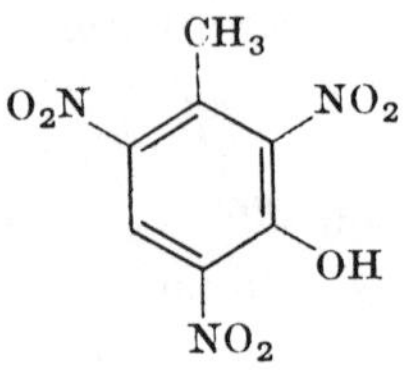          (**43**) 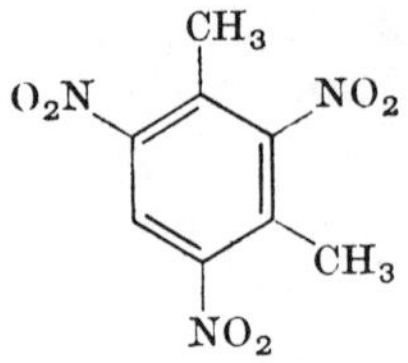          (**44**)

Binary melting-point diagrams have been determined for a number of aromatic hydrocarbons with picryl chloride, picramide, and styphnic acid.[127]

Two trisubstituted trinitrobenzenes have been studied by Hammick and Hellicar.[136] These are trinitromesitylene and trichlorotrinitrobenzene, in both of which the planes of the nitro groups are likely to be nearly perpendicular to the planes of the rings. The former compound does not form complexes with naphthalene or hexamethylbenzene, whereas the latter forms yellow, incongruently melting 1:2 molecular compounds with both donors, and also a yellow liquid phase.

(iii) General conclusions. General conclusions about the complexing ability of substituted polynitrobenzenes have been proposed by Sinomiya,[92,93,95] as follows:

The tendency of isomeric polynitro compounds to form molecular compounds is:

trinitrobenzenes: sym > unsym > vic.

dinitrobenzenes: *p* or *m* > *o*.

trinitrotoluenes: 2,4,6- > 2,3,4- or 2,4,5-.

dinitrophenols and dinitrotoluenes: 2,4-, 2,5- or 3,5- > 2,6- > 2,3- or 3,4-.

tetranitrobenzenes: 1,2,4,6- > 1,2,4,5.

These comparisons indicate that *ortho*-substitution hinders the formation of molecular compounds.

The effectiveness of other substituents in the polynitrobenzene ring is as follows:

1,3,5-trinitrobenzenes: $H > OH > NH_2 > CH_3 > Cl > OCH_3$.

2,4-dinitrobenzenes: $CO_2H > Cl > OH > Br > CH_3 > H > OCH_3$.

nitrobenzene: $CO_2H > CHO > Cl > Br$ for the same position; *p*- > *m*- > *o*- for the same group.

*b. Polynitronaphthalenes.* The molecular compounds formed by some polynitronaphthalenes with aromatic hydrocarbons, arylamines, phenols, and unsaturated compounds were studied by Sudborough, Picton, and Karve,[137] who found that only arylamines formed molecular compounds with 1,3,5-, 1,3,8-, or 1,4,5-trinitronaphthalene, whereas 1,3,6,8-tetranitronaphthalene formed molecular compounds with both aromatic hydrocarbons and arylamines. The factors of importance seem to be the number of nitro groups (the more the better), their position in the ring (*meta* is best), and minimum deviation from molecular planarity (*peri*-nitro groups will be roughly perpendicular to the naphthalene ring). On this basis 1,3,5,7-tetranitronaphthalene (**29**) should be the most effective electron–acceptor, and further study of these systems seems necessary.

The only other polynitro aromatic hydrocarbon that appears to have been studied as a "complexing agent" is 1,3,6,8-tetranitropyrene (**31**) (Bavin[138]), which has the advantages of low solubility and easy availability. Naphthalene and indole form 2:1 molecular compounds, while the other donors used (fluorene, carbazole, phenanthrene, anthracene, 9,10-dimethylphenanthrene, benz[*a*]anthracene, pyrene, benzidine, 4,4″-diamino-*p*-terphenyl) formed 1:1 molecular compounds. These molecular compounds crystallized as needles of various colours, but no crystallographic information has been reported. The tetranitropyrene molecular compounds are generally not suitable for identification or isolation of aromatic hydrocarbons as they are formed only in the presence of a large excess of hydrocarbon.

*c. Benzotrifuroxans and related compounds.* Bailey and his co-workers[139–141] have suggested that benzotrifuroxan (26) (BTF) should be an effective electron-acceptor because the furoxan rings are constrained to be coplanar with the central benzene ring. About fifty molecular compounds have been prepared, even with such weak electron-donors as 1,2,3-trimethylbenzene, mesitylene, 1,2,3,4-tetramethylbenzene, styrene, biphenyl, diphenylmethane, tetralin, 6-methyl-, 5-ethyl-, and 5,7-dimethyl-tetralin, 3,4-dihydro-1,7-dimethylnaphthalene, 2-*sec*-butyl- and 2-*n*-octyl-naphthalene, 1-hexyl- and 1-phenyl-naphthalene, and indene. Thus BTF appears to be at least as effective an electron-acceptor as TNB.

Replacement of the furoxan rings by one or more substituents of another kind, such as nitro or cyano, reduces the ability to form molecular compounds. Changing the furoxan to furazan rings gives benzotrifurazan (27), which forms molecular compounds with a wide variety of partners (Bailey and Evans[141]), including phosphate esters. The one structure that has been reported (triethyl phosphate–benzotrifurazan[142]) shows that the latter molecular compounds do not fall within the ambit of the present Review.

*d. Polycyanobenzenes and related compounds.* There is a striking contrast between the complexing abilities of polynitro- and polycyano-benzenes. TNB is a powerful acceptor, but 1,2,4,5-tetranitro- and hexanitro-benzene appear not to form charge-transfer compounds, presumably because the nitro groups cannot be coplanar with the benzene rings. Although 1,3,5-tricyanobenzene (20) is a poor electron-acceptor, both 1,2,4,5-tetracyano- (22) (TCNB) and hexacyano-benzene (23)[143] form crystalline molecular compounds. Bennet and Wain[114] showed that 1,3,5-tricyanobenzene formed 1:1 molecular compounds with benzidine and 1-naphthylamine, but not with anthracene (the latter result was confirmed by Bailey *et al.*[144]). Tricyanomesitylene (45) forms a crystalline molecular compound with pyrene.[143] Bailey *et al.* also studied the complex-forming ability of TCNB, tetrachlorophthalonitrile, and 2,4,6-tricyano-*s*-triazine (21); all three gave crystalline equimolar molecular

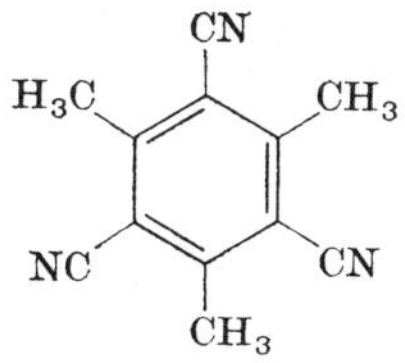

(45)

compounds. Their conclusions can be summarized as follows: A nitro group enhances electron affinity more than a cyano group; an anhydride group enhances electron affinity more than two adjacent cyano groups; a hetero atom in the ring enhances the electron affinity; and the shape of the acceptor is important, and when different acceptors are compared the same donor and solvent should, if possible, be used.

The crystal structures and spectroscopic properties of some TCNB molecular compounds will be discussed below.

*e. Anhydrides and imides.* Neither maleic anhydride* nor phthalic anhydride forms crystalline molecular compounds with aromatic hydrocarbons, but the effectiveness of tetrachlorophthalic anhydride (**24**, X = Cl) (TCPA) as complex-former points to the importance of appropriate substituents in particular instances. Early studies by Pratt and Perkins[147] have been extensively supplemented by Jacquignon and Buu-Hoï,[148–152] who compared tetra-chloro-, tetrabromo-, and tetraiodo-phthalic anhydride and the corresponding tetrahalophthalimides (**46**, R = H) and their *N*-aryl derivatives.

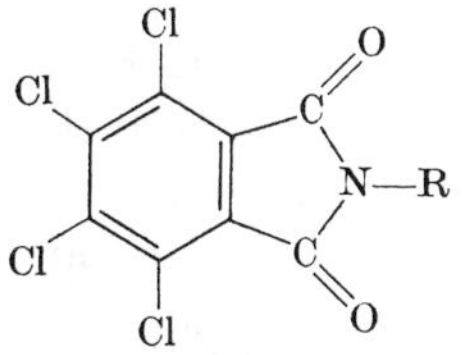

(46) 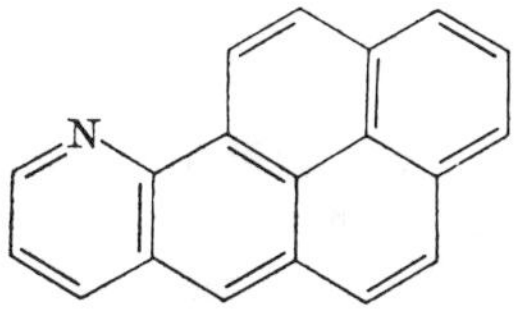          (47)

The electron donors used included polycyclic hydrocarbons and their oxygen and sulphur analogues, as well as indole and carbazole derivatives and naphthalene derivatives with an oxygen function.

* Dichloromaleic anhydride forms crystalline π-molecular compounds with durene and HMB.[145]

TCPA was the most effective of the electron-acceptors studied, with tetrabromophthalic anhydride only slightly less so. The most stable molecular compounds in this group should be formed by tetra-fluorophthalic anhydride,[153] which has not yet been studied as an electron-acceptor. The crystal structure of TCPA has been reported by Rudman.[154]

Buu-Hoï and Jacquignon[152] have exploited the ability of TCPA to form molecular compounds by using it as a substrate for chromatographic separation of aromatic hydrocarbons.

Ethylenetetracarboxylic acid dianhydride (furo[3,4-c]furantetra-one] (**2**) should be a powerful electron-acceptor but has not been synthesized.[44] Pyromellitic dianhydride (**28**, X = H) (PMDA) forms many crystalline compounds with substituted aromatic hydrocarbons;[155,156] early preparative work with this anhydride has been supplemented by synthetic and crystallographic studies by Boeyens and Herbstein,[157] and synthetic and spectroscopic studies by Nakayama, Ichikawa, and Matsuo.[158] Most of these addition compounds are equimolar in composition, but o-xylene, tetralin, and indene were reported[158] to have donor:acceptor ratios of 2:1. The crystallographic results will be discussed in more detail below.

Ferstandig *et al.*[156] checked whether a number of compounds related to pyromellitic dianhydride formed molecular compounds with p-xylene. These compounds included pyromellitic acid and its 1,4-dimethyl and tetramethyl esters. Only pyromellitoyl chloride showed any signs of complexing, but to a much smaller extent than pyromellitic dianhydride.

The addition compounds formed by mellitic dianhydride and mellitic trianhydride (**25**) were studied by Meyer and Raudnitz,[159] who isolated a number of coloured crystalline molecular compounds; similar results were obtained later by Mustafin[160] for mellitic trianhydride. Naphthalene and anthracene, respectively, form 1:1 orange-red and deep blue crystalline molecular compounds; molecular compounds are also formed with pyridine (colourless), di-bromostyrene (yellow), and m-dinitrobenzene (yellow), but it is not clear whether these belong to the same structural family. Molecular compounds were not formed with aliphatic hydrocarbons. The electron affinity of mellitic trianhydride is 0.31 ev higher than that of PMDA.[48]

Some studies[161,162] have been made of the acceptor properties of substituted naphthalic anhydrides (**30**) including the hexachloro-, hexabromo-, 3-nitro-, 4-nitro-, 4-chloro-, 4-bromo-, and 4,5-dinitroderivatives. The 4-halonaphthalic anhydrides formed unstable molecular compounds with 3-methoxypyrene, 3-pyrenol, and pyrenoline (**47**), but the other acceptors all gave reasonably stable products.

The effectiveness of hexachloro- and hexabromo-naphthalic anhydride as electron-acceptors is noteworthy because of their probable deviations from planarity [cf. octachloronaphthalene (Gafner and Herbstein[163])].

Maleic anhydride and maleimide form coloured crystalline molecular compounds with quinol[164,165] and resorcinol.[166] These molecular compounds appear to be similar to the quinhydrones (Section VI-H) in that both charge-transfer and hydrogen bonding contribute to their stabilization. No structural result has been reported.

*f. Other substituted benzenes that act as electron-acceptors.* Trimesic acid trichloride has been shown to form equimolar, yellow molecular compounds with naphthalene, anthracene, phenanthrene, acenaphthene, fluorene, and dimethylaniline (Bennett and Wain.[114]). Presumably these are charge-transfer compounds, although this has not been checked by physical methods.

*g. Other possible electron-acceptors, of uncertain significance.* There are a number of other substances that form molecular compounds with aromatic hydrocarbons and may behave as electron-acceptors. Despite some lack of direct information it is worthwhile recalling the most relevant results. Physical measurements and structure determinations are needed to establish the nature of the interaction between the components.

Our first example is the benzene:2,2-diphenyl-1-picrylhydrazyl (DPPH) equimolar compound. Williams[168] has shown that the benzene molecules simply fit between the irregularly shaped DPPH molecules; the shortest intermolecular C–C distance is 3.52 Å and there is no crystallographic evidence for donor–acceptor interaction.

4-Nitro- and 4,4′-dinitro-biphenyl form equimolar compounds with $N,N,N',N'$-tetramethylbenzidine (Rapson, Saunder, and Stewart[169]). The diffuse reflectance spectrum of tetramethylbenzidine:4,4′-dinitrobiphenyl has been measured[170] and interpreted as indicating some charge-transfer. However, charge-transfer was also inferred from the diffuse reflectance spectra of the tetramethylbenzidine:(4,4′-dinitrobiphenyl)$_4$ complex, but neither the crystal structure (Saunder[171]) nor the polarized single-crystal absorption spectrum of the analogous biphenyl:(4:4′-dinitrobiphenyl)$_3$ complex (Nakamoto[172]) supports this conclusion.

Complex formation between benzene and certain optically active polynitrobiphenyl derivatives has been exploited[173,174] in order to resolve mixtures of the components. Our present interest is in the crystalline complexes themselves, which have 1:2 (benzene:biphenyl derivatives) or 1:1 compositions. A possibly related example is the brown toluene:$(2,2',4,4',6,6'$-hexanitrobiphenyl)$_2$ (**48**) molecular compound.[175] Phase-diagram studies show that naphthalene forms a 1:1 molecular compound with $2,2',4,4'$-tetranitrobibenzyl; crystals were not isolated.

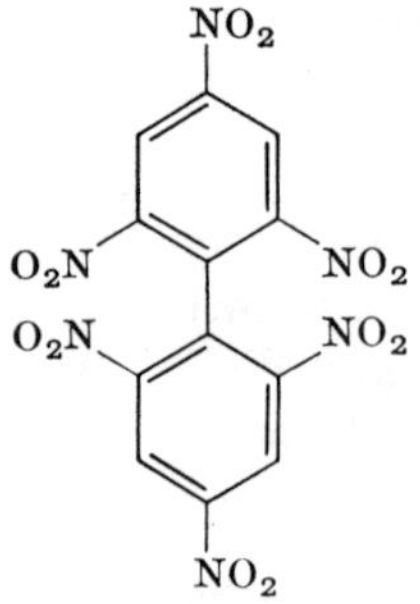

(48) 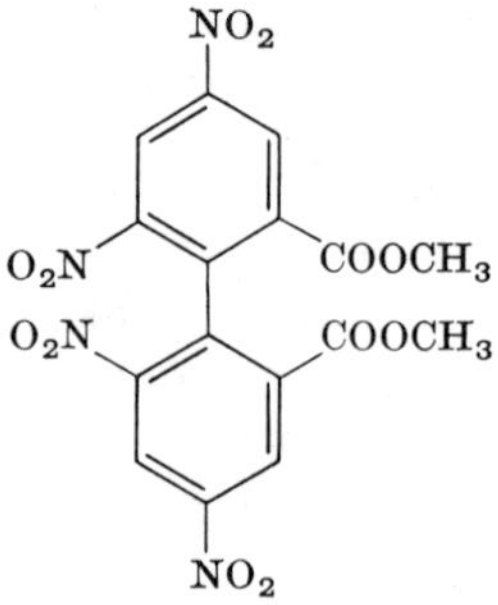                                    (49) 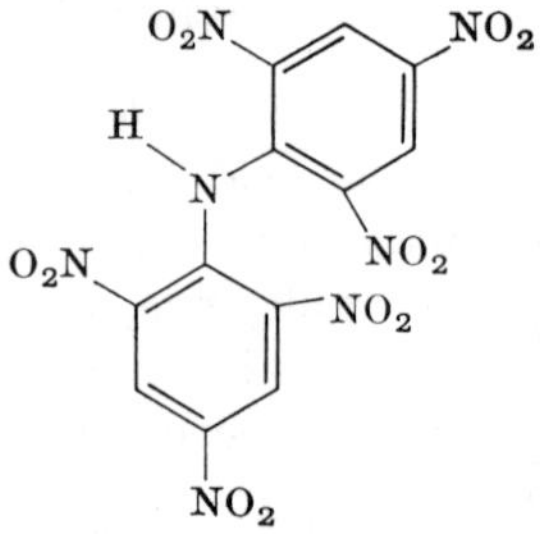                                    (50)

Methyl $4,4',6,6'$-tetranitrodiphenate (**49**) forms 1:2 (hydrocarbon: ester) molecular compounds with benzene, toluene, xylenes, naphthalene, anthracene, nitrobenzene, *m*-dinitrobenzene, TNB, 1-nitronaphthalene, and $2,2'$-dinitrobiphenyl (Hammick and Sixsmith[176]). These molecular compounds are pale yellow and the catholic variety of partners suggests that some type of clathration is involved. There is also a 1:1 complex with acenaphthene; as this is deep yellow it seems possible that some donor–acceptor interaction is involved. Another potentially interesting molecular compound is (naphthalene)$_2$:$2,2',4,4',6,6'$-hexanitrodiphenylamine (**50**), which crystallizes in yellow prisms.[177]

Cyanuric cyanide (2,4,6-tricyano-*s*-triazine), (**21**) has been shown to act as an electron-acceptor in molecular compounds (Bailey *et al.*[144,146]). It also forms 1:3 and 3:1 molecular compounds with 2,4,6-tris(dimethylamino)-*s*-triazine (Das, Shaw, and Smith[178]). These have high melting points (231°, 221°) but their structures are not known; Das *et al.*[178] suggest that they are $n:\sigma$ complexes similar to the homocomplex formed by cyanuric chloride itself (Hassel and Rømming[179]). It has been suggested that the molecular compounds

formed by hexamethylbenzene with a number of chlorocyclophos-
phazenes are donor–acceptor compounds, with hexamethylbenzene
acting as donor.[180] Apparently analogous molecular compounds are
formed by various dialkylamino- with chloro-cyclophosphazenes,[180]
so inference of some sort of donor–acceptor interaction seems justi-
fied. A difficulty is that hexamethylbenzene also forms molecular
compounds with octakis(dimethylamino)cyclotetraphosphazatetra-
ene (**51**) and dodecakis(dimethylamino)cyclohexaphosphazahexaene
(**52**), which are unlikely to behave as electron-acceptors.

(**51**)  (**52**)

**4. Unsubstituted and substituted quinones.**—The complexes of *p*-
benzoquinone (**8**) and its derivatives with a number of phenols,
aromatic amines, and hydrocarbons were studied by Kremann *et al.*,[181]
who determined the binary phase diagrams. It is worth noting that
addition compounds were not formed with naphthalene, phenanthrene, acenaphthene, fluorene, or aniline. Laskowski[182,183] showed
later that *p*-benzoquinone formed molecular compounds with pyrene
and dibenz[*a,h*]anthracene but not with other aromatic hydrocarbons. Only the pyrene : benzoquinone compound has been studied
further; its composition is 1:1,[184] and not 2:1 as first stated.[185]
TMPD : *p*-benzoquinone has been prepared and shows a charge-
transfer band in its (solid-state) spectrum (Foster and Thomson[186]).
Tetramethylbenzidine forms a 1:2 molecular compound with *p*-
benzoquinone while benzidine forms a 2:1 molecular compound;
the polarized absorption spectra of single crystals have been
measured.[187] The first of this pair appears to have the usual mixed-
stack structure but that of the second may be more complicated.

The most comprehensive studies of the complex-forming abilities
of aromatic quinones are those of Laskowski,[182,183,188] who used
the microscope fusion method to test for formation of crystalline
molecular compounds. The electron-acceptors examined included
unsubstituted benzoquinones, naphthoquinones, and anthraquinones

and their methylated derivatives; the related compounds vitamin K$_1$ (**53**), α-tocopherolquinone (**54**), and Q-coenzymes (**55**) were included because of their biological importance. The donors studied (97 compounds in all) included polycyclic hydrocarbons, nitrogen heterocyclics, aromatic amines, aromatic azo-compounds, and a number of miscellaneous compounds. Because of the possible role of

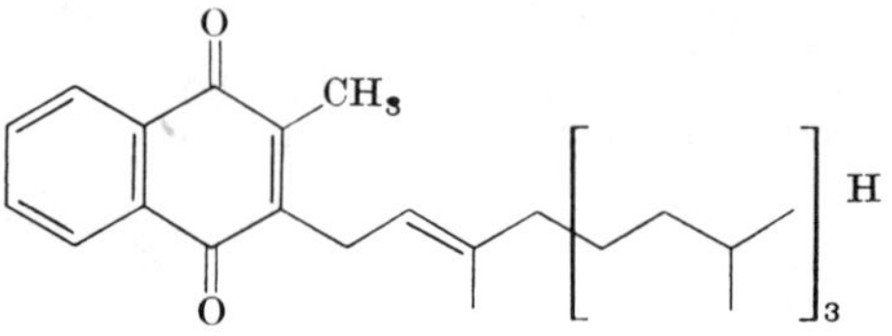

(**53**)

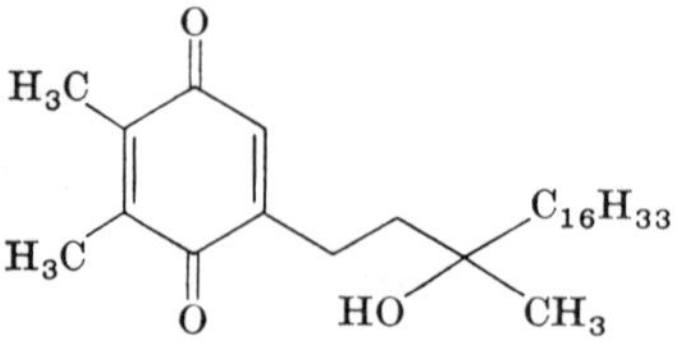

(**54**)

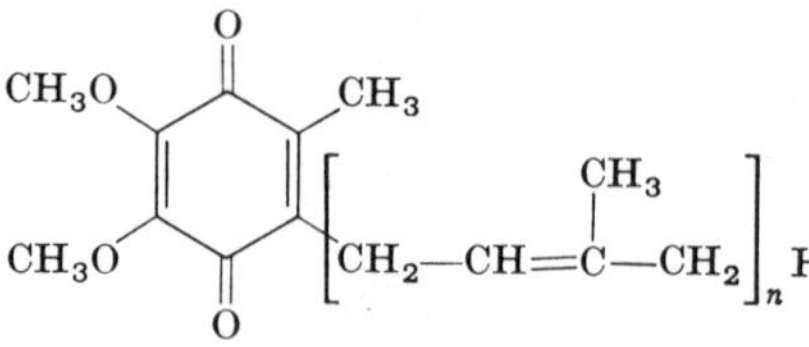

(**55**)

$n = 6$–$10$

charge-transfer in carcinogenesis some carcinogenic compounds were included to detect any correlation between complex-forming ability and carcinogenic properties; current opinion[189] is against correlation between donor or acceptor properties of organic molecules and their carcinogenic activity. The order of reactivity found for quinones of comparable degree and position of methylation was benzoquinones > naphthoquinones ≫ anthraquinones. The benzoquinones react with all categories of donors while the naphthoquinones react primarily with polycyclic hydrocarbons and aromatic amines. More than half of the solid compounds formed involved 2,3-dimethyl-, 2,3,5-trimethyl-, or 2,3,5,6-tetramethyl-*p*-benzoquinone (duroquinone) (**9**, X = CH$_3$). Vitamin K$_1$, α-tocopherolquinone, and the Q-coenzymes formed coloured melts but no solid compound was obtained. In the naphthoquinone series methoxyl and hydroxyl groups drastically reduced activity, while amino and alkylthio groups eliminated activity. In agreement with the tendency shown in earlier work, biphenyl, naphthalene, anthracene, naphthacene, and picene did not form compounds with any of the electron-acceptors studied. The order of reactivity of 3- or 4-ring unsubstituted aromatic hydrocarbons is *peri*-condensed > angular *cata*-condensed ≫ linear *cata*-condensed.

*p*-Benzoquinone and some other quinones form coloured addition products with polyhydroxy derivatives of aromatic hydrocarbons. The 1:1 composition is most common (the well-known quinhydrones), but the 2:1 composition is also found in many systems. The redox reactions involved in the formation of quinhydrones have been studied, particularly by Michaelis and his collaborators,[190] while synthetic work, summarized in Table 4, has been reported by Jackson and Oenslager,[191] Meyer,[192] and Siegmund.[194] Crystal structures have been determined for some of the molecular compounds listed in Table 5 and these are discussed below, especially in Section VI-H.

Schlenk[195] has suggested that analogous molecular compounds can be prepared by replacing quinol by *p*-phenylenediamine, and *p*-benzoquinone by its diimine. Some of the possible combinations have been prepared and found to give deeply coloured crystals which are rather unstable. Later work suggests that these molecular compounds have ionic ground states; particular attention has been paid

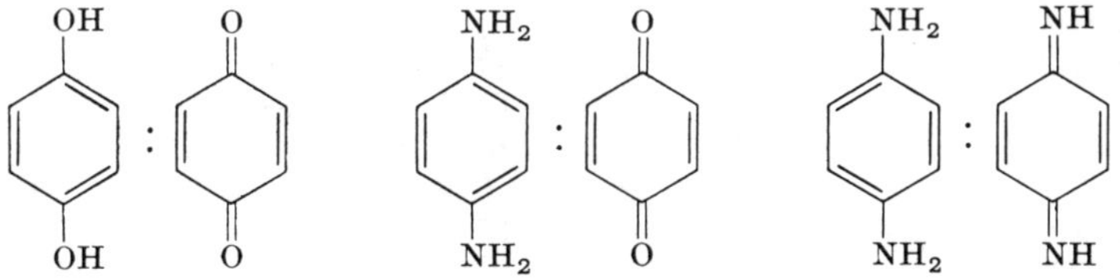

Quinhydrone       Quinone:diamine

to TMPD:chloranil, which has been studied crystallographically and by a number of other physical techniques. The 1:2 molecular compounds between *p*-phenylenediamine or benzidine and fluorenone[195] are possibly analogous to the quinone:diamines.

Indole and carbazole (also some *N*-substituted carbazoles) form 1:1 and 2:1 molecular compounds with 1,4-naphthoquinone and with

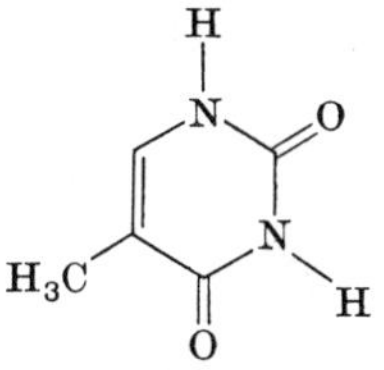

(**56**)

quinizarin (1,4-dihydroxyanthraquinone).[196] These could have structures analogous to those of the quinhydrones.

*p*-Benzoquinone and thymine (5-methyluracil) (**56**) form a 1:1 molecular compound (Sakurai and Okunuki[197]) whose crystal structure is described in Section VI-H.

Substituted *o*- and *p*-benzoquinones form addition compounds with both unsubstituted and substituted aromatic hydrocarbons, but not with aliphatic hydrocarbons. Pfeiffer *et al.*[198] showed that

Table 5. Crystalline molecular compounds[a] of polyhydroxy aromatic compounds with quinones[b] (and some possibly related examples).

| Donor (D) | Acceptor (A) | Ratio D:A | Remarks | Reference |
|---|---|---|---|---|
| Phenol | *p*-BQ | 2:1 | "Phenoquinone" | 181 |
| *p*-Chlorophenol | *p*-BQ | 1:1 | Orange needles | 192 |
| | | 2:1 | Dark red crystals | 192 |
| *p*-Bromophenol | *p*-BQ | | {Results similar to those for *p*-chlorophenol | |
| *p*-Nitrophenol | *p*-BQ | 1:1 | | 181 |
| Hydroquinone monomethyl ether | *p*-BQ | 2:1 | | 193 |
| 1-Naphthol | *p*-BQ | 1:1 | Dark red plates | 181,192 |
| | | 2:1 | Dark brown needles | |
| 2-Naphthol | *p*-BQ | 1:1 | Black plates | 192 |
| | | 2:1 | | 181,191 |
| 1-Naphthol | NQ | 1:1 | Red needles | 192 |
| 1-Naphthol | PQ | 1:1 | Red plates | 192 |
| *p*-Toluidine | *p*-BQ | 1:1 | | 181 |
| 2-Naphthylamine | *p*-BQ | 1:1 | | 181 |
| | | 1:2 | | 181 |
| Hydroquinone | *p*-BQ | 1:1 | Quinhydrone | 181,192 |
| Durohydroquinone | DQ | 1:1 | Brown needles | 193 |
| Resorcinol | *p*-BQ | 1:2 | | 181 |
| | | 1:1 | Red prisms | 192,193 |
| Pyrocatechol | *p*-BQ | 1:2 | | 181 |
| | | 1:1 | Dark green needles | 192 |
| | | 2:1 | Red needles | 194 |
| 2,3-Naphthalenediol | *p*-BQ | 1:1 | | 194 |
| Pyrogallol | *p*-BQ | 1:3 | | 181 |
| Phloroglucinol | *p*-BQ | 1:2 | Red-brown crystals | 193 |

[a] Kremann *et al.*[181] inferred the existence of molecular compounds from the appearance of the phase diagrams; the other investigators[191–194] isolated the molecular compounds reported.

[b] *p*-BQ = *p*-benzoquinone; NQ = 1,4-naphthoquinone; PQ = phenanthraquinone; DQ = duroquinone.

electron-donating substituents such as $CH_3$, $CH_3O$, and HO enhanced the complex-forming power of the donor molecules but reduced that of the acceptor molecules; halogen substituents had the opposite effects. Mono-, di-, and tetra-chlorobenzoquinone form molecular compounds with benzene, toluene, *p*-xylene, and hexamethylbenzene. Only HMB:*p*-chloranil is stable in air. Some aromatic hydrocarbon:*p*-chloranil molecular compounds have been studied by crystallographic methods (Sections VI-G,I). Pfeiffer *et al.*[198] prepared similar compounds with *o*-chloranil (15, X = Cl) as electron-acceptor; these seemed less stable than the *p*-chloranil molecular compounds. The solid-state spectra (UV and visible) of TMPD:*o*-chloranil and TMPD:*o*-bromanil molecular compounds have been studied by Foster and Thomson,[186] and magnetic measurements have been made by Kainer and Überle.[52] Perylene:*o*-chloranil has been studied spectroscopically by Eastman, Andros and Calvin,[199] and preliminary crystallographic data are also given in their paper. Some di- and tri-methylpyrroles form crystalline molecular compounds, mainly 1:1 in composition, with various halogenated *p*-benzoquinones (including chloranil) and 1,4-naphthoquinones;[196] their structures are not known.

The electron-acceptor with the highest electron affinity to date (Briegleb[47]) is 2,3-dichloro-5,6-dicyanobenzoquinone (DDQ) (12) (first synthesized by Thiele and Gunther[200]). Coloured solutions are formed when this compound is added to solutions of various aromatic compounds and some crystalline molecular compounds have been reported, *e.g.*, by Ottenberg, Brandon, and Brown.[201] Coloured crystalline compounds have also been obtained between pyrene and 2,3,5-trichloro-6-cyano-*p*-benzoquinone[202] and tetracyano-*p*-benzoquinone (cyanil) (9, X = CN).[203] The electron-withdrawing power of the cyano groups is so large that tetracyano-*p*-hydroquinone behaves as an electron-acceptor and forms red $\pi$-compounds with anthracene and pyrene.[203]

Suitably substituted *p*-benzoquinones can act either as electron-acceptors or electron-donors, depending on the nature of the second component.[204] Thus 2,5-diethoxy-*p*-benzoquinone (57) (DEQ) and 2,5-bis(methylamino)-*p*-benzoquinone (BAQ) (58) give charge-transfer bands in methylene chloride solution in the presence of donors such as benzidine or acceptors such as DDQ or TCNE. Coloured crystalline molecular compounds DEQ:(DDQ)$_2$ and (BAQ)$_2$:DDQ have been prepared, and the crystal structure of the former is said to be compatible with $\pi$–$\pi^*$ interaction between the

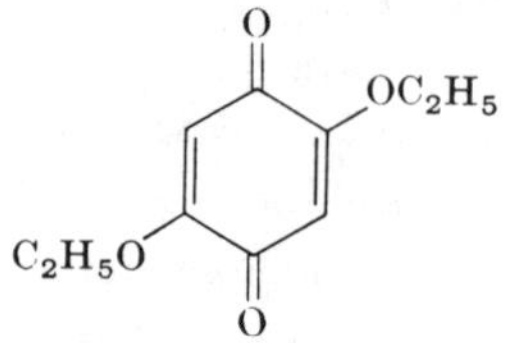

(57) 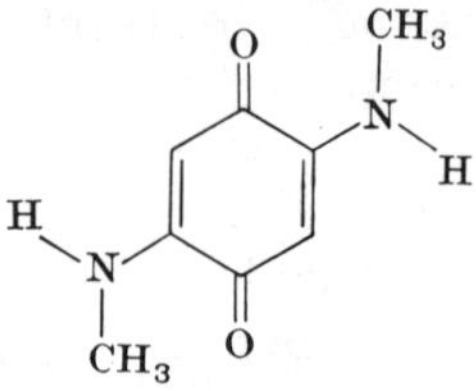                              (58)

components.[205] Both molecular compounds give strong ESR signals and thus presumably have ionic ground states.

*p*-Benzoquinonetetracarboxylic dianhydride (**11**) (Hammond[206]) is rather unstable both in moist air and in solution but forms coloured complexes in solution with polymethylbenzenes. The molecular compounds were not obtained crystalline by Hammond but there is no reason to doubt their existence.

Although hexachloro-1,4-naphthoquinone (**13**) is unlikely to be planar (in view of the demonstrated non-planarity of octachloronaphthalene[163]), it nevertheless forms many 1:1 molecular compounds with aromatic hydrocarbons and related substances (Willems[207,208]).

Various polynitro quinones have proved valuable electron-acceptors. An early example is 2,7-dinitroanthraquinone (**10**), introduced by Fritzsche[209] and investigated later by Schmidt[210] and by Bornstein *et al.*[211,212] It forms relatively insoluble, coloured, crystalline, 1:1 molecular compounds with anthracene, certain polymethylanthracenes, phenanthrene, stilbene, chrysene, and fluorene, but not with naphthalene or biphenyl. Preliminary crystallographic results have been reported for fluorene:2,7-dinitroanthraquinone (Hertel and Römer[213]). Chrysaminic acid (2,4,5,7-tetranitrochrysazin) (**59**) has also been reported by Coffey and van Alphen[214] to form molecular compounds, but we have not found any reference to original papers; this molecule meets the requirement of having nitro groups *meta* to

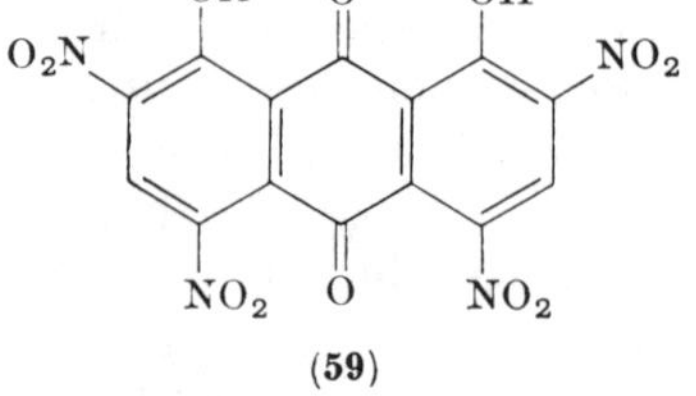

(59)

one another; however, on one side of the molecule, the *peri*-nitro groups must cause considerable distortion from planarity. Pfeiffer *et al.*[198] have reported a 1:1 compound of hexamethylbenzene with phenanthraquinone, but the behaviour of other aromatic hydrocarbons is not known. Both 1,8- and 3,6-dinitrophenanthraquinone (**16a,b**) form equimolar coloured molecular compounds with acenaphthene, and the latter forms also violet and red 2:1 (hydrocarbon:quinone) molecular compounds with anthracene and fluorene, respectively (Hertel and Kurth[215]). The melting points of the last two compounds, 250° and 270°, respectively, are higher than usual.

**5. Substituted cyclopentadienone systems.**—The most important electron-acceptors of this type are based on fluorenone. 2,6,7-Trinitrofluorenone gives molecular compounds with stilbene and anthracene.[215] The effectiveness of these polynitrofluorenones as electron-acceptors is somewhat surprising in view of the vicinal nitro groups. However, although *o*-dinitrobenzene does not give molecular compounds, other molecules with *ortho*-nitro groups act as acceptors, particularly when other groups are present (Sinomiya[92,94]). Nevertheless, one would expect the *meta*-relationship of nitro groups to be more effective than *ortho*, as shown by the introduction of 2,4,7-trinitro-9-fluorenone (TNF) (**33a**) by Orchin *et al.*[216,217]

The effectiveness of TNF has been explored by Laskowski *et al.*[218] Formation of molecular compounds with, *inter alia*, biphenyl, stilbene, *p*-aminoazobenzene, and 1,3,5-triphenylbenzene was demonstrated by microscopic fusion analysis. Twelve equimolar TNF molecular compounds have been prepared[219] with different methylbenz[*a*]anthracenes and six with different methylbenzo[*c*]phenanthrenes. Free energies of formation of the TNF molecular compounds *in solution* were measured by a spectrophotometric method. Cell dimensions and space groups have been reported[220] for TNF molecular compounds of three aromatic amines. The only molecular compounds of this group for which full crystal structures have been determined are 1,12-dimethylbenzo[*c*]phenanthrene:5-bromo-2,4,7-trinitrofluorenone and hexahelicene:5-bromo-2,4,7-trinitrofluorenone.[31]

2,4,5,7-Tetranitrofluorenone (TENF) (**33b**) is a sufficiently powerful electron-acceptor to form a complex with tetralin (Ray and Francis)[222] as well as with various aromatic hydrocarbons (Hertel and Kurth).[215] Newman and Lutz[223] have been particularly successful in using TENF to make molecular compounds with overcrowded non-planar aromatic hydrocarbons such as 4,5-dimethylphenanthrene and 1,12-dimethylbenzo[*c*]phenanthrene. Two interesting variations have been made to the basic TNF and TENF structures. First, Newman and Lutz[223] have synthesized 2-[(2,4,5,7-tetranitro-9-fluorenylidene)aminooxy]propionic acid (34) and shown that it can be resolved into its enantiomers. In this way they have separated the enantiomers of hexahelicene. Secondly, Mukherjee and Levasseur[224] have synthesized 2,4,7-trinitro-9-fluorenylidenemalonitrile (5b), which is an even more powerful electron-acceptor than TNF and forms stable ion-radical salts with, for example, lithium and triethylammonium cations. These authors prepared ten equimolar charge-transfer molecular compounds with various donors. 2,4,5,7-Tetranitro-9-fluorenylidenemalonitrile (5c) has also been synthesized, its electron affinity has been measured, and solid ion-radical salts have been prepared.[49]

Hexachloro-1-indenone (32) is another substituted cyclopentadienone system which forms molecular compounds (composition 1:1) with unsubstituted and substituted aromatic hydrocarbons.[167,208] A somewhat related molecule is 1,2,3-indenetrione (60). Its electron

(60)

affinity has been estimated at 1.07 ev from the wavelengths of the charge-transfer bands in its solution spectra with various electron-donors; this value suggests that it should be a good acceptor. The indenetrione itself is sensitive to water, and no crystalline π-molecular compound has been prepared from it. By analogy with the greater complex-forming ability of TCPA than of phthalic anhydride, tetrachlorindenetrione is likely to be a more powerful electron-acceptor than the parent compound.

**6. Heterocyclic acceptors.**—The only examples of heterocyclic molecules known to act as electron-acceptors are 2,4-[225] and 2,5-dinitrothiophene.[226] Both give yellow 1:1 molecular compounds with naphthalene; in addition, 2,4-dinitrothiophene gives a red 1:1 molecular compound with benzo[*b*]thiophene (thionaphthene).[225]

## E. Quasi-acceptors

Two groups of molecular compounds are known that are structurally analogous to the π-donor–acceptor molecular compounds but do not show any charge-transfer bands in their spectra. Thus, there is doubt as to the type of interaction between the electron-donor and the second partner; many workers consider the interaction to be due to dipole or polarization effects. It is convenient to consider these groups together with the $\pi$-donor–acceptor molecular compounds; however, we shall refer to the second partner as a quasi-acceptor because it does not have all the properties of a conventional acceptor.

**1. Polyfluorinated aromatic hydrocarbons.**—There is evidence for weak interaction between hexafluorobenzene (and related compounds) and various $\pi$- and $n$-donors in the liquid state, as indicated by excess volumes of mixing,[227] enthalpies of mixing,[228] refractive indices, and dielectric constants.[229] On the other hand, UV,[230] IR, and $^2$H- and $^{19}$F-NMR spectra[231] of liquid mixtures are simply the sums of the spectra of the components, emphasizing that the interaction cannot be large.

Crystalline molecular compounds reported are listed in Table 6; phase diagrams have been determined for some combinations,[227,230,233]

Table 6. Crystalline 1:1 molecular compounds prepared from $\pi$- and $n$-donors and quasi-acceptors related to hexafluorobenzene.

| Quasi-acceptor | $\pi$-Donor | $n$-Donor |
|---|---|---|
| $C_6F_6$ | Benzene,[227,232] naphthalene,[157,233] pyrene,[157] perylene,[157] phenanthrene,[235] triphenylene,[235] toluene,[233] $p$-xylene,[227] fluorobenzene,[238] mesitylene,[227,232,239] durene,[232] biphenyl,[233] 2-methylnaphthalene[238] | Aniline,[234] DMA,[235,236] TMPD,[235,236] $N,N'$-dimethyl-$m$-toluidine,[237] $N,N'$-dimethyl-$p$-toluidine[237] |
| $C_6F_5CN$ | | TMPD,[236] DMA[236] |
| $C_6F_5Cl$ | Benzene,[239] toluene,[239] mesitylene[239] | Aniline,[239] triethylamine[237] |
| $1,3,5\text{-}C_6F_3Cl_3$ | Mesitylene,[239] HMB[239] | |
| $C_{10}F_8{}^a$ | Benzene[233] | |
| $C_{12}F_{10}{}^b$ | Benzene[233] | |

$^a$ Perfluoronaphthalene.  $^b$ Perfluorobiphenyl.

and congruent melting points were obtained for all molecular compounds except biphenyl:$C_6F_6$.[233] Crystallographic studies (Boeyens and Herbstein[157]) have shown that mixed stacks are formed in naphthalene: $C_6F_6$, pyrene: $C_6F_6$, and perylene: $C_6F_6$ and presumably they exist in the other molecular compounds as well. Some work has been done on the structure of mesitylene: $C_6F_6$, which is disordered at room temperature.[240] The crystalline molecular compounds all lose $C_6F_6$ on exposure to the atmosphere. Wide-line NMR studies[232] show that molecular reorientation takes place in the crystals down to temperatures of about 100°K. The $^{35}$Cl-NQR spectra of some $C_6F_3Cl_3$ molecular compounds show broad lines indicative of disorder; the frequency shifts are at the limits of those characteristic of lattice effects, and their amounts were related to the geometries of the donors but not to their electronic structures.[239]

Perfluorotriphenylene forms crystalline molecular compounds with aromatic hydrocarbons (Smith and Massey[241]). The 1:1 compound with triphenylene is more stable than the hexafluorobenzene molecular compounds; this is shown by its high melting point (250–252° in a sealed tube, with dissociation) and by its furnishing a mass spectrum from the vapour at 80° corresponding to $(C_{18}H_{12} + C_{18}F_{12})$. The structure is not known. Similar results were obtained[233] for $C_6H_6$: perfluoronaphthalene (m.p. 70°) and $C_6H_6$: perfluorobiphenyl (m.p. 65°).

Spectroscopic measurements show that π-π* charge-transfer interaction is very weak in aromatic hydrocarbon–$C_6F_6$ solutions but that n-π* charge-transfer does occur in amine–$C_6F_6$ solutions.[235,236]

The band at 297 mμ in the absorption spectrum of tetrafluoro[2,2]-paracyclophane has no counterpart in the spectra of octafluoro-[2,2]paracyclophane or [2,2]paracyclophane itself and has been ascribed to transannular charge-transfer interaction between the electron-donating phenyl ring and the electron-attracting tetra-fluorophenyl ring.[242] This result suggests that charge-transfer interaction is possible, but it should be remembered that the two phenyl rings are appreciably closer together in the [2,2]paracyclophane molecule than in the crystals of any $C_6F_6$ molecular compound.

The overall evidence is that the interaction between quasi-donor and quasi-acceptor is rather weak and that the contribution of charge-transfer to this interaction is small; nevertheless, the occurrence of mixed stacks in the crystalline molecular compounds suggests that these molecular compounds are not just packing complexes.

**2. 1,3-Dimethylalloxazine, purines, and pyrimidines.**—The three related molecules, 1,3-dimethylalloxazine (**61**), 1,3,7,9-tetramethyl-uric acid (TMU) (**62**), and caffeine (**63**), form molecular compounds with aromatic and heterocyclic hydrocarbons, but there are distinct differences between the molecular compounds of 1,3-dimethylalloxazine, on the one hand, and those of TMU and caffeine on the other. Matsunaga[243] has shown that 1,3-dimethylalloxazine forms molecular compounds with pyrene, TCNE, and DDQ [ratio (other component):(dimethylalloxazine) = 1:2]. Thus it appears that 1,3-dimethylalloxazine, like certain *p*-benzoquinone derivatives (Section IV-D4), can behave both as an acceptor (*vis-à-vis* pyrene) and as a donor (*vis-à-vis* TCNE and DDQ). The diffuse reflectance spectra show evidence of charge-transfer bands; the IR spectra indicate that

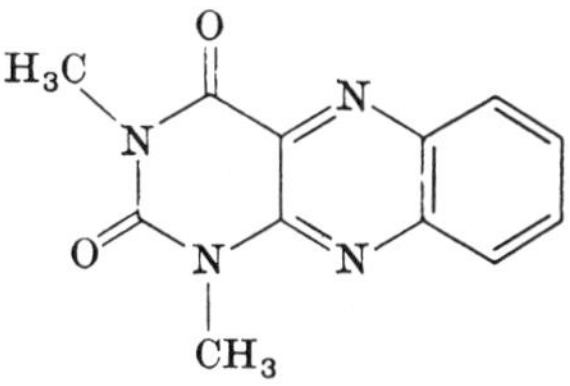

ground states of the molecular compounds are non-ionic. No crystallographic study has been reported for this group of compounds.

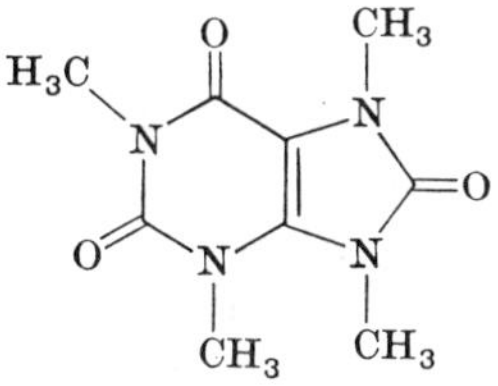

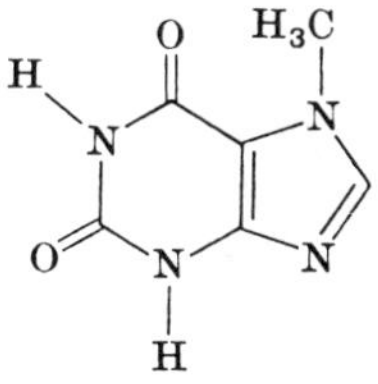

The enhanced solubility of aromatic hydrocarbons in aqueous solutions containing caffeine or TMU was studied by Weil-Malherbe,[244] who also showed that TMU formed crystalline molecular compounds with pyrene (1:1), benzo[*cd*]pyrene (benzopyrene:TMU = 1:2), and coronene (1:2) but not with phenanthrene, anthracene, chrysene, 20-methylcholanthrene, or dibenz[*a,h*]anthracene. Molecular compounds of aromatic heterocycles with caffeine and with TMU have also been prepared (Booth and Boyland[245]). The supposition that enhanced solubility and molecular-compound formation are due to similar interactions, and the biological importance of such solubilization, have stimulated further study of the molecular compounds.

There is no new band in the solution spectra of aromatic hydrocarbons with purines or pyrimidines (Booth, Boyland, and Orr;[246] Bergmann[247]) nor are the crystalline molecular compounds coloured; these results, together with studies of fluorescence spectra by Van Duuren,[248] indicate that charge-transfer does not occur. Calculations[247,249] in which dispersion interactions are assumed show that TMU interacts more strongly with benzo[*cd*]pyrene than do other purines and pyrimidines. Also the hierarchy of interaction energies of various aromatic hydrocarbons with TMU agrees well with the experimental order of solubilization by TMU.[250]

The results for the molecular compounds of purine and pyrimidine with aromatic hydrocarbons are similar to those obtained for the aromatic hydrocarbons:$C_6F_6$ molecular compounds. There is virtually no spectroscopic or other evidence for donor–acceptor charge-transfer, but the crystal structures of the molecular compounds (Section VI-M) are analogous to those of π-molecular compounds; thus TMU behaves as a quasi-acceptor. Little is known of the molecular compounds of aromatic heterocycles with purines and pyrimidines beyond that they exist.

**3. Flavins.**—Various isoalloxazine derivatives have been reported to form molecular complexes of different types with electron-donors such as phenols (the chemistry and biochemistry of flavins have been succinctly reviewed by Beinart and Hemmerich[251]), and the molecular complexes have been discussed by Tollin.[252]

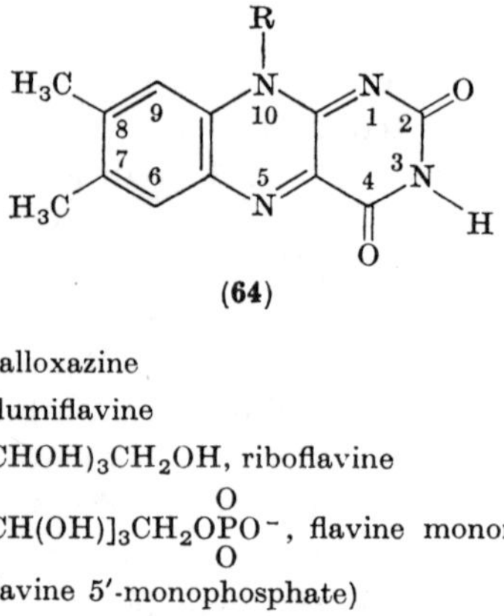

**(64)**

(a) R = H, isoalloxazine

(b) R = $CH_3$, lumiflavine

(c) R = $CH_2(CHOH)_3CH_2OH$, riboflavine

(d) R = $CH_2[CH(OH)]_3CH_2OPO^-$, flavine mononucleotide (FMN)

(riboflavine 5′-monophosphate)

(e) R = $CH_2[CH(OH)]_3CH_2$—O—P—O—P—O—CH$_2$

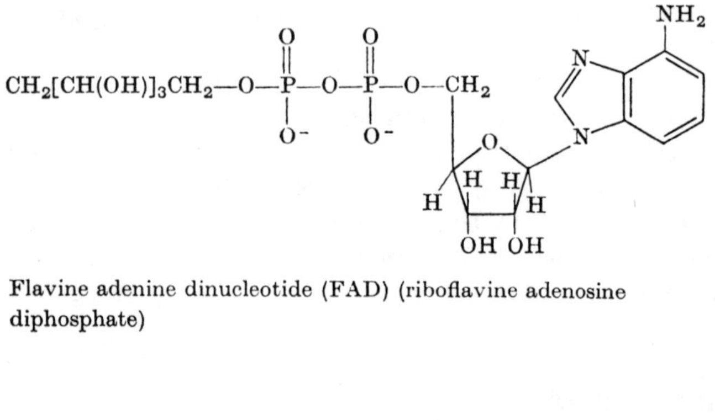

Flavine adenine dinucleotide (FAD) (riboflavine adenosine diphosphate)

In (64) the flavine nucleus is shown in the fully oxidized (flavo-quinone) form, stable under aerobic conditions, but reversibility of oxidation–reduction interconversions allow the flavins to be charac-terized as true quinonoid systems. In each oxidation state, cationic, neutral, and anionic species can be obtained, depending on the pH of the system (see Kosower[253] and Tollin[252] for details). Neutral and charged flavoquinone complexes have been obtained, as well as charged flavosemiquinone complexes. Most studies to date are based on visible and ESR spectra of solutions but some crystalline complexes have been prepared. For convenience, neutral and charged species will be discussed together. The neutral crystalline complexes are generally orange to orange-red, while the protonated complexes range from green through deep red to black. Charge-transfer inter-action between donor and flavin (acceptor) moieties has been sug-gested for certain complexes, although Kosower[253] has pointed out that this may not have been appropriate in some cases.

Neutral crystalline molecular complexes of 2,3- and 2,7-naphth-alenediol with riboflavin were prepared by Fleischman and Tollin.[254] Chemical analysis showed that the phenol:flavin ratio was 2:1, whereas spectroscopic analysis showed 1:1 complexes in solution. Spectroscopic study of mineral-oil suspensions of powdered 2:1 crystals showed an extra band, which was ascribed to charge-transfer interaction. Equimolar neutral crystalline complexes have been pre-pared between lumiflavin and various indoles, such as 5-methylindole (65), carbazole (66), and tryptophan (67). The corresponding riboflavin complexes did not crystallize. Acid complexes such as lumi-flavin–tryptophan–HCl–$H_2O$ have also been obtained. The crystal-line complexes are more highly coloured than the parent flavins, but diffuseness of the additional bands in the spectra makes it difficult to decide whether charge-transfer interaction is really present (Pereira and Tollin[255]).

(65)     (66)     (67)

Crystalline complexes have also been prepared with compositions 1:1:1 flavin:phenol:HCl. The flavins are represented by riboflavin, lumiflavin, and 9-methylisoalloxazine, while the phenols include hydroquinone, 4-chloropyrocatechol, resorcinol, and 1,2- and 1,4-naphthalenediol (Fleischman and Tollin[256]). The spectra suggest charge-transfer interaction between the components.

Finally, mention should be made of the paramagnetic semi-quinone salts and complexes first prepared by Kuhn and Strobele[257] and studied later by Fleischman and Tollin.[258] The latter authors showed that isoalloxazine derivatives in concentrated hydriodic acid were reduced to the semiquinone state and that some crystalline flavin hydriodides could be obtained; ESR measurements showed that these had 100% unpaired spins. Riboflavin hydriodide (no analysis given) gave pink crystalline platelets, and lumiflavin hydriodide (composition $FH_2 \cdot 2HI$) was a dark-brown solid, which could not be recrystallized. Black crystals (red in thin section) were obtained from saturated solutions of riboflavin in hydriodic acid to which various phenols (*e.g.*, hydroquinone, 2-naphthol, 2,3- 1,7-, 2,7-, 1,5-, and 1,4-naphthalenediol) had been added. A typical composition was $FH_3{}^+ \cdot I^- \cdot 2,7$-naphthalenediol. There were no indications of charge-transfer interactions in the spectra (Tollin[252]).

Kuhn and Strobele[257] have shown that a fairly complicated series of equilibria exists between the fully oxidized and fully reduced forms of flavins, each intermediate being characterized by "excellent crystallizing power, vivid colours, and sharply defined composition." The best results were obtained with riboflavin where equilibria can be represented as:

Flavin $\rightleftharpoons$ Verdoflavin $\rightleftharpoons$ Chloroflavin $\rightleftharpoons$
$\qquad\qquad\qquad\qquad\qquad\qquad$ Rhodoflavin $\rightleftharpoons$ Leucoflavin

All the crystalline intermediates were paramagnetic (bulk suscepti-bility measurements), ionic ($Na^+$ or $Cl^-$ ions were necessary), and supposed to be composed of different mixtures of the various flavin oxidation states.

Crystallographic studies have been briefly reported (Kierke-gaard[259]) of a number of oxidized and reduced forms of some iso-alloxazine derivatives. The oxidized forms are planar, but in the reduced form the isoalloxazinium moiety is folded. Brief reports have also appeared of crystallographic studies of some naphthalenediol complexes (Fritchie, Trus, and Langhoff[260]). In 10-methylisoalloxaz-inium bromide sesqui-(2,7-naphthalenediol) hydrate, $(C_{10}H_{10}N_4O_2)^+$ $Br^- \cdot (C_{10}H_8O_2)_{3/2} \cdot H_2O$, a donor–acceptor complex is formed be-tween flavin and one naphthalenediol molecule, while the remaining diol molecule is disordered and serves merely as a molecule of solvation (see also Addenda).

## F. Classification of Donors

**1. Aromatic hydrocarbons and heterocycles.**—*a. Unsubstituted aromatic hydrocarbons.* Benzene forms molecular compounds with most acceptors, but unless the products are kept in closed vessels benzene generally evaporates rapidly at room temperature. The composition is mostly 1:1, but both TNB and picryl chloride give 2:1 crystals, as shown by phase diagrams[261] (however, Hertel and Bergk[262] report equimolar compositions for benzene:TNB and toluene:TNB). Naphthalene is prolific in its ability to form molecular compounds; about 90 different types are listed in Elsevier's *Encyclopedia of Organic Chemistry.* Anthracene forms many molecular compounds, but tetracene and pentacene have been reported to form molecular compounds only with $SbCl_4$ and $SnCl_5$.

The condensed aromatic hydrocarbons, pyrene, perylene, and coronene, all form molecular compounds with apparent ease, pyrene, in particular, forming stable and well crystalline materials. Perylene is noteworthy in that its monopicrate[263] is reported to be dark violet-blue while its dipicrate is red;[264] usually the colour of molecular compounds between two particular components does not change with composition.

Acenaphthene, acenaphthylene, fluorene, and fluoranthene all form molecular compounds with most acceptors, but molecular compounds of biphenyl are comparatively rare and it yields isolable molecular compounds only with picryl chloride,[265] TCNB, and BTF;[139,157] PMDA gives only a broad maximum in the phase diagram,[157] while TNB gives an incongruently-melting equimolar compound.[266] 9,10-Dihydroanthracene forms a 1:2 compound with TNB,[101] but not a picrate.

Aromatic hydrocarbons that are not planar nevertheless form $\pi$-molecular compounds. This was shown by Orchin,[267] using a picric acid derivative as acceptor; the most dramatic example is surely hexahelicene (Newman, Darlak, and Tsai;[268] Ferguson *et al.*[31]), which forms $\pi$-molecular compounds with a number of acceptors. Among other non-planar donors are 9,10-dihydroanthracene (with TNB, 1:2[101]), and phenothiazine (with TNB;[269] see Section VI-E), and 9,10-dihydro-9-methylacridine (**68**).[270]

(68)

TNB and picric acid compounds of various *cis*- and *trans*-stilbenes have been prepared,[103] as well as of many polyenes.[104] The latter are all 1:2 compounds, suggesting that each acceptor molecule interacts more or less independently with the phenyl groups of the polyene. Chloranil, PMDA, and TCNE compounds of various *cis*- and *trans*-stilbenes have been prepared and their diffuse reflectance spectra measured.[105,106]

Sinomiya[94] found the following order for the ability of aromatic hydrocarbons to form molecular compounds with polynitrobenzenes: naphthalene, pyrene, fluoranthene > acenaphthene > phenanthrene, benzene > fluorene > anthracene, chrysene.

*b. The effects of substitution.* Alkyl substituents in general (if not too bulky), and methyl groups in particular, enhance the stability of molecular compounds. The converse applies to hydroxyl and ether substituents, although many molecular compounds are known where the donors carry hydroxyl or other groups. 1-Naphthol:picric acid does not have a charge-transfer band in its UV or visible spectrum, nor is there any evidence for hydrogen bonding between the components.* This has led Rastogi and Singh[271] to conclude that dipole–dipole interaction makes the major contribution to the stability of the crystalline molecular compound. There may be resemblances here to the molecular compounds of $C_6F_6$ or TMU.

Donor–acceptor compounds of brominated aromatic hydrocarbons with a number of electron-acceptors have been studied spectroscopically in solution by Spotswood.[272] Although crystalline samples were not prepared, there is no reason to doubt that this can be done. Crystalline molecular compounds of halogenated donors (or acceptors) are potentially convenient subjects for crystal structure analysis (cf. ref. 31).

Two series of qualitative studies have been made of the effects of substitution in aromatic hydrocarbons on their capacity to form

---

* 1-Naphthol:picric acid has been reported to be isomorphous with naphthalene: picric acid, etc. (see Table 19).

molecular compounds. Sinomiya[95] studied in particular the effects of substitution in positions 1 and 2 of naphthalene on compound formation with various polynitrobenzenes, and Baril and Hauber[120] considered picric acid compounds of polymethyl- and polyhydroxy-benzenes and -naphthalenes and of some aromatic hydrocarbons. Both the nature of the substituent and its position must be taken into account. Sinomiya gives the following sequence of decreasingly favourable effect of substituent (in a given position in naphthalene) on molecular-compound formation: $NH_2 > CH_3 > OH > C_2H_5 > OCH_3 > Cl > Br > OC_2H_5 > H > COOH > COOCH_3 > OC_6H_5 > CN > COOC_6H_5 > NO_2$. Substitution in position 1 tends to favour molecular-compound formation while substitution in position 2 sometimes hinders it. For benzene derivatives the tendency to compound formation is greater the larger the number of methyl substituents and the more symmetrical their disposition in the ring. Baril and Hauber[120] found that methyl groups were more effective in a sidechain than in the benzene ring; unsaturation in the sidechain produced very explosive picrates.

The presence of amino groups undoubtedly enhances the ability of a substituted hydrocarbon to form stable molecular compounds with electron-acceptors; the effect of the dimethylamino group is even greater. These strongly electropositive substituents lower the ionization potential of the hydrocarbon and thus enhance its donor power. Participation of nitrogen lone pairs is also possible. Thus arylamines form molecular compounds with acceptors that do not form isolable molecular compounds with the parent hydrocarbons; this has been illustrated for various polynitronaphthalenes.[137] In the limit, some arylamines may transfer an electron entirely to sufficiently powerful electron-acceptors and so form molecular compounds with ionic ground states. The aromatic amines with the most powerful electron-donor properties are shown in formulae (**69–75**).

TDAE[273] and TMPD are perhaps the most extensively studied of this group of strong donors. The chemistry of TDAE forms an interesting contrast to that of TCNE, while TMPD can be contrasted with TCNQ; the skeleton is similar in both pairs, but the substituent groups impart very different properties to the molecules. TDAE melts at $-4°$ and is easily oxidized to the monocation and the more stable dication. The ionization potential of TDAE is about 6.5 ev and it forms a charge-transfer compound with TNB of composition $TDAE:(TNB)_2$, which is fairly stable in air. Solid donor–acceptor molecular compounds of TDAE with chloranil, bromanil, and

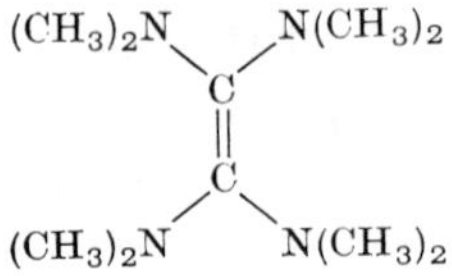

(69)

Tetrakis(dimethylamino)-
ethylene (TDAE)

(70)

*p*-Phenylene-
diamine (PD)

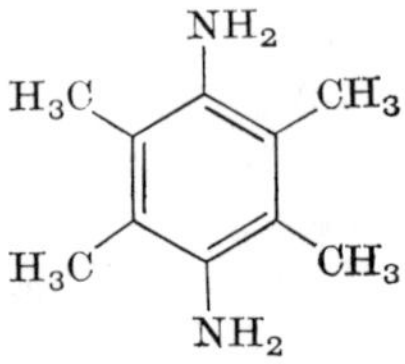

(71)

Durenediamine
(DAD)

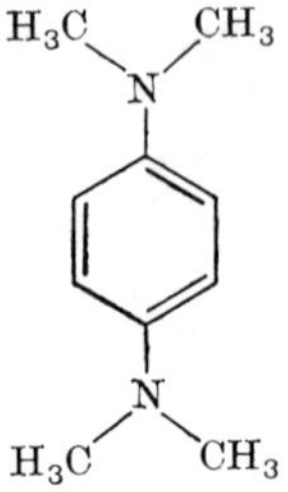

(72)

*N,N,N',N'*-
Tetramethyl-*p*-
phenylenediamine
(TMPD)

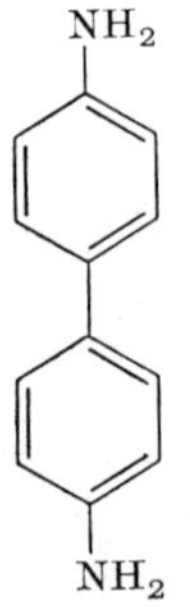

(73)

Benzidine

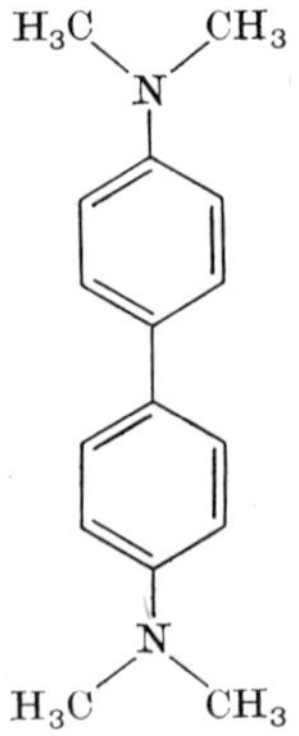

(74)

*N,N,N',N'*-
Tetramethylbenzidine

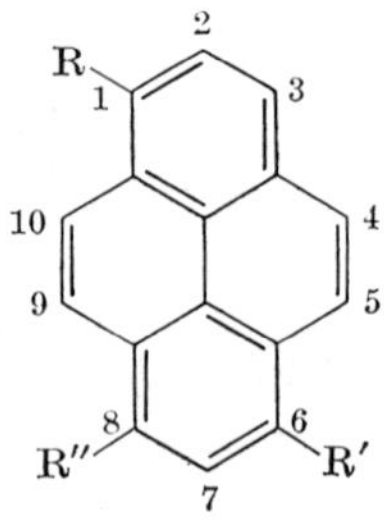

(75)

(a) R = R' = NH₂; R″ = H
(b) R = R″ = NH₂; R′ = H

(a) $R = R' = NH_2$; $R'' = H$
(b) $R = R'' = NH_2$; $R' = H$

3,4,5,6-tetrabromo-3,5-cyclohexadiene-1,2-dione have been prepared.[274] The ion-radical salt $(TDAE)^{2+}(TCNE)_2^-$ is formed with
TCNE;[274] the ionic nature of its ground state has been demonstrated
by UV, visible, and IR spectroscopy and by ESR.

Hammond and Knipe[275] suggest that in neutral TDAE the
planes of the dimethylamino groups are rotated well out of the
$N_2C\!=\!CN_2$ plane. The molecule will therefore not be planar and this
may somewhat hinder charge-transfer. As the available electrons are
associated largely with the nitrogen atoms, it was suggested that
TDAE is essentially an *n*-type donor but differs from the usual *n*-type
donors in that the "lone pairs" are delocalized in much the same
sense as in a *π*-type donor. Spectra in solution give some evidence
that TDAE is such a powerful donor that aromatic hydrocarbons

such as anthracene and pyrene act as electron-acceptors in its presence.[275]

A large number of molecular compounds of arylamines and other nitrogen-containing compounds with TNB have been prepared by Sudborough and Beard.[276] Primary amines of the benzene, naphthalene, and other aromatic hydrocarbon series* were all found to give black or deep brown equimolar molecular compounds, the naphthalene derivatives being appreciably more stable than the benzene derivatives. Secondary amines form molecular compounds with TNB provided that at least one of the groups attached to nitrogen is an aryl group; sometimes the composition is 2:1, as with diphenylamine. Tertiary amines with only one aryl group attached to nitrogen form molecular compounds with TNB; when two aryl groups are present the stability is appreciably reduced. Thus *N*-methyldiphenylamine (**76**) does not give a molecular compound whereas *N*-methyldi-2-naphthylamine (**77**) does. Triphenylamines and similar substances

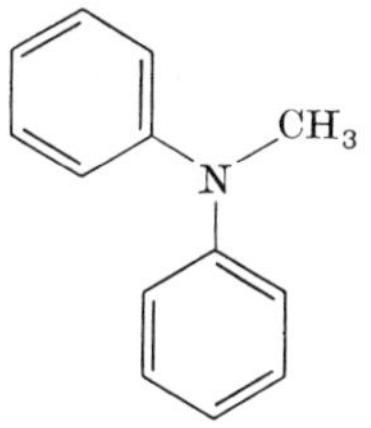

(**76**) 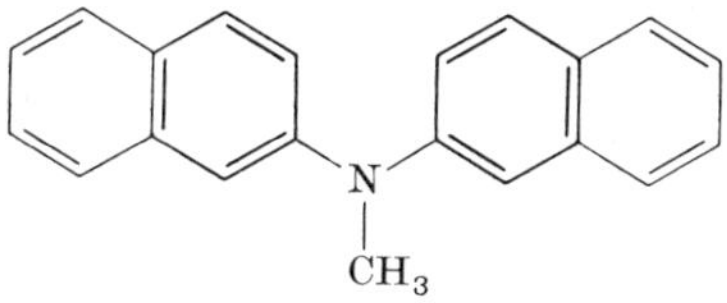     (**77**)

do not give molecular compounds with TNB or picric acid; this is presumably because the three phenyl groups are not coplanar. Aromatic amines and alkylarylamines with the amino group attached to a sidechain do not generally give crystalline TNB molecular compounds although they give intense red solutions in the presence of TNB.

Some other molecules containing nitrogen external to aromatic rings form molecular compounds with TNB,[276] but it is not known whether the nitrogen plays any special role in the donor–acceptor interaction. The following are some examples:

*c. Heteroaromatic molecules as electron-donors.* Replacement of carbon in an aromatic or quasi-aromatic ring system by nitrogen, oxygen, or sulphur does not usually affect greatly the ability of these molecules to function as electron-donors; however, the possibility of donation of lone-pair electron (*n–π** compounds) instead of,

---

* Nitrogen attached directly to aromatic ring.

$$C_6H_5CH_2NHC_6H_5$$

N-Benzylaniline:TNB
(1:1; deep red)

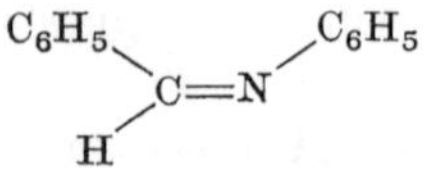

N-Benzylideneaniline:TNB
(1:2; yellow)

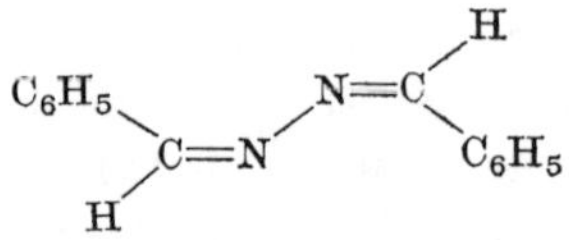

Benzylideneazine:TNB
(1:1; pale yellow)

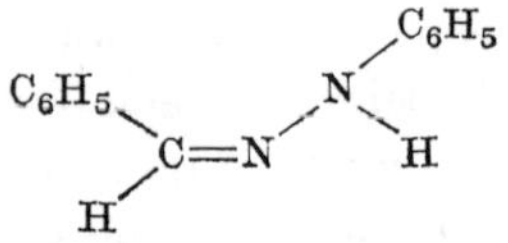

benzylidenephenylhydrazine: TNB
(1:1; deep red)

or in addition to, π-electrons of the aromatic rings must now be taken into account. Too few relevant physical measurements have been made for the importance of this effect to be apparent as yet.

(i) Nitrogen heterocycles. Most of the molecular compounds of nitrogen heterocycles that have been studied have TNB as electron-acceptor. The donors that form 1:1 compounds with TNB include quinoline and tetrahydroquinoline,[276] phenazine, acridine, and 9,10-dihydro-9-methylacridine,[270] indole, and carbazole.[118] Molecular compounds have also been prepared from many pyrroles.[196] However, molecular compounds are not formed if the nuclear heteroatom is basic as, for example, in benz[a]acridine.

The capacity of 1,2,3,4-tetrahydro-1-oxocarbazole (**35**) to give molecular compounds virtually equals that of naphthalene. Many polynitrobenzene molecular compounds have been prepared, including those of *m*- and *p*-dinitrobenzene, TNB, and picric acid (Kent[100]). The study has been extended to the monomethyl derivatives of the carbazole ketone (**35**) (Kent and McNeil[277]). Similar studies have been made of the polynitrobenzene molecular compounds of methyl-carbostyrils (cf. **78**) (Kent, McNeil, and Cowper[278]).

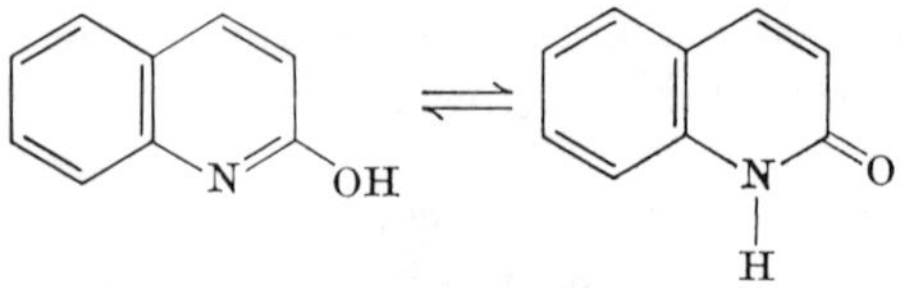

Carbostyril (**78**)

lactim form          lactam form

The *s*-triazine ring system behaves similarly to benzene in that substituted compounds behave either as electron-acceptors (Cl or

CN substituents) or electron-donors (dimethylamino substituent), depending on the nature of the substituent. 2,4,6-Tris(dimethylamino)-*s*-triazine forms 1:1 molecular compounds with a number of electron-acceptors[279,280] (Table 7), and the TNB molecular com-

Table 7. Molecular compounds in which substituted *s*-triazines and cyclophosphazenes act as electron-donors (status of Group III doubtful). Results taken from refs. 178 and 279.

| Molecular compound | | Temp. (°C)[a] |
|---|---|---|
| *Group I* | | |
| $N_3C_3(NMe_2)_3$ | Picric acid | 160–162 |
| $N_3C_3(NMe_2)_3$ | TNB | 155–175[b] |
| $N_3C_3(NMe_2)_3$ | TCNE | 190–195 |
| $N_3C_3(NMe_2)_3$ | Chloranil | 120–130; 150–165 |
| *Group II* | | |
| $N_3P_3(NMe_2)_6$ | Picric acid | 195–197 |
| $N_3P_3(NMe_2)_6$ | "$C_6H_3(NO_2)_2OH$" | 44.5–46; 112 |
| $N_3P_3(NMe_2)_6$ | TCNE | 113–114 |
| $N_3P_3(NHMe)_6$ | (Picric acid)$_2$ | 190–192,210–212 |
| $N_4P_4(NMe_2)_8$ | Picric acid | 112–113 |
| *Group III* | | M.p.(°C) |
| $3N_3P_3(NMe_2)_6$ | $N_3P_3Cl_3$ | 114 |
| $3N_3P_3(NHPr^i)_6$ | $N_3P_3Cl_3$ | 173 |
| $2N_3P_3Cl_2(NHBu^t)_4$ | $N_3P_3Cl_4(NHBu^t)_2$ | 108 |
| $2N_4P_4(NMe_2)_8$ | $N_4P_4Cl_8$ | 116 |
| $N_4P_4(NMe_2)_8$ | $4N_4P_4Cl_8$ | 122 |

[a] Temperatures are those at which a characteristic change can be observed on the hot stage of the polarizing microscope.
[b] Crystal structure has been determined.[281]

pound has been shown by crystal structure analysis[281] to have the same type of component arrangement as in other π-molecular compounds (see Section VI-D).

Various dialkylaminocyclophosphazenes form 1:1 and 1:2 molecular compounds with electron-acceptors (Group II in Table 7). Several other donor–acceptor molecular compounds (Group III in Table 7) have more complicated component ratios and presumably quite different structures. It is not clear from the original papers whether single crystals of these materials have been prepared or whether these is other evidence for donor–acceptor interaction.

(ii) Oxygen heterocycles. Ketones, *e.g.*, benzophenone, do not form π-molecular compounds, nor do molecules such as xanthone (**79**) and dimethylpyrone (**80**) where the oxygen atom and carbonyl group belong to the same ring system. However, oxygen hetero-cyclics such as dibenzofuran (**81**) and benzofuran (**82**) do form

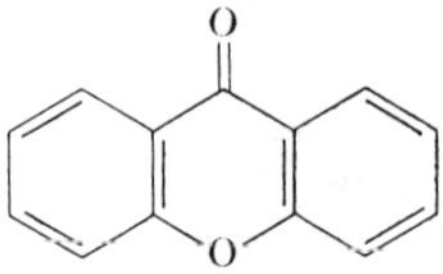

(79)

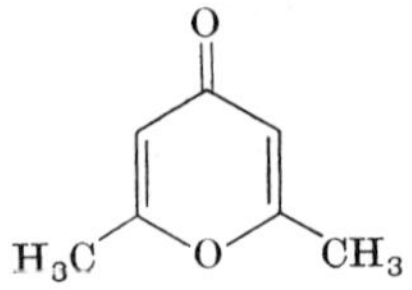

(80)

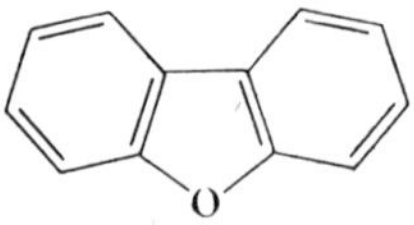

(81)

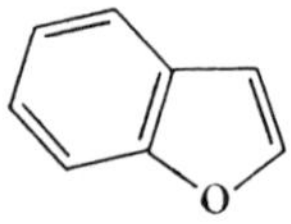

(82)

isolable TNB compounds; and hydroquinone (**83**) (1:1), fluorenone (2:1), and coumarin (**84**) (2:1) compounds with TNB have been identified by phase diagrams.[282]

(83)

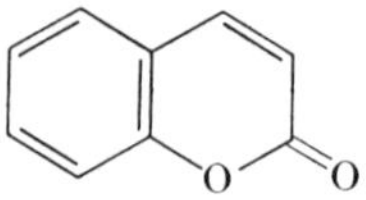

(84)

(iii) Sulphur heterocycles. Kraak and Wynberg[107] have prepared crystalline 1:1 TNB molecular compounds of a number of derivatives of 2,2′-bithiophene (**85**) and cyclopenta[*bb*′]dithiophene (**86**).

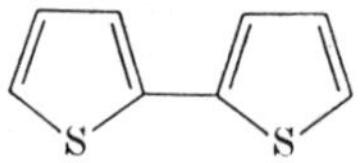

(85)

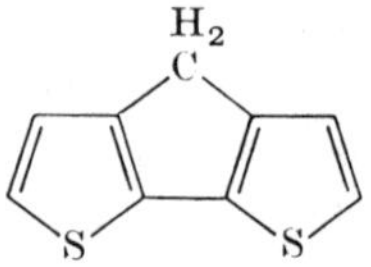

(86)

Benzotrifuroxan (BTF) molecular compounds have been prepared with benzo[*b*]thiophene (**87**) and 13,14-dithiatricyclo[8.2.1.1$^{4,7}$]tetra-deca-4,6,10,12-tetraene (**88**); the crystal structure of the molecular compound of the latter has been determined[283] (see also Section VI-I). Matsunaga[284] has prepared molecular compounds of the tetrathiotetracene derivative (**89**) with *o*-chloranil and *o*-bromanil

(3:1) and TCNE (3:2). These substances have ionic ground states and low electrical resistivities. Molecular compounds of phenothiazine (**90**) have also been prepared;[119,269,285] some have non-ionic ground states[119,269] and others have ionic ground states[285] (see Section VI-A).

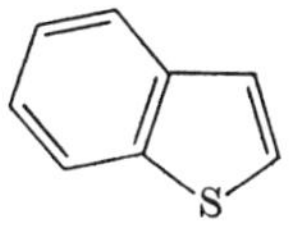

(87)

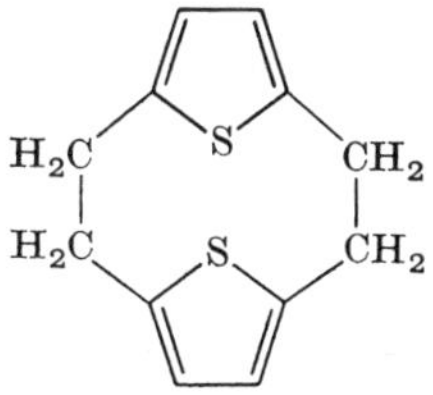

(88) (DDDT)

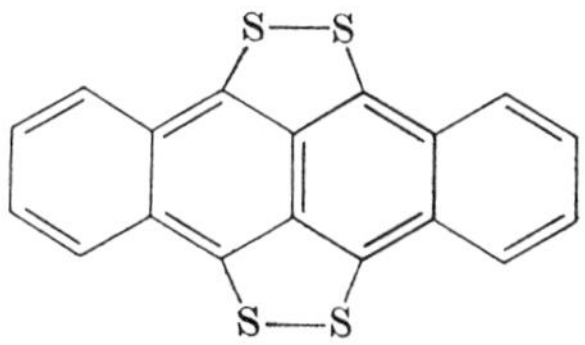

(89)

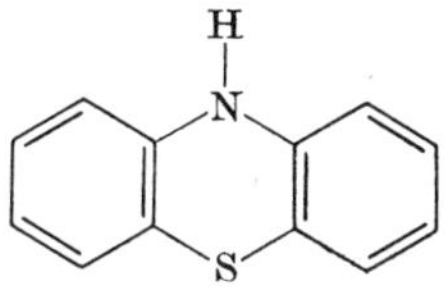

(90)

**2. Metallocenes.**—The molecular compounds that have been reported between various metallocenes and electron-acceptors are listed in Table 8, together with some of their properties. The stability in air of crystals of most of the cobaltocene and nicklelocene molecular compounds (or salts) contrasts with the oxidative instability of the parent metallocenes. The crystal structure of ferrocene:TCNE[288] has been determined[289] (see Section VI-C), confirming its status as a π-molecular compound with non-ionic ground state.[290]

Dibenzenechromium interacts with TNB, *p*-benzoquinone, chloranil, and TCNE in benzene to give deeply coloured solutions from which equimolar solids were obtained (Fitch and Lagowski[291]), but crystallinity of the solids was not reported. Spectroscopic studies (UV, visible, IR, and ESR), mainly on the solutions, suggest that complete electron-transfer occurs and that the solids are radical salts. The crystal structures of 1:1 and 1:2 ditoluenechromium:TCNQ have been determined (see Section VI-K).

**3. Carbonylarene complex.**—Tricarbonylchromium anisole:TNB has been prepared and its crystal structure determined (Carter,

Table 8. Summary of experimental results for metallocene–acceptor molecular compounds.

| Donor (D) ($E°$ in v) | Acceptor[a] (A) | Compd. (D:A)[b] | Product type[c] | Spins / Formula wt. | Remarks | Ref. |
|---|---|---|---|---|---|---|
| Ferrocene (0.30) | TCNB | | | | Melt or solution only | 84 |
| | TCPA | | | | | 84 |
| | p-Benzoquinone | | | | | 287 |
| | Chloranil | 1:1 | π | | | 287 |
| | TCNE[d] | 1:1 | π | | Green needles | 286,287,288 |
| | DDQ[d] | 1:1 | I | | Black crystals | 287 |
| | TCNQ | 1:2 | I | | Black needles $\rho$ 0.24 ohm cm | 286 |
| Bis(tetrahydroindenyl)iron (+0.05) | TCNE[d] | 1:2 | I | | Black ppt. | 287 |
| | DDQ | 1:1 | I | | Black crystals | 287 |
| Nickelocene (-0.10) | Chloranil | 1:2 | I | $10^{-7}$ | Dark-brown solid | 286 |
| | TCNQ | 1:2 | I | | | 84 |
| Cobaltocene (-1.16) | Chloranil | 1:1 | I | 1 | Green powder | 286,287 |
| | | 1:2 | I | 0.02–0.2 | Olive-green | e |
| | TCNE[d] | 1:1 | I | | Dark blue ppt. ($O_2$-sensitive) | 287 |
| | DDQ[d] | 1:1 | I | | Black crystals | 287 |
| | TNB | 1:1 | I | | Brown solid | 286 |
| | TCNQ | 1:2 | I | | Black scale $\rho$ 6.5 ohm cm | 84 |

[a] For electron affinities of acceptors see Table 1.
[b] Products stable in air unless stated otherwise.
[c] The nature of the ground state ($\pi$ = non-ionic; $I$ = ionic) was identified from IR spectra, where the acceptor-ion contribution was decisive because of the weakness of the absorption of the metallocinium ion, and from ESR spectra.
[d] Resistivities of these compounds are in the range $10^8$–$10^{10}$ ohm cm.
[e] Some confusion may have occurred between the 1:1 and 1:2 samples (see ref. 287).

McPhail, and Sim[292]) (see Section VI-C). There does not appear to be any unusual feature.

**4. Porphyrins.**—Eighty-nine different molecular compounds of various porphyrins with various electron-acceptors have been prepared by Treibs[293] and their powder spectra determined. Most samples were obtained as powders (degree of crystallinity not known), but some single crystals were reported. The products were usually coloured. The electron-acceptors included picric, styphnic, picrolonic, and flavianic acid, and so most of the products presumably have a salt-like character. The molecular ratios were usually 1:1 or 1:2, but some 1:3 complexes were formed and other ratios were found on occasion (*e.g.*, mesoporphyrin ester gives a red complex with 2,6-dinitrobenzoic acid, the ratio being 1:5).

More recently Hill, MacFarlane, and Williams[294] have reported solid complexes (powders) of Co(II) ætioporphyrin, of Co(II) mesoporphyrin dimethyl ester, and of ætioporphyrin with acceptors such as TNB, 3,5-dinitrobenzonitrile, and TNF. There is some spectroscopic evidence for charge-transfer interaction.

**5. Coordination complexes.**—$\pi$-Molecular compounds of 8-quinolinol (**91**) or its Cu(II), Pd(II), and Ni(II) chelates (metal oxinates) with

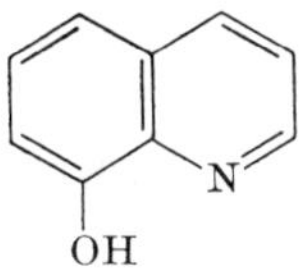

(**91**)

various electron-acceptors have been prepared and their reflection spectra measured.[295] The molecular compounds are either 1:1 or 1:2 (Table 9) but the spectra show no significant difference between the different compositions. Electrical conductivities of single crystals have been measured.[296] Several crystal structures in this series (discussed below) indicate that the metal atoms do not contribute importantly to the cohesion between the components.

## G. Thermodynamic Parameters of Crystalline $\pi$-Molecular Compounds

The free energy, enthalpy, and entropy of formation of crystalline donor–acceptor compounds have been measured for only a limited number of substances. The relevant reaction is:

$$D_{cryst} + A_{cryst} = (DA)_{cryst}$$
$$\Delta X_f^\circ = X_f^\circ(DA_{cryst}) - [X_f^\circ(D_{cryst}) + X_f^\circ(A_{cryst})]$$
$$(X = G, H, \text{ or } S)$$

Various methods have been used, based on solubility,[297] EMF,[298] and depression of the freezing point.[299] Thermodynamic parameters measured for molecular compounds in solution should not be used for discussion of the properties of the crystalline compounds because the usually rather small values may be influenced as much by environment as by differences in structure or interaction between components.

Table 9. Compositions of various metal 8-quinolinolate–acceptor molecular compounds.[a]

| | Donor | | |
| Acceptor | Cu salt | Pd salt | 8-Quinolinol |
| --- | --- | --- | --- |
| BTF | 1:1 | 1:1 | 2:1 |
| Benzotrifurazan | 1:2 | — | 1:1 |
| 4,6-Dinitrobenzofuroxan | 1:1 | 1:1 | — |
| TNB | 1:2 | — | 1:1 |
| TNF | 1:1 | 1:1 | 2:1 |
| TCNB | 1:2 | 1:1 | 1:1 |
| Chloranil | 1:1 | 1:1 | 2:1 |
| DDQ | 1:1 | 1:1 | — |
| Picryl azide | 1:2 | — | 1:1 |

[a] Bis-(8-quinolinolato)-copper(II) or -palladium(II). The compositions are expressed as donor–acceptor molecular ratios; thus, 1:1 for a metal salt is equivalent to 2:1 for 8-quinolinol.

Most of the available results are summarized in Tables 10 and 11. The measured free energies of formation can be compared with the sequences of donor strengths proposed by Dimroth and Bamberger[131] (from semiquantitative solubility measurements) and by Sinomiya.[94] Both sequences fit the experimental free energies of formation fairly well although there are some discrepancies. Melting points of molecular compounds are sometimes used as measures of their stability; for example, Hertel[62] noted that molecular compounds of TNB, picric acid, and picramide with a given donor usually have similar melting points, whereas the melting points of the analogous TNT and picryl chloride molecular compounds are considerably lower. The results in Table 10 also suggest some correlation between thermodynamic stability and melting point. As the melting point is

Table 10. Measured free energies of formation of crystalline $\pi$-molecular compounds ($\Delta G_f°$ in cal/mole at 298°K unless stated otherwise).[a]

| Donor | Acceptor TNB | | Acceptor picric acid | | Dimroth–Bamberger sequence for picric acid compound | Acceptor TNT | | Sinomiya donor sequence |
|---|---|---|---|---|---|---|---|---|
| | $\Delta G_f°$ | M.p. | $\Delta G_f°$ | M.p. | | $\Delta G_f°$ | M.p. | |
| Fluoranthene | −2910 | 204° | | | | −1570 | 132° | 1 |
| Phenanthrene | −2410 | 164° | | 133° | | −1340 | 102° | 3 |
| Naphthalene | −2070 | 152° | −2070 | 150° | 1 | −1300 | 97° | 1 |
| Acenaphthene | | 161° | | 161° | | −1300 | 111° | 2 |
| Fluorene | −890 | 105° | | 84° | 4 | −830 | 89° | 4 |
| Indene | | | | | 2 | | | |
| Anthracene | −840 | 162° | −490 | 138° | 3 | | | 5 |
| Benzene | | | | | 5 | | | 3 |
| 1-Methylnaphthalene | | 147° | | | 2 | −1690 | 74° | 1 |
| 2-Methylnaphthalene | −2480 | 123° | | | 1 | −1300 | 97° | 2 |
| Naphthalene | −2080 | 152° | | | 3 | | | 3 |
| Pentamethylbenzene | −1590 | 119° | | | | −660 | 81° | 2 |
| Hexamethylbenzene | −1340 | 173° | | | | −280 | 123° | 1 |
| Durene | −840 | 101° | | | | | | 3 |

[a] Compositions are 1:1 except for fluorene:TNB = 3:4. Measurements are by Hammick and Hutchison[297] except for naphthalene and anthracene:picric acid (see Table 11). M.p.s are from various sources.

Table 11. Thermodynamic parameters for some molecular compounds at 297°K.[a]

| Molecular compound | $\Delta G_f{}^\circ$ (cal/mole) | $\Delta H_f{}^\circ$ (cal/mole) | $\Delta S_f{}^\circ$ (cal mole$^{-1}$ deg$^{-1}$) |
|---|---|---|---|
| Naphthalene:TNB[b] | − 2070 | − 1120 | 3.2 |
| Anthracene:TNB[c] | − 840 | − 341 | 1.7 |
| Naphthalene:picric acid[d] | − 2070 | − 880 | 4.0 |
| Anthracene:picric acid[e] | − 490 | − 2210 | − 5.8 |
| Quinhydrone[f] | − 3670 | − 5390 | − 6.3 |

[a] Other values (all cal/mole) reported are as follows: $\Delta H_f{}^\circ$ for benzene:TNB[300] − 700 ± 260 and 4-bromo-1-naphthylamine:2,6-dinitrophenol[64] yellow polymorph − 3200, red polymorph 100; also the following:

[b,c] $\Delta G_f{}^\circ$ from ref. 297, $\Delta H_f{}^\circ$ from ref. 300.

[d] $\Delta G_f{}^\circ = -1920, \; -2050,[298] \; -2150,[298] \; -2083,[299] \; -2070;[130] \; \Delta H_f{}^\circ = -880$ (direct calorimetric measurement).[298]

[e] $\Delta G_f \, (297°) = -490, \; \Delta G_f \, (273°) = -629;[131] \; \Delta H_f{}^\circ$ and $\Delta S_f{}^\circ$ (both assumed independent of temperature) calculated from these $\Delta G_f$ values.

[f] Quinhydrone, see Section VI-H.

the temperature at which solid and liquid are in equilibrium, its use as a criterion of thermodynamic stability implies that the properties of the various molten molecular compounds are similar. Another factor here is surely the similarity between the structures of the donors and of the acceptors considered in Table 10.

The thermodynamic parameters that are most directly informative about structure are the standard enthalpy and entropy of formation; available values are summarized in Table 11. The molecular compounds listed there fall into two groups according to whether the entropy of formation is positive or negative. In the former group the entropy change on compound formation makes an important contribution to stability, whereas in the latter the entropy change diminishes the stability of the molecular compound. Since the entropy of formation of crystalline naphthalene:picric acid is positive (perhaps owing to disorder in the crystals, see Section V-H), this compound becomes more stable (with respect to its components) as the temperature is raised, as Brønsted[298] pointed out. The converse applies to anthracene:picric acid. Of course, it is not to be expected that naphthalene:picric acid and anthracene:picric acid will dissociate into their components on cooling or heating, respectively, but rather that phase changes will occur. Kofler[109] has shown that naphthalene:picric acid is stable from room temperature to the melting point, whereas anthracene:picric acid has a phase change at 85°. No study below room temperature has been reported for either of these molecular compounds.

In addition to the major qualitative difference between the two groups regarding the sign of $\Delta S_f{}^\circ$, there are smaller quantitative differences within the groups. For example naphthalene:TNB and naphthalene:picric acid are isomorphous and have the same free energy of formation; nevertheless, the small differences in their values of $\Delta H_f{}^\circ$ and $\Delta S_f{}^\circ$ must reflect differences in the interactions between the pairs of components.

## H. Stoichiometry of π-Molecular Compounds

Most π-molecular compounds are equimolar in composition, and this type has been the most intensively studied, both in regard to crystallography and to physical properties. There are a number of examples where other, less stable, molecular compounds appear in the phase diagrams—*e.g.*, for perylene:pyromellitic dianhydride both 1:1 and 2:1 compositions are found (Boeyens and Herbstein[301]), and for pyrene:pyromellitic dianhydride (Figure 4) three compositions

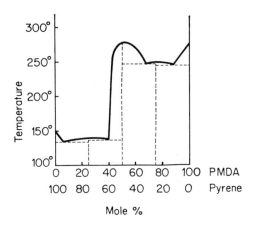

Figure 4. Melting point diagram for the system pyrene:PMDA. The full lines were determined experimentally, the broken lines were inferred. [Reproduced with permission from reference 292].

(2:1, 1:1, and 1:2 or 1:3) are found (Ilmet and Kopp;[302] Herbstein and Snyman[303]).

There are some systems where the 2:1 or 1:2 molecular compound is the most stable species (*e.g.*, the 2:1 molecular compounds and benzo[c]pyrene with TMU).[244] Unfortunately the binary phase diagrams for these systems have not been determined. It is not known if there is a system where both 2:1 (say) and 1:1 molecular compounds occur, with the former more stable than the latter. A remarkable phase diagram has been reported by Sinomiya[93] for the 2-naphthol:1,2,4-trinitrobenzene system (Figure 5). Here the 1:1

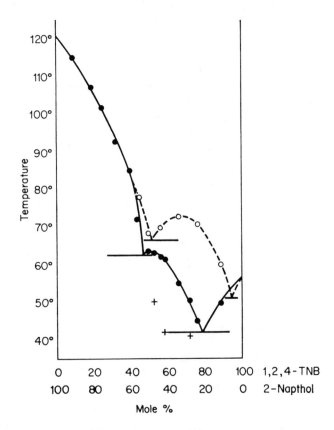

Figure 5. Melting point diagram for the system 2-naphthol:1,2,4-trinitrobenzene.[93] Full lines show the stable system and broken lines the metastable system.

and 1:2 molecular compounds appear to form independent systems; the 1:1 was obtained as an orange-red powder (m.p. 63.5°) and the 1:2 as red-brown crystals (m.p. 73°). Further study of this system seems desirable.

There are also a few examples where other ratios are said to occur. These reports should be viewed with some suspicion, particularly if they are based only on chemical analysis of crystals with possible volatile components; for example, Bailey and Case[139] report a 1:4 composition for benzene:BTF, whereas the crystallographic results of Boeyens and Herbstein[157] show a 1:1 composition. One authenticated example is the 3:4 composition for fluorene:TNB, first reported by Sudborough[101] and checked crystallographically by Hertel and Römer[304] and by Herbstein and Regev[305] (who obtained the same unit-cell volume and density as Hertel and Römer but found the crystals to be triclinic and not monoclinic). Other examples, not yet checked crystallographically, are given in Table 12.

Table 12. Some unusual compositions reported for $\pi$-molecular compounds.[a]

| Donor (D) | Acceptor (A) | D:A | Ref. |
|---|---|---|---|
| Bromodurene | BTF | 3:2 | 140 |
| Tetralin | Nitrobenzodifuroxan | 1:3 | 139 |
| Pyrene | PMDA | 1:3 | 303 |
| $N,N$-Dibenzyl-$m$-toluidine | TNB | 3:2 | 278 |
| Triphenylmethanol | TNB | 3:2 | 306 |
| Phenanthrene | $p$-Dinitrobenzene | 3:1 | 99 |
| Fluorene | TNB | 3:4 | 101,304,305 |
| 1-Naphthylamine | 2,3-Dinitrophenol | 3:2 | 93 |
| 2-Naphthylamine | 2,3-Dinitrophenol | 3:2 | 93 |
| Phenanthrene | 1,2,4,6-Tetranitrobenzene | 2:3 | 94 |
| Naphthalene | 1,2,4,6-Tetranitrobenzene | 3:2 | 92 |
| 1-Naphthol | Tetryl | 3:2 | 92 |
| $p$-Phenylenediamine | $p$-Benzoquinone | 2:5 | 195 |
| 2-Naphthol | "Dinitroaniline" | 3:2 | 90 |
| Naphthalene | "Dinitroaniline" | 3:1 | 90 |
| Dibenzo[$c,d$]phenothiazine | DDQ | 3:2 | 285 |
| Benzene | $o$-Chloranil | 3:1 | 198 |
| 1,4-Diphenylbutadiene | TNF | 3:1 (also 1:2) | 216 |
| Tetrathiotetracene (89) | $o$-Chloranil, $o$-bromanil | 3:1 | 284 |
|  | TCNE | 3:2 |  |

[a] The 5:4 composition reported for 2-aminofluorence:TNB from ethanol (Sudborough[101]) appears to be incorrect as 1:1 crystals of nearly the same colour and melting point were also reported by this author as being obtained from benzene; the 5:4 composition reported for an acetyl derivative of 2-aminofluorene with TNB is also suspect. Orchin *et al.*[216] report that analysis of the TNF complex with benzo[$c$]fluorene does not correspond to any simple molar proportion of hydrocarbon to TNF.

## I. Ternary π-Molecular Compounds

There are a number of molecular compounds which contain three components; does the third component participate in the charge-transfer interaction in the same way as the other components or is it essentially inert in this respect? Structural or spectroscopic studies could settle this point but none has been made, apart from a few preliminary X-ray studies.

**1. Non-interacting third component.**—A number of examples are known where the third component does not appear to affect the physical properties of the molecular compound and is thus probably present as "solvent of crystallization" without participating in any donor–acceptor interaction.

(i) Phenanthrene:TNB crystallizes as such from ethanol but contains an unspecified amount of benzene when crystallized from benzene (Kofler[109]). The two samples gave similar diffraction patterns,[305] and thus the benzene is presumably present as solvent of crystallization.

(ii) The yellow phenanthrene:2-[(2,4,5,7-tetranitro-9-fluorenyl-ideneamino)oxy]propionic acid compound contains one molecule of acetic acid (the solvent) for each donor–acceptor pair (Newman and Lutz).[223] The molecule of acetic acid may be hydrogen-bonded to the propionic acid portion of the acceptor molecule.

(iii) The azulene:BTF molecular compound contains one-half of a propionic acid molecule per formula unit.[140]

(iv) Tetrabenzonaphthalene:picric acid crystallizes with one molecule of ethanol per formula unit (Herbstein and Regev[305]). The unit-cell dimensions of the solvated and unsolvated crystals are very similar, but there are some differences in the intensities of corresponding reflections.

(v) 2,2′,4,4′,5,5′-Hexamethylstilbene forms a molecular compound with two molecules of picric acid and one of benzene (Elbs[307]). Nothing is known of the structure or of the interaction among the components.

(vi) 1-Naphthol:hexachloro-1-indenone crystallizes with one-half molecule of benzene or acetic acid per donor–acceptor pair (Pfeiffer *et al.*[167]). Similar results were reported by these workers for the analogous 2-naphthol molecular compounds and benzene. The structures are not known.

**2. Interacting third component.**—There are three examples of ternary donor–acceptor molecular compounds where there seems to be some likelihood of the third component's interacting with the other two. One is 1-naphthylamine:pyridine:picric acid (1:1:1). Kofler[109] has shown that 1-naphthylamine forms both 1:1 and 1:2 molecular compounds with picric acid, as well as the ternary pyridine compound. Since the three compounds have somewhat different colours it is possible that the pyridine also participates in charge-transfer interaction. The second example is benzidine:TCNQ which crystallizes in 1:1 proportion from $CHCl_3$ or $CH_3CHCl_2$, whereas from $CH_2Cl_2$, $(CH_3)_2CO$, or $CH_3CH_2Br$ an additional 1–1.5 moles of solvent are present. There are differences in the IR spectra of the binary and ternary molecular compounds, and the electrical conductivities of the ternary compounds are $10^4$–$10^6$ times as large as those of the binary compounds.[308, 481] The third example is benzidine:2,4-dinitrophenol which occurs as brown crystals (m.p. 144°), whereas a yellow monohydrate [m.p. 138° (dec.)] is obtained from water (Buehler and Heap[98]). Participation of water in a charge-transfer interaction seems much less likely than for pyridine.

**3. Molecular compounds of organic salts with trinitrobenzene.—** There are a number of examples where both a particular donor, and also its potassium salt, form molecular compounds with TNB. Some examples are given in Table 13. No information is available about their structure or physical properties.

Table 13. Examples of TNB molecular compounds formed by a particular donor and by its potassium salt.[a]

| Donor | TNB molecular compound | |
| --- | --- | --- |
| | Crystals | M.p. |
| *o*-Aminobenzoic acid | Orange needles | 192–193° |
| *m*-Aminobenzoic acid | | — |
| *p*-Aminobenzoic acid | Red | 151° |
| 1-Anthrol | Red-brown plates | 161° (dec.) |
| | TNB molecular compound of K salt | |
| | Crystals | M.p. |
| *o*-Aminobenzoic acid | Deep red needles | 114° |
| *m*-Aminobenzoic acid | Red-brown needles | 118–119° |
| *p*-Aminobenzoic acid | Red needles | 115° (dec.) |
| 1-Anthrol | Black needles | 275° |

[a] The anthrol potassium salt molecular compound has composition $C_{14}H_9OK$: $(C_6H_3(NO_2)_3)_2$.[102] All the other compounds are 1:1.[276]

## V. CRYSTAL CHEMISTRY OF π-MOLECULAR COMPOUNDS

### A. The Information Sought

We consider here what information about the chemical and physical nature of donor–acceptor compounds can be obtained from crystallographic studies at various levels of sophistication. Determination of cell dimensions and space group is an essential preliminary step and sometimes allows general deductions to be made about component arrangement. However, a full structure determination is necessary to decide whether any special donor–acceptor interaction exists and to allow correlation between structure and physical properties (*e.g.*, polarized absorption spectra, electrical-conductivity tensor) of the crystals.

The most characteristic feature of crystalline donor–acceptor π-molecular compounds is an alternating arrangement of donor and acceptor molecules in mixed stacks. The interplanar distance and degree of parallelism of the component molecules gives information about the nature of the donor–acceptor interaction. Although the observed mutual orientation of adjacent component molecules within a stack can be compared with theoretical predictions, such a comparison is meaningful only for the polymorphic form stable at absolute zero, because the theoretical calculations (based on interaction energies and not free energies) refer to this temperature. Knowledge of the polymorphism of the system and of order–disorder transformations is thus particularly important. Results from disordered crystals are liable to be especially misleading since they refer to an average of two (or more) different orientations.

Molecular dimensions, if of sufficient accuracy, should give an indication of the effects of complex formation on the components and, in particular, show whether the components are present as charged or uncharged species. However, molecular dimensions are rather insensitive to changes in molecular charge, and conclusions should be checked against the results of other physical measurements. It is worth noting that the precision of atomic coordinates derived from measurements on disordered crystals is likely to be much higher than their accuracy, and so again erroneous conclusions may be drawn if the presence of disorder is not recognized. Differences in the shapes of one or both components in different molecular compounds may provide useful information in some instances (see Section VI-A).

Phase changes and order–disorder transformations in crystalline

$\pi$-molecular compounds are of considerable interest in themselves as these crystals provide well-defined examples of two-component systems. Thus, establishment of the occurrence of disorder is as important for the crystal physics of these systems as it is for their crystal and molecular chemistry.

## B. The Available Results

The crystal structures of over forty different $\pi$-molecular compounds have been actually determined (*i.e.*, atomic positions have been obtained by one of the standard crystallographic techniques, and not simply inferred from cell dimensions, as was done in certain earlier work). These crystal structures are distributed rather unevenly among the various donors and acceptors (see Table 14), so that general conclusions drawn from the available samples will necessarily be tentative. Not many results are available for compositions other than equimolar, but structural similarities are evident between 1:1 and *some* 2:1 (or 1:2) crystals.

The available results can be treated in a number of different ways but no single scheme has been found for discussion of all the information currently available. Comparisons will be made here of the molecular compounds of various donors with a given acceptor, and of the converse situation. In addition, it is found that many molecular compounds can be grouped together into families of quasi-isomorphous crystals. Component arrangements in molecular compounds with neutral and ionic ground states can also be usefully compared.

## C. Packing Densities in $\pi$-Molecular Compounds and in Crystals of Their Components

The difference between the volume of a formula unit in a crystalline molecular compound and the sum of the molecular volumes of the components is a useful but limited index of the comparative packing densities of the combined and the separated components. The limitations arise mainly from the inaccuracy of measured unit-cell dimensions, the desirability of comparing volumes measured at the same reduced temperatures (*i.e.*, at the same fractions of the absolute melting points of the respective substances), and the difficulties of defining adequately the molecular volumes of the components because of the occurrence of polymorphism or special interactions in their crystals (*e.g.*, chloranil[309]). A reduction of volume on formation of a molecular compound would indicate tighter binding (or perhaps

Table 14. Electron-donors and -acceptors occurring in the molecular compounds whose crystal structures have been reported. The number of crystal structures in which each donor and acceptor was reported up to early 1970 is given in parentheses.

| Donors | Acceptors |
|---|---|
| *Aromatic hydrocarbons*<br>Naphthalene (3), anthracene (3), pyrene (3), perylene (3), benzo[c]pyrene (1), coronene (1), hexahelicene (1) | *Polynitroaromatics, etc.*<br>1,3,5-Trinitrobenzene (8), picryl azide (1), 1-bromo-3,6,8-trinitrofluorenone (2), benzotrifuroxan (2) |
| *Substituted aromatic hydrocarbons*<br>Hexamethylbenzene (2), 1,12-dimethylbenzo[c]phenanthrene (1), p-iodoaniline (1), N,N,N′,N′-tetramethylphenylenediamine (3), benzidine (1), hydroquinone (2), phenol (1), p-chlorophenol (2), resorcinol (1) | *Cyano compounds*<br>Tetracyanoethylene (4), 1,2,4,5-tetracyanobenzene (6), tetracyanoquinodimethane (6) |
| *Non-benzenoid aromatic hydrocarbons*<br>Azulene (1), acepleiadylene (1) | *Aromatic quinones*<br>p-Benzoquinone (5) |
| *Aromatic heterocycles*<br>Indole (1), skatole (1), 8-quinolinol (1), 2,4,6-tris-(dimethylamino)-s-triazine (1) | *Substituted anils*<br>Fluoranil (1), chloranil (5) |
| *Coordination compounds*<br>Bis-8-quinolinolatocopper(II) (4), bis-8-quinolinolatopalladium(II) (2) | *Aromatic anhydrides*<br>Pyromellitic dianhydride (3) |
| *Organometallics*<br>Ferrocene (1), anisoletricarbonylchromium (1), ditoluenechromium (2) | *Various*<br>1,3,7,9-Tetramethyluric acid (3), 3,5-dinitrobenzoic acid (1) |
| *Various*<br>13,14-Dithiatricyclo[8.2.1.4$^{4,7}$]-tetradeca-4,6,10,12-tetraene (1), phenothiazine (2). | |

only more efficient packing) in the molecular compound than in the crystalline components. Such results are found for the quinhydrones (Table 22, page 312), but most other groups of molecular compounds show small contractions and expansions about equally distributed (Table 20, page 253).

Two illustrations of these difficulties may be given. Perylene occurs in two polymorphic forms; for the molecular volume values of 304.3,[310] 312,[311] and 308 Å$^3$ [312] have been given for the first polymorph, and 319 Å$^3$ [313] for the second polymorph. The molecular volume of PMDA is 214 Å$^3$ [157] and that of perylene:PMDA is 528.2 Å$^3$.[301] These results could indicate a small amount of either expansion or contraction on molecular-compound formation, depending on which perylene polymorph is taken as basis. The molecular volume of the anthracene:TNB[314] compound is 460 Å$^3$ at 300°K and 440 Å$^3$ at175°K; the sum of the molecular volumes of anthracene (237 Å$^3$)[315] and TNB (212 Å$^3$)[316] at 300°K is 449 Å$^3$; the results at 300°K yields a +2.5% volume change on molecular-compound formation; this is considerably less than the +8.7% volume change claimed by Skraup and Eisemann,[317] whose older results should generally be viewed with some caution.

Volume changes on formation of molecular compounds *in solution* have been determined by Ewald[318] using the pressure-dependence of the charge-transfer spectra. The values found for $\Delta V$ varied between $-3$ and $-20$ Å$^3$ per formula unit (see Table 20, page 253, groups 1, 2, 4b). Ewald's value for anthracene:TNB is $\Delta V = -8$ Å$^3$ per formula unit, compared with the small positive value of $\Delta V$ derived from crystallographic results. The general agreement of Ewald's values with the crystallographic results is rather varied, presumably because the two sets of measurements refer to different physical circumstances.

## D. The Occurrence of Mixed Stacks

**1. Equimolar compounds.**—The basic feature of the crystal chemistry of the equimolar donor–acceptor molecular compounds is the arrangement of the components in mixed stacks of alternating donor (D) and acceptor (A) molecules:

$$------ D A D A D A D A D A -------$$

The essentially planar component molecules are usually inclined to the stack axis at an angle of between 20° and 30°, and the perpendicular distance between the molecular planes varies from 3.2 to 3.5 Å. There are two lines of evidence suggesting that there are stronger interactions between adjacent molecules within a stack than between molecules in different stacks. The more direct evidence comes from measurements of elastic constants of crystals of $\pi$-molecular compounds. By use of the forced resonance vibration method, Young's modulus (= stress/strain) in the direction of the needle axis was measured for various aromatic hydrocarbons and hydrocarbon:TNB molecular compounds.[319] The results (Table 15) show that Young's modulus along the stack (needle) axis is about

Table 15. Values of Young's modulus ($E$) for some aromatic hydrocarbon and π-molecular-compound crystals.[319]

| Substance | $E$ ($10^{-10}$ dyn/cm$^2$) |
|---|---|
| Triphenylene | 1.19 |
| Dibenzo[a,c]anthracene | 2.81 |
| Coronene | 1.10 |
| Coronene:TNB | 9.99 |
| Anthracene:TNB | 12.4 |
| Perylene:TNB | 7.77 |

ten times greater for the molecular compounds than for the aromatic hydrocarbons. One can roughly equate the intermolecular interactions in the aromatic hydrocarbons to those between the stacks in the molecular compounds. The experimental results thus show that the intra-stack interactions are appreciably stronger than the inter-stack interactions. Measurement of the complete set of elastic constants for members of the two groups of compounds would put these conclusions on a much sounder basis.

The second method relies on a comparison of the measured interplanar distance between molecules within a stack and the sum of the thicknesses of the adjacent donor and acceptor molecules. Care is needed in applying this method because of the lack of standard thicknesses for donor and acceptor molecules. For example, the interplanar distances in pyrene[320] and coronene[321] are 3.53 and 3.40 Å, respectively, and this range is typical of aromatic hydrocarbons; however, the only parallel information for acceptors is the interplanar distance of 3.45 Å in TCNQ.[322] (There is no molecular stack in TCNE, so that an appropriate molecular thickness cannot be extracted from its known crystal structure.[323]) If one may generalize from these few values one concludes that measured interplanar distances within a stack of less than about 3.35 Å are evidence for an interaction operating roughly perpendicular to the molecular planes in addition to the usual dispersion forces.

A component thickness of approximately 3.5 Å leads to a periodicity along the stack axis of about 7 Å, which is an indication, but not a proof, of the existence of mixed stacks; in ditoluenechromium:TCNQ the stack axis is 7 Å, but segregated and not mixed stacks are found in this substance (Section VI-K[324]). Even in molecular compounds with mixed stacks there may be two cell dimensions close to 7 Å and resort must then be made to crystal morphology or diffuse scattering to identify the stack axis. It has so far been possible to interpret distances between molecules in different stacks solely in terms of van der Waals interactions.

The evidence for a stronger interaction within the mixed stacks than between them is the justification for describing these crystals in terms of stacks instead of choosing another mode of description, such as in terms of planes of molecules which would be equally valid from a geometrical point of view but would lack any special physical significance. The stacks themselves have a roughly cylindrical shape and are arranged in the crystal in a quasi-hexagonally close-packed fashion. (Figure 33, page 296, shows the component arrangement in pyrene:PMDA.) In some instances the sequence along the stack axis is more complicated, as different orientations of donor and/or acceptor molecules may occur within a given stack; the resulting periodicity will then be an integral multiple of 7 Å. So far, structures of two crystals with 14 Å periodicities along the stack axes have been determined: these are anthracene:TNB[314] and the ordered form of pyrene:PMDA.[303] Other crystals with 14 Å periodicities along the stack axes are known, but their crystal structures (Table 16) have not been determined. The 1:1 HMB:picryl chloride, bromide, and iodide molecular compounds (Powell and Huse;[325] Bernstein and Herbstein[326]) form an isostructural group where the stack axis is an integral multiple of 14 Å, but complications are introduced by diffuse reflections due to disorder effects not yet understood in detail (Section VI-D).

Description of the crystal structures in terms of isolated mixed stacks appears to be valid for all structures reported hitherto, except for the 1:1 bis-8-quinolinolatocopper(II):TCNQ molecular compound (Williams and Wallwork[327]). Here the familiar plane-to-plane arrangement is found, but the acceptor molecules act as bridges between donor molecules in different stacks (see Section VI-C).

**2. Molecular compounds with component ratio different from unity.**—Change in stoichiometry of $\pi$-molecular compounds from 1:1 to another ratio may or may not be accompanied by changes in structure and properties. There is one type of $\pi$-molecular compound

Table 16. Crystallographic results for 1:1 π-molecular compounds with periodicities other than *ca.* 7 Å along the stack axis.[a]

| Molecular compound | $a$ (Å) | $b$ (Å) | $c$ (Å) | $\beta$ (deg) | $Z$ | Space group | Volume of formula unit[b] measured (Å³) | additive (Å³) | Ref. |
|---|---|---|---|---|---|---|---|---|---|
| Benzo[c]phenanthrene:DDQ | 18.59 | 16.11 | **13.82** | 96.3 | 8 | $B2_1/c$[c] | 514 | — | 305 |
| Pyrene:PMDA (at 110°K) | 13.667 | 9.130 | **14.404** | 91.5 | 4 | $P2_1/n$ | 450 | 476 | 303 |
| Anthracene:TNB (at 173°K) | 11.35 | 16.27 | **13.02** | 133.2 | 4 | $C2/c$ | 440 | 449 | 314 |
| HMB:picryl chloride[d] | **14.0** | 9.0 | 15.4 | | 4 | $A2_1am$, $Ama2$, or $Amam$ | 485 | 490 | 325 |
| | | | | | | | | | 326 |
| Guaiacol:picric acid (yellow polymorph) | 25.35 | 8.57 | **13.7** | 96 | 8 | $P2_1/a$ | 370 | — | 305 |
| Phenanthrene:TNB | 7.7 | 17.13 | **13.7** | 101.5 | 4 | $P2_1/c$ | 451 | 456 | 305 |
| Resorcinol:$p$-benzoquinone | **14.63** | 5.976 | 11.529 | — | 4 | $Pnca$ | 252 | 273 | 328 |
| Benzene:picric acid[e] | 12.85 | 14.13 | 14.82 | — | 8 | $Pcm2_1$ or $Pc2m$ | 336 | 336 | 330 |

[a] The values for the stack axis (or needle axis for unknown structures) are in bold type. This facilitates comparisons without requiring confusing reorientation of unit cells.

[b] Unless noted otherwise, both "measured" and "additive" volumes are room temperature values.

[c] Non-standard space group for convenient comparison with other results.

[d] Diffuse scattering ignored in these results. The HMB:picryl bromide and HMB:picryl iodide molecular compounds give similar but more complicated results.

[e] Benzene:TNB has been studied by C. H. MacGillavry (personal communication). The crystals are triclinic with the stack axis [001] ~ 7.2 Å; there are 4 formula units per cell.

where the 1:2 ratio of components results from a disparity in size between donor and acceptor molecules. When the donor molecule is appreciably larger than the acceptor, two acceptor molecules can be sandwiched between each pair of donor molecules in a mixed stack. A number of examples are known (Table 17); the periodicity along the stack axis remains at about 7 Å, and physical properties are similar to those of analogous 1:1 molecular compounds. The converse situation, with size of donor and acceptor reversed, does not appear to have been encountered.

In a second type II in Table 17 there is a definite change in structure with the stacking sequence becoming:

$$-----A D A A D A A D A A D A A D A A D A A D------$$

for a donor–acceptor ratio of 1:2. The converse situation (D:A = 2:1) is found in this group. The periodicity along the stack is about 10.5 Å. Despite the change in stacking sequence, the properties of these $\pi$-molecular compounds do not differ markedly from those of comparable 1:1 compounds; the detailed studies needed to establish expected minor differences are lacking.

One may guess that other compositions (3:2, 4:3, etc.) have analogous stacking sequences with appropriate periodicities, but there is no evidence supporting this suggestion; the cell dimensions of $(fluorene)_3:(TNB)_4$ are not compatible with it.[305]

The third type in Table 17 is as yet represented by only two examples, which serve as a warning against premature generalization. The segregated stacks of $TMPD:(TCNQ)_2$ and ditoluenechromium:$(TCNQ)_2$ are similar to those in $Cs_2(TCNQ)_3$.[329] However, TMPD:TCNQ (Hanson[340]) has mixed stacks, whereas ditoluenechromium:TCNQ[324] has segregated stacks.

## E. Relation between Crystal Structures of Molecular Compounds and Those of Their Constituent Components

A molecular compound and its two constituent components form three separate phases and no relationship is required between the molecular arrangements in the three phases. In general, no relationship is found; but there are at least three examples where a rather close resemblance is found between the cell dimensions of the $\pi$-molecular compound and those of one of its components (Table 18), and in one of these (thymine:$p$-benzoquinone, Section VI-H-2, p. 321) a close structural resemblance has been found. Structures have not been reported for the other two examples. There are some

Table 17. Crystallographic results for 2:1 and 1:2 π-molecular compounds.[a]

| Type | Molecular compound | $a$ (Å) / $\alpha^\circ$ | $b$ (Å) / $\beta^\circ$ | $c$ (Å) / $\gamma^\circ$ | $Z$ | Space group | Ref. |
|---|---|---|---|---|---|---|---|
| I | Bis-(8-quinolinato)Cu(II):(TCNB)$_2$ | **7.37** / 98.9 | 7.70 / 96.3 | 14.29 / 97.3 | 1 | $P\bar{1}$ | 331 |
| | Bis-(8-quinolinato)Cu(II): (picryl azide)$_2$ | 16.14 / — | **6.90** / 105.6 | 30.93 / — | 4 | $A2/a$ | 332 |
| | *trans*-Stilbene:(TNB)$_2$ | 12.7 / 102.3 | 15.4 / 85.5 | **7.7?** / 87.6 | 2 | $P1$ or $P\bar{1}$ | 333 |
| | 9,10-Dihydroanthracene:(TNB)$_2$ | 25.7 / — | **7.34** / 92.0 | 7.41 / — | 2 | $P2_1/a$ | 305 |
| II | Benzo[$c$]pyrene:(TMU)$_2$ | 9.33 / 119.6 | **10.59** / 113.3 | 10.84 / 87.3 | 1 | $P\bar{1}$ | 334 |
| | Coronene:(TMU)$_2$ | 9.36 / 118.3 | **11.18** / 113.9 | 10.51 / 85.6 | 1 | $P\bar{1}$ | 335 |
| | Phenoquinone[b] [(Hydroquinone)$_2$:$p$-benzoquinone] | **11.152** / — | 5.970 / 100.0 | 11.499 / — | 2 | $P2_1/c$ | 336,337 |
| III | TMPD:(TCNQ)$_2$ | 7.782 / 93.52 | 15.020 / 102.77 | **6.488** / 82.97 | 1 | $P\bar{1}$ | 338 |
| | Ditoluenechromium:(TCNQ)$_2$ | 8.25 / 94.7 | **7.76** / 92.3 | 13.77 / 112.5 | 1 | $P\bar{1}$ | 339 |

[a] The stack axis (or needle axis for unknown structures) is in bold type to avoid confusing changes of orientation.
[b] (ClC$_6$H$_4$OH)$_2$:C$_6$H$_4$O$_2$ and (BrC$_6$H$_4$OH)$_2$:C$_6$H$_4$O$_2$ are isomorphous with phenoquinone. This probably holds also for (CH$_3$C$_6$H$_4$OH)$_2$:C$_6$H$_4$O$_2$.

Table 18. Resemblances between the unit cell dimensions of molecular compounds and those of their components. Structural resemblances have been demonstrated for the last three examples.

| Substance | $a$ (Å) | $b$ (Å) | $c$ (Å) | $\beta$ (deg) | $Z$ | Space group | Ref. |
|---|---|---|---|---|---|---|---|
| Acenaphthene | 8.29 | 14.00 | 7.225 | — | 4 | $Pcm2_1$ | 341 |
| 4,6-Dinitro-1,3-xylene | 11.5 | 5.49 | 7.2 | 98 | 2 | $P2_1/m$ | 43 |
| 1:1 Molecular cpd. | 18.5 | 14.2 | 7.25 | 103 | 4 | $P2_1/a$ | 43 |
| 9-Isopropylcarbazole | 18.01 | 7.963 | 16.82 | — | 8 | $Ic2a$ | 342 |
| Picryl chloride | 11.10 | 6.83 | 12.62 | 102.5 | 4 | $P2_1/c$ | 343 |
| 1:1 Molecular cpd. | 18.12 | 6.962 | 16.70 | 100.1 | 8 | $P2_1/c$ | 342 |
| Thymine | 12.87 | 6.83 | 6.70 | 105 | 4 | $P2_1/c$ | 344 |
| $p$-Benzoquinone | 7.005 | 6.795 | 5.767 | 101.5 | 2 | $P2_1/a$ | 345 |
| 1:1 Molecular cpd. | 12.427 | 6.859 | 12.639 | 90 | 4 | $Pbmn$ | 346 |
| Tetrabromoethylene | 14.193 | 4.140 | 12.164 | 112.0 | 4 | $P2_1/c$ | 347 |
| Pyrazine | 9.316 | 3.815 | 5.911 | — | 2 | $Pmmn$ | 348 |
| 1:1 Molecular cpd. | 11.803 | 4.161 | 12.094 | 113.1 | 2 | $P2_1/c$ | 349 |
| Tetraiodoethylene | 15.10 | 4.45 | 13.00 | 109 | 4 | $P2_1/c$ | 349 |
| Pyrazine | 9.316 | 3.815 | 5.911 | — | 2 | $Pmmn$ | 348 |
| 1:1 Molecular cpd. | 12.37 | 4.47 | 12.68 | 117.6 | 2 | $P2_1/c$ | 349 |
| 1,2,4,5-Tetrabromobenzene | 10.32 | 10.71 | 4.02 | 102.4 | 2 | $P2_1/a$ | 350 |
| Hexabromobenzene | 15.38 | 4.002 | 8.38 | 92.7 | 2 | $P2_1/a$ | 351 |
| 1:1 Molecular cpd. | 17.80 | 4.01 | 14.42 | 111.2 | 2 | $P2_1/a$ | 352 |

cases (not of π-molecular compounds) where crystal-structure analysis has confirmed that resemblances in cell dimensions between a component and the molecular compound do result from similarities in molecular arrangements. These are also given in Table 18.

## F. Mixed-donor π-Molecular Compounds

Appreciable mutual solid solubility of certain pairs of molecular compounds has been found (Table 19). This means that a fraction

Table 19. Mutual miscibility of some π-molecular compounds (mixed-donor systems).

| Mixed donors | Acceptor | Mutual miscibility | Crystallographic information available |
|---|---|---|---|
| Naphthalene/1-naphthol<br>Naphthalene/2-naphthol<br>Naphthalene/1-bromo-<br>naphthalene | Picric acid | Complete[a] | Only for naphthalene: picric acid |
| Anthracene/phenanthrene[b]<br>Phenanthrene/azulene<br>Perylene/pyrene<br>Anthracene/acridine<br>Stilbene/azobenzene<br>Naphthalene/anthracene | TNB | Partial[c] | For both components (different structures)<br>Only for anthracene: TNB and stilbene:TNB<br>For both components (different structures) |
| Naphthalene/fluorene<br>Naphthalene/chrysene<br>Anthracene/fluorene | Picric acid | None[a] | Only for naphthalene: picric acid<br>Only for anthracene: picric acid |

[a] Rheinboldt and Senise.[353]    [b] Phase diagram in Figure 6.    [c] Lower.[354]

of donor molecules of one kind in a given molecular compound can be replaced by donor molecules of another kind. Rheinboldt and Senise[353] showed that certain pairs of picric acid molecular compounds exhibit a complete range of solid solubility, whereas others were immiscible in the solid state. Similar studies have been made by Lower,[354] especially for the anthracene/phenanthrene:TNB mixed-donor system (Figure 6). The results apply essentially only to the temperature region in which the measurements were made and not necessarily to room temperature. However, the phase diagrams obtained are in all instances compatible with the available (room-temperature) crystallographic data. Incomplete miscibility in the solid state must be found whenever the pure end-members of the system have different crystal structures. The complete solid-state miscibility of the picric acid molecular compounds of naphthalene, 1-naphthol, 2-naphthol, and 1-bromonaphthalene shows that, in these examples, the substituents have only minor influence on the

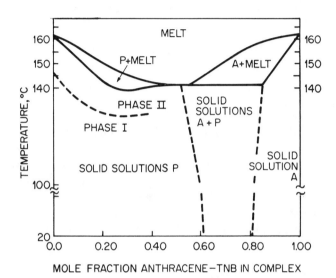

Figure 6. Melting point–composition diagram of the anthracene/ phenanthrene:TNB system. Solid phases P and A show X-ray powder patterns identical with those of pure phenanthrene:TNB and pure anthracene:TNB, respectively. The two phases of solid P could not be distinguished at anthracene:TNB concentrations exceeding 40 mole %. [Reproduced with permission from reference 354.]

crystal structure of the molecular compounds. The wide range of solid solubility found in the anthracene/phenanthrene:TNB system suggest that those systems where no solid solubility was reported should be reinvestigated.

Hertel and Bergk[262] have shown that, although naphthacene does not form molecular compounds with TNB, it nevertheless goes into solid solution, to the extent of about 1%, in the 1:1 benzene:TNB and toluene:TNB molecular compounds.

Attempts to prepare mixed-acceptor molecular compounds based on TNB/TENF mixtures gave glasses rather than crystalline melts (Lower[354]).

## G. Quasi-isomorphous Groups of Crystal Structures

Simplification of the classification of crystalline molecular compounds will result if we can find groups with similar crystal structures (isomorphism) or at least related crystal structures (quasi-isomorphism). The requirements for quasi-isomorphism are less rigid than those for strict isomorphism, and so, despite the wide variety of molecules functioning as donors and acceptors, it is possible to find resemblances in the crystal structures of certain molecular compounds that extend beyond the fundamental alternating arrangement in mixed stacks. The crystallographic results which serve as a basis for the classification into groups* are given in Table 20. Individual structures are discussed at greater length in the text below. Table 20 does not give a complete list of all the π-molecular compounds for which crystallographic results have been reported, as many compounds do not fall into recognizable quasi-isomorphous groups.

Quasi-isomorphism is sometimes a consequence of disorder and does not apply to the ordered crystals stable at lower temperatures. For example, the various molecular compounds that crystallize in Group 1 at room temperature transform to different ordered arrangements on cooling (Bernstein and Herbstein;[326] see Table 22, page 312).

## H. Polymorphism and Disorder

**1. Polymorphism.**—Studies with the hot-stage microscope (Kofler;[109] Laskowski, Grabar, and McCrone[218]) and X-ray analyses have shown that polymorphism and/or order–disorder transformations occur in many π-molecular compounds. The microscopical observations refer only to temperatures above room temperature, while the X-ray diffraction studies cover isolated points in the range between 100°K and the melting point. Determination of specific heat as a function of temperature would be of considerable interest; the only molecular compound so far studied in this way is quinhydrone[371,372] which does not show any transformation below room temperature. Above room temperature many molecular compounds show one or more transformations, and (fluorene)$_3$:(TNB)$_4$ is even said to have four phases between 25°C and the melting point of 106° (Kofler[109]). It is tempting to assume that the various polymorphic structures differ only in the arrangement of the mixed stacks and that the sequence within the stacks does not differ between two polymorphs.

---

* Group is here defined as "an assemblage of objects having some relationship, resemblance or common character (Webster)."

Table 20. Groups of quasi-isomorphous crystals (results refer to room temperature unless stated otherwise). $V_{mc}$ = volume in molecular compound; $V_{add}$ = sum of component volumes.

| Molecular compound | a (Å) | b (Å) | c (Å) | β (deg) | Volume of formula unit $V_{mc}$ (Å³) | Volume of formula unit $V_{add}$ (Å³) | Ref. |
|---|---|---|---|---|---|---|---|
| *1. Space group C2/m, 2 formula units per cell, stack axis [001].* | | | | | | | |
| Naphthalene:TCNE[a] | 7.26 | 12.69 | 7.21 | 94.4 | 339 | 343 | 355 |
| Naphthalene:TCNB | 9.39 | 12.66 | 6.87 | 107.2 | 390 | — | 356 |
| Naphthalene:PMDA | 9.19 | 13.0 | 6.81 | 104 | 395 | 395 | 157 |
| Anthracene:$C_6F_6$ | 9.03 | 12.2 | 7.26 | 95 | 398 | — | 157 |
| Pyrene:$C_6F_6$ | 9.88 | 13.5 | 6.98 | 113 | 429 | — | 157 |
| TMPD:TCNQ | 9.88 | 12.71 | 7.72 | 97.3 | 481 | — | 340 |
| Anthracene:TCNQ | 11.48 | 12.95 | 7.00 | 105.4 | 502 | 492 | 357 |
| *2. Space group C2/m, 2 formula units per cell, stack axis [010].* | | | | | | | |
| TMPD:chloranil | 16.32 | 6.57 | 8.81 | 112 | 438 | — | 358 |
| *3a. Space group $P2_1/a$, 2 formula units per cell, stack axis [001].* | | | | | | | |
| Pyrene:TCNE | 14.33 | 7.24 | 7.98 | 92.4 | 403 | 424 | 359 |
| Pyrene:PMDA | 13.9 | 9.25 | 7.24 | 94 | 464 | 476 | 303 |
| Perylene:fluoranil | 18.5 | 7.49 | 6.97 | 112.3 | 447 | — | 360 |
| Pyrene:chloranil[b] | 13.83 | 9.04 | 7.65 | 96 | 476 | 470 | 220 |
| HMB:chloranil[c] | 15.26 | 8.64 | 7.30 | 106 | 463 | 461 | 361 |
| HMB:TCNB | 14.90 | 8.92 | 7.41 | 104 | 478 | — | 362 |
| *3b. Space group $P2_1/a$, 4 formula units/cell, stack axis [001].* | | | | | | | |
| s-TAB:TNB | 14.063 | 15.079 | 6.982 | 103.5 | 358 | — | 363 |
| Anthracene:picric acid | 19.3 | 12.83 | 7.23 | 91 | 434 | 451 | 330 |
| Ditoluenechromium:TCNQ[a] | 22.45 | 15.45 | 7.00 | 115 | 552 | — | 324 |

Table 20 *(continued)*

| Molecular compound | $a$ (Å) | $b$ (Å) | $c$ (Å) | $\beta$ (deg) | Volume of formula unit | | Ref. |
| --- | --- | --- | --- | --- | --- | --- | --- |
| | | | | | $V_{mo}$ (Å³) | $V_{add}$ (Å³) | |
| 4a. *Space group P2₁/a, 2 formula units per cell, stack axis [010].* | | | | | | | |
| Perylene:TCNE | 15.70 | 8.28 | 7.31 | 96.1 | 471 | 470 | 364 |
| Perylene:PMDA | 17.0 | 7.16 | 10.13 | 121 | 528 | 522 | 301 |
| Perylene:C₆F₆^e | 17.6 | 7.73 | 7.31 | 106 | 478 | — | 157 |
| Benzene:PMDA | 12.9 | 6.7 | 7.68 | 96 | 330 | 331 | 157 |
| 4b. *Space group P2₁/a, 4 formula units per cell, stack axis [010].* | | | | | | | |
| Azulene:TNB (at −95°C) | 16.39 | 6.66 | 13.77 | 96.1 | 374 | — | 28 |
| Skatole:TNB (at −140°C) | 16.76 | 6.61 | 13.45 | 95.6 | 371 | — | 365 |
| Indole:TNB (at −140°C) | 15.87 | 6.58 | 13.47 | 94.8 | 350 | — | 365 |
| Naphthalene:TNB^f | 16.19 | 6.97 | 14.5 | 97.6 | 405 | 393 | 366 |
| Naphthalene:picric acid | 16.14 | 6.80 | 14.34 | 97 | 390 | 395 | 330 |
| Perylene:BTF | 19.2 | 6.95 | 15.6 | 98 | 515 | — | 157 |
| 9-Isopropylcarbazole:pioryl chloride | 16.7 | 6.96 | 18.12 | 100 | 514 | 535 | 342 |
| 4c. *Space group P2₁/a, 8 formula units per cell, stack axis [010].* | | | | | | | |
| Benzo[c]phenanthrene:picric acid | 31.4 | 7.0 | 20.0 | 114 | 503 | 514 | 305 |
| Benzo[c]phenanthrene:TNB | 32.1 | 7.0 | 19.9 | 114 | 507 | 512 | 305 |
| 5a. *Space group Pbcn, 4 formula units per cell, stack axis [100].* | | | | | | | |
| Phenothiazine:TNB | 7.01 | 15.14 | 17.15 | | 455 | 457 | 269 |
| Biphenyl:BTF | 7.04 | 15.1 | 16.8 | | 447 | 437 | 157 |
| 5b. *Space group P2₁2₁2₁, 4 formula units per cell, stack axis [001].* | | | | | | | |
| Perylene:TNB | 31.5 | 9.6 | 7.2 | | 545 | 520 | 262 |
| Anthracene:BTF | 15.8 | 17.3 | 6.8 | | 465 | 457 | 157 |

| Molecular compound | a (Å) | b (Å) | c (Å) | α (deg) | β (deg) | γ (deg) | Volume of formula unit | | Ref. |
|---|---|---|---|---|---|---|---|---|---|
| | | | | | | | $V_{mc}$ (Å³) | $V_{add}$ (Å³) | |
| 6a. *Triclinic, space group P1̄. 1 formula unit per cell, stack axis* [001]. | | | | | | | | | |
| Anthracene:PMDA | 7.6 | 10.0 | 7.3 | 105 | 115.5 | 101 | 454 | 451 | 301 |
| TMPD:TCNB | 7.654 | 8.041 | 7.462 | 96.7 | 85.9 | 101.3 | 447 | — | 367 |
| Ferrocene:TCNE | 7.77 | 7.87 | 6.78 | 113.6 | 96.7 | 77.0 | 370 | 367 | 289 |
| 6b. *Triclinic, space group P1̄, 2 formula units per cell, stack axis* [001]. | | | | | | | | | |
| Acepleiadylene:TNB | 8.79 | 16.055 | 6.515 | 92.9 | 102.2 | 99.0 | 442 | — | 27 |
| Pyrene:TNB | 8.4 | 16 | 6.7 | 84 | 77 | 87 | 436 | 474 | 262 |
| Dibenzo[g,p]chrysene:TNB^g | 9.97 | 20.0 | 7.0 | 92 | 77 | 113 | 617 | 631 | 305 |
| Dibenzo[g,p]chrysene:picric acid^g | 9.97 | 20.0 | 7.0 | 92 | 77 | 113 | 617 | 633 | 305 |
| Phenothiazine:3,5-dinitrobenzoic acid | 10.00 | 13.92 | 7.54 | 105.8 | 105.2 | 101.8 | 473 | — | 119 |
| Piperidine:picric acid | 8.8 | 12.1 | 6.92 | 92 | 93 | 108 | 349 | — | 68 |
| 6c. *Molecular compounds with metal 8-quinolinolates as donors; triclinic, space group P1̄, 1 formula unit per cell, stack axis* [100]. | | | | | | | | | |
| Bis-8-quinolinolatoPd(II):chloranil | 8.17 | 8.18 | 9.69 | 99.5 | 77.8 | 66.0 | 546.5 | — | 368 |
| (8-quinolinol)₂:chloranil | 7.88 | 8.03 | 11.30 | 129.9 | 116.7 | 68.2 | 552.5 | — | 369 |
| Bis-8-quinolinolatoPd(II):TCNB | 8.73 | 7.97 | 9.38 | 109.6 | 72.8 | 107.6 | 572.2 | — | 370 |
| Bis-8-quinolinolatoCu(II):(TCNB)₂ | 7.37 | 7.70 | 14.29 | 98.9 | 96.3 | 97.3 | 788.5 | — | 331 |
| Bis-8-quinolinolatoCu(II):TCNQ | 7.12 | 12.00 | 7.54 | 96.8 | 112.5 | 88.8 | 591.2 | 595.1 | 327 |

[a] $\Delta V$ in solution is $-7$ Å³ per formula unit.[318]

[b,c] $\Delta V$ in solution is $-8$ Å³/formula unit for b and $-18$ Å³/formula unit for c.[318]

[d] This crystal has segregated and not mixed stacks.

[e] No proof of stack-axis direction.

[f] $\Delta V$ in solution is $-5$ Å³/formula unit.[318]

[g] Dibenzo[g,p]chrysene ≡ tetrabenzonaphthalene.

This is correct for the two polymorphs of quinhydrone (Section VI-H), the only molecular compound where the structures of two polymorphs are known. However, it might be dangerous to generalize from one example where both hydrogen bonding and charge-transfer interactions are important.

**2. Disorder.**—There are grounds for believing that two different types of disorder occur in π-molecular compounds. The first type is a true thermodynamic disorder, the equilibrium degree of order varying with temperature (and pressure). The degree of order increases with falling temperature until a transformation occurs from a state of short-range to one of long-range order. The degree of long-range order is also temperature-dependent and perfect order is attained only at absolute zero, although the ordering may be virtually complete at considerably higher temperatures. The second type of disorder is non-equilibrium in nature and is probably due to growth faults which are "frozen-in" during crystallization; this type of disorder does not change with temperature.

**3. Diagnostic tests for the presence of disorder.**—*a. Specific heat–temperature curves.* This classical method of demonstrating the occurrence of phase and/or order–disorder transformations[373] has not yet been used for molecular compounds. Measurement of the absolute entropy of the crystalline compound, and of the entropy changes accompanying phase transformations, should be useful in determining whether a particular phase is disordered and would then indicate the most appropriate temperatures for crystal structure analysis.

*b. Various crystallographic tests.* (i) Evidence from space group. Direct evidence of a disordered arrangement of one or both components may sometimes be obtained from an incompatibility between the site-symmetry demanded by the space group of the molecular compound and the point-group symmetry of one or both components. For example, the molecule 9,10-dihydroanthracene has a butterfly shape, but in its 1:2 molecular compound[101] with TNB (Table 17) it has an apparent centre of symmetry. This presumably results from disorder as it is unlikely that the molecule changes shape on formation of the molecular compound.

(ii) Diffuse reflections on X-ray diffraction photographs. Diffuse scattering of X-rays can arise from thermal vibrations, equilibrium-type disorder, or non-equilibrium-type disorder. These factors can be distinguished to some extent by careful analysis of the variation in shape and intensity of the diffuse reflections with temperature (*e.g.*, Jagodzinski,[374] Wooster[375]). Diffuse scattering from perylene:$C_6F_6$ at room temperature[157] results from equilibrium-type disorder, while that from HMB:picryl chloride[325] is caused by non-equilibrium disorder. Quantitative study of the diffuse scattering is the key to the detailed description of disorder in crystals.

(iii) *Debye–Waller factors.* Many papers on the determination of the crystal structures of $\pi$-molecular compounds contain a statement about the high degree of apparent thermal motion in the crystals at room temperature as demonstrated by the rapid decline in intensity with increasing $\sin \theta/\lambda$. However, as Powell and Huse[325] pointed out, melting points of most molecular compounds are not very different from the means of the melting points of their components and thus a much higher degree of thermal vibration than in the components is not to be expected. If, however, the room-temperature structure of the molecular compound lacks long-range order in some respects, then the Debye–Waller factors will be larger than normal because they will include a contribution from static displacements of the atoms from their mean positions. This explanation assumes that the two (or more) possible orientations for the molecules are not too different. As illustration we can compare the values of the individual isotropic Debye–Waller factors (B) found in anthracene (m.p. 216°) at room temperature with those found in pyrene:PMDA (m.p. 260°) at room temperature. For anthracene Cruickshank[376] found B values ranging from 2.83 to 4.00 Å$^2$, whereas for pyrene:PMDA Herbstein and Snyman[303] found values ranging from 3.1 to 7.0 Å$^2$.

More illuminating results can often be obtained if B values for the various atoms in the structure are plotted against their distances from the centres of gravity of the individual components. This has been done in Figure 7 for naphthalene:TCNE, whose room-temperature structure was determined by Williams and Wallwork.[355] The Debye–Waller factors of the atoms of the naphthalene molecule are so large that they cannot correspond to a physically reasonable amount of thermal vibration but must indicate disorder, whereas the B values of the atoms of the TCNE molecule are of reasonable magnitude for carbon and nitrogen atoms in a crystal that melts at 99°. The approximately parabolic dependence of B on distance from the centre of the naphthalene molecule can be explained by assuming that the naphthalene molecule takes up two different orientations, about 15° apart, in the molecular stacks. The same conclusion was reached by Williams and Wallwork,[355] using Fourier methods, as shown below. Thus any statement in a description of the results of a structure analysis that one component has an appreciably larger apparent amplitude of thermal vibration than the other component

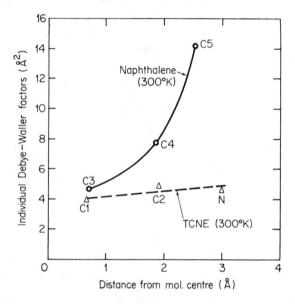

Figure 7. The equivalent isotropic Debye–Waller factors of the atoms in naphthalene:TCNE at room temperature.[355] Separate curves are shown for the two molecules. [Reproduced with permission from reference 303.]

should be taken as *prima facie* evidence of disorder in the crystals. In our discussion of results we shall encounter a number of examples where such disorder has passed unrecognized by the original authors.

A clear-cut distinction between the distance-dependence of the B values of the atoms of the two components is not always found, as *e.g.*, for pyrene:PMDA [303] and pyrene:TMU [377] at room temperature. In these examples, evidence of disorder is provided by the large absolute values of B. A note of warning is necessary here: large values may be obtained for Debye–Waller factors for reasons unconnected with intrinsic disorder in the arrangement of the components; for example, poor or impure crystals, or progressive decomposition during the measurements, may lead to artificially high values for B—*caveat emptor*. The converse decision—that a crystal is not disordered—is often easier: for example, Kamenar *et al.*[368] comment that their crystals of bis-8-quinolinolatopalladium(II):chloranil were "remarkable in that the intensity and extent of the Bragg scattering is such as might be expected from an ionic, or highly hydrogen-bonded crystal."

(iv) Electron-density and difference syntheses. Analysis of the distance-dependence of Debye–Waller factors of the two components is rapid and requires little extra work; computation of electron-density and difference syntheses in appropriate molecular planes, although more time-consuming, gives more information. Electron-density syntheses in the planes of the naphthalene and TCNE molecules of the naphthalene:TCNE molecular compound are shown in Figures 8 and 9,* as well as a difference synthesis in the naphthalene molecular plane. It is clear that the naphthalene molecules are

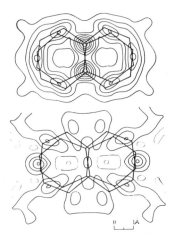

Figure 8. Electron-density and difference syntheses in the plane of the naphthalene molecule in naphthalene:TCNE at room temperature.[357] The contours of electron density are at intervals of 1 eÅ$^{-3}$ and start at 0 eÅ$^{-3}$; the contours of difference density are at intervals of 0.5 eÅ$^{-3}$ (thick lines zero and positive contours, thin lines negative contours). [Reproduced with permission from reference 303.]

disordered and that the electron-density and difference syntheses can be explained, as noted above, by a random superposition of two different orientations for the naphthalene molecules; on the other

* These are more convenient representations of results already given by Williams and Wallwork.[355]

hand, the TCNE molecules take up only one orientation in a particular stack. Disordered naphthalene molecules have also been detected in naphthalene:TCNB by use of electron-density syntheses (Kumakura, Iwasaki, and Saito[356]).

Difference syntheses in molecular planes are particularly useful when the probability of the two (or more) orientations differs widely,

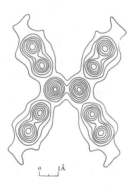

Figure 9. Electron-density synthesis in the plane of the TCNE molecule in naphthalene:TCNE. The difference synthesis is featureless and is not reproduced. For other details see caption to Figure 8. [Reproduced, with permission, from reference 303.]

instead of being equal, as in the examples cited above. For example, Hanson[28] refined the crystal structure of azulene:TNB at $-95°$ on the assumption of a particular orientation of the azulene molecules. Although apparently satisfactory convergence was obtained, a difference synthesis in the azulene molecular plane showed ten small peaks which were explained by postulating that $7\%$ of the azulene molecules were in the alternative orientation. As we shall see below, the real situation in azulene:TNB is probably even more complicated.

(v) Behaviour on cooling. Many crystalline molecular compounds undergo solid-state transformations on cooling. In one particular instance—pyrene:PMDA—determination of the crystal structures above and below the transformation temperature ($\sim 200°$K) has shown that the two structures are related by a disorder-to-order transformation (Herbstein and Snyman[303]). These results are discussed in greater detail in Section VI-E. Transformations on cooling have also been found in several of the molecular compounds that crystallize in space group $C2/m$ at room temperature (Table 21). It is striking that three different low-temperature structures have

Table 21. Phase transformations of some π-molecular compounds on cooling.

| Molecular compound | Temp. (°K) | a (Å) | b (Å) | c (Å) | β (deg) | Space group | Ref. |
|---|---|---|---|---|---|---|---|
| Naphthalene:TCNE | 300 | 7.26 | 12.69 | 7.21 | 94.4 | C2/m | 326 |
| | <150 | | | Transforms to triclinic | | | |
| TMPD:chloranil | 300 | 16.32 | 6.57 | 8.81 | 112 | C2/m | 358 |
| | <250 | | | Transforms to triclinic | | | |
| Anthracene:C$_6$F$_6$ | 300 | 9.03 | 12.2 | 7.26 | 95 | C2/m | 326 |
| | 150 | 9.0 | 12.0 | 14.5 | 95 | I2/a | |
| Pyrene:C$_6$F$_6$ | 300 | 9.88 | 13.5 | 6.98 | 113 | C2/m | 326 |
| | 120 | 9.8 | 13.4 | 6.9 | 113 | P2$_1$/a | |
| Pyrene:PMDA | 300 | 13.9 | 9.25 | 7.24 | 94 | P2$_1$/a | 303 |
| | 110 | 13.67 | 9.13 | 14.4 | 91.5 | P2$_1$/n | |
| Perylene:PMDA[a] | 300 | 14.61 | 7.16 | 10.13 | 94.7 | P2$_1$/n | 301 |
| TDT:TNB | 300 | 6.72 | 16.06 | 9.59 | 105.2 | P2$_1$/m | 281 |
| | <140 | | | Transforms to triclinic | | | |

[a] At 100°K the crystals shatter; the structure is unknown.

been found, although the compounds are quasi-isomorphous at room temperature. The general nature of the low-temperature structures can be predicted from the changes in cell dimensions and symmetry with respect to the room temperature structures, but no detailed analyses have been made. In these transformations the overall crystallographic and molecular orientations in the two phases appear to be maintained, but ordering of orientations presumably occurs; the low-temperature crystals are often twinned as the transformations can take place in two related ways.

Thus occurrence of a transformation on cooling is a strong indication that the high-temperature phase is disordered. However, some discretion is necessary in using this criterion; for example, perylene:PMDA undergoes a violent phase transformation on cooling to about 100°K, but the room-temperature structure appears to be ordered (Boeyens and Herbstein[301]).

## VI. MORE DETAILED DISCUSSION OF CRYSTALLOGRAPHIC RESULTS FOR VARIOUS GROUPS OF RELATED π-MOLECULAR COMPOUNDS

### A. Introduction

In the discussion that follows we shall pay particular attention to four areas where comparisons among the different groups of molecular compound are possible. These are:

(1) The relative arrangements of the molecular stacks in the crystal, or other packing arrangement in the rare instances where stacks are not obvious.

(2) Presence or absence of disorder at the temperature at which the crystal structure was determined. Phase transformations are included.

(3) The mutual arrangement of component molecules within a stack. Appreciable deviations from parallelism of the donor and acceptor molecules suggest that some localized interaction is strong enough to tilt the molecules from the densest packing arrangement. The various distances between atoms in adjacent molecules will give a measure of the interactions between them. If the components are essentially parallel then the average interplanar distance is one suitable measure of donor–acceptor interaction. Finally, the mutual orientation of donor and acceptor molecules projected on to their average plane* may indicate how strong the charge-transfer forces

---

* We shall call these "overlap diagrams."

are. Conclusions here will be reinforced if we can find a mutual orientation which occurs often, particularly if a satisfactory theoretical explanation can also be supplied. Since the charge-transfer interaction between the components is only one of a number of roughly equal contributions, attention should be concentrated on general resemblances in mutual orientation rather than dissipated in trying to account for differences between generally similar orientations.

(4) Comparison of shapes and dimensions of the component molecules in various environments, in the hope of establishing how molecular shape is influenced by the formation of molecular compounds. This may provide another means of distinguishing between Mulliken- and Weiss-type $\pi$-molecular compounds. For example, neutral phenothiazine has a butterfly shape,[378] while the cation-radical has been reported to be planar.[379-381] Thus phenothiazine:TNB[269] and phenothiazine:3,5-dinitrobenzoic acid[119] have neutral ground states on the basis of the shape of phenothiazine in these crystals. The next step should be to determine the crystal structures of phenothiazine:$p$-benzoquinone and phenothiazine:DDQ, which have ionic ground states according to Matsunaga.[285]

We first consider molecular compounds where the overlap (and hence presumably the main charge-transfer interaction) between donor and acceptor appears to be confined to a "benzene-ring" portion of the donor. Here we include various donor:TCNE compounds, the quinhydrones, and donor:chloranil compounds. Next we consider examples where there is overlap between the acceptor (here required to be a substituted benzene such as PMDA, TNB, and TCNB) and a "naphthalene" portion of the donor. We conclude with a discussion of the miscellaneous groups of molecular compound where a clear-cut classification is not yet possible.

For convenience we have summarized the crystallographic results for each of the molecular compounds at the head of the discussion. Much of this information is also given in Table 20; if so, the group in Table 20 is included in the summary.

## B. Theoretical Studies of the Interaction between Components

The interaction energy between a pair of parallel donor and acceptor molecules has been calculated for a number of examples. The most stable arrangements were obtained by mutually translating and rotating the pair of molecules, the interplanar distance usually being taken from experiment. Because of the model used, one would expect the results to be comparable with measurements made on the molecular compounds in the gas phase rather than in the crystalline state. Because of the paucity of gas-phase results, the theoretical results are sometimes compared with measurements made on solutions.

Phenomenological calculations were made by Mantione[381] for a number of aromatic hydrocarbon:TCNE combinations. Four contributions were included: the attractive electrostatic, polarization and dispersion energies, and a repulsive term. The first three were calculated by using the "monopole-bond polarizability" interaction described by Claverie,[382] while the repulsive term was assumed to be exponential and similar to that used by Kitaigorodskii[383] in his calculations of the lattice energies of molecular crystals. Apart from azulene:TCNE, fair correlation was obtained between calculated interaction energies and observed free energies of formation of the molecular compounds in solution (a more appropriate comparison would have been with the enthalpies of formation). However, there is no agreement between the calculated most stable arrangements and those actually found in the crystal-structure analyses of naphthalene:TCNE, pyrene:TCNE, and perylene:TCNE. Interaction energies, but not preferred arrangements, have been calculated for some methylbenzene:TCNE compounds (Mantione[384]).

Quantum-mechanical calculations were made for molecular compounds of TCNE, TNB, and some other electron-acceptors. Kuroda *et al.*[385] estimated charge-transfer interactions in the isolated 1:1 complexes naphthalene:TCNE and pyrene:TCNE. They considered only the π-electrons but took into account possibilities of charge-transfer to all unfilled TCNE orbitals. Thus in naphthalene:TCNE fifty electronic configurations were included in the calculations, and eighty in pyrene:TCNE. The minimum-energy arrangement of donor and acceptor molecule centres agrees with the results of the crystal-structure analysis of naphthalene:TCNE (see overlap diagram in Figure 11a, page 268), the depth of the energy minimum being $-6.3$ kcal/mole, compared with an experimental $\Delta H_f^\circ$ (in solution) of $-4.1$ kcal/mole. For pyrene:TCNE an energy minimum was found with the TCNE molecule over the centre of each type of six-membered ring of the pyrene molecule. The respective charge-transfer interaction energies were $-6.42$ and $-6.28$ kcal/mole, compared with $\Delta H_f^\circ$ of $-4.1$ kcal/mole (in solution). The lower of the two calculated interaction energies corresponds to the actual position in the crystal (Figure 11c, page 268); however, the difference of 0.14 kcal/mole is probably too small to be significant. The energy barrier hindering rotation of naphthalene or pyrene relative to TCNE is only about 1 kcal/mole, so the influence of charge-transfer interactions on mutual orientation may be much less than on mutual position.

All valence electrons (both $\sigma$ and $\pi$) were taken into account in another set of calculations,[386] and two aspects of the intermolecular interaction were studied in detail: relative position and orientation of donor and acceptor molecules, and the relative magnitudes of the $\pi$- and $\sigma$-interactions. The most striking feature of these calculations is the small contribution of $\pi$–$\pi$* interaction to the total stabilization energy, for example, 1.6% for pyrene:TCNE (in contrast, the interactions involving $1s$ orbitals of hydrogen of methylated compounds contribute about 60% of the perturbation energy). Despite the small contribution of $\pi$–$\pi$* interactions, the energy minima were found at the same relative positions of naphthalene:TCNE and pyrene:TCNE as reported by Kuroda *et al.*[385] and, in agreement with their results, essentially free relative rotation of the two partners was found.

The extended Hückel scheme of Hoffman[387] has been used to calculate minimum energy conformations for some molecular compounds and ion-radical salts (Wold;[388], Chesnut and Moseley[389]). Binding was obtained for perylene:fluoranil, $Cs_2(TCNQ)_3$, and $N$-methylphenazinium:TCNQ, but not for naphthalene:TCNE, TMPD:TCNQ, anthracene:TNB, or skatole:TNB,[389] and also not for benzene:TCNE.[388] The crystal structure of the last compound is not known, but comparison with experiment is possible for the other molecular compounds and for the ion-radical salts. The energy minima obtained after translating and rotating the two molecules were in fair agreement with the experimental crystallographic results. The largest shifts ($\sim 1$ Å) between calculated and observed structures were obtained for perylene:fluoranil and for anthracene:TNB.

The effect of environment is clearly important and two major discrepancies between theory and experiment emerge from these studies. First, the calculated pairwise interaction energies are appreciably smaller than the enthalpies of formation of the crystalline molecular compounds from the gaseous components. Thus the appropriate comparison of theory is with the gas-phase molecular compounds (see, *e. g.*, refs. 41 and 42); also allowances for long-range interactions are needed in the calculations before the calculated values can be usefully compared with the experimental results for the crystalline molecular compounds. Secondly, while the calculated barriers to mutual rotation of the components are small, the experimental results show that very similar overlap diagrams are found for analogous molecular compounds, suggesting an appreciable barrier to mutual rotation (see, for example, Figure 11). It is not clear whether this difference between theory and experiment is due to an underestimate of the orientation-dependence of the charge-transfer energy in the calculations or because the restraining effects of surrounding stacks have not been taken into account.

### C. Molecular Compounds between Donors containing "Benzene" Regions and TCNE or TCNQ

Of the donors considered in this Section, only hexamethylbenzene (HMB) is truly a benzene derivative, although ferrocene has a certain resemblance. However, the experimental results suggest that TCNE interacts primarily with a single benzene ring of the other aromatic donor systems considered, as do each of the two halves of TCNQ with bis-8-quinolinolatocopper. Therefore the classification has been based on interaction of TCNE (or TCNQ) with a "benzene" portion of a (generally larger) donor.

#### 1. Naphthalene : TCNE
$[a = 7.26, b = 12.69, c = 7.21 \text{ Å}, \beta = 94.4°, C2/m, Z = 2; \text{group 1}]$

The crystal structure of naphthalene:TCNE at room temperature was determined by Williams and Wallwork[355] (Figure 10). The overlap diagram is shown in Figure 11a. The arrangement of molecules along the stack axis is similar to that found in most other π-molecular compounds. The naphthalene molecules are disordered, taking up two orientations about 15° apart (see Figures 7–9). Cooling the crystals to about 150°K causes the monoclinic high-temperature phase to change to a twinned triclinic structure.[326] The crystal structure of the triclinic phase has not been determined, nor is there much hope of so doing, because the crystals are damaged in passing through the transformation.

#### 2. Pyrene : TCNE
$[a = 14.33, b = 7.24, c = 7.98 \text{ Å}, \beta = 92.4°, P2_1/a, Z = 2; \text{group 3a}]$

The crystal structure of pyrene:TCNE at room temperature has been determined by three-dimensional methods (Ikemoto and Kuroda[359]) (Figure 12; overlap diagram in Figure 11c). There does not appear to be any disorder in the position or orientation of either molecule*. The usual mixed stacks are obtained in the usual quasi-

---

* [Added in proof:] This structure has been refined further[478] (diffractometer measurements at 105°K, $R = 5.5\%$). There is no phase change on cooling.

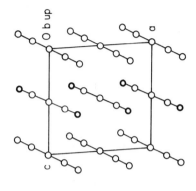

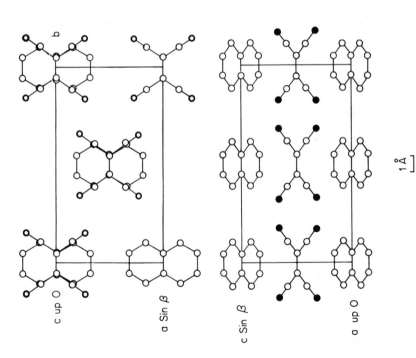

Figure 10. Naphthalene : TCNE. The crystal structure[355] in projection down the three crystallographic axes. The average (disordered) positions of the naphthalene molecules are shown. Naphthalene and TCNE molecules have been omitted in the lower portion of the projection down [001].

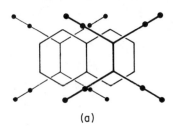

(a)

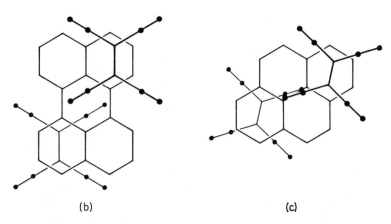

(b)                                    (c)

Figure 11. Overlap diagrams. (a) Naphthalene:TCNE; the averaged and not the actual orientation of the naphthalene molecules is shown. (b) Perylene:TCNE. (c) Pyrene:TCNE. The mutual arrangement of the two components in (b) and (c) is that found experimentally; there is no indication of disorder in these two crystals.

hexagonal packing, the stack axis being [001]. Any one stack is surrounded by two identical stacks and four other stacks derived from the reference stack by the *a* glide plane. There is no relative displacement along the stack axis.

The molecular planes are very nearly parallel (angle 2°); the distance between them is 3.32 Å. The mutual arrangement of a TCNE molecule to a "naphthalene region" of pyrene is remarkably similar to that in "ordered" naphthalene:TCNE.

The polarized absorption spectra of single crystals (Figure 13) have been reported by Kuroda, Kunii, *et al.*[390] The first and second charge-transfer bands are at 12.7 and 20 kK, frequencies close to those found in solution spectra. The first band has been assigned to

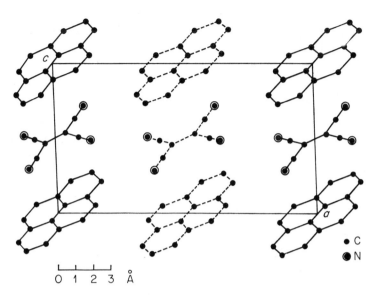

Figure 12. Pyrene:TCNE. Projection down [010] showing mixed stacks along [001]. [Reproduced with permission from ref. 359.]

charge-transfer from the highest occupied orbital of pyrene to the lowest vacant orbital of TCNE, and the second band to charge-transfer from the second highest orbital of pyrene to the lowest vacant orbital of TCNE. The band at 30 kK is due to an overlap of a third charge-transfer band (predicted theoretically by Kuroda, Ikemoto, and Akamatu[391]) and the $p$-band of pyrene, while the $\beta$-band of pyrene appears beyond 33 kK.

### 3. Perylene:TCNE

[$a = 15.70$, $b = 8.28$, $c = 7.31$ Å, $\beta = 96.1°$, $P2_1/a$, $Z = 2$; group 4a]

The crystal structure of perylene:TCNE at room temperature has been determined by Ikemoto and Kuroda,[364] using two-dimensional methods. No disorder was detected, but the individual isotropic Debye–Waller factors are rather high (4.1 to 8.2 Å$^2$ for perylene, 4.5 to 9.2 Å$^2$ for TCNE), so it would not be too surprising if a transformation were found at lower temperature.*

With the stack axis along [010], a reference stack is surrounded by two identical stacks and four others, derived from the reference stack by action of the $a$ glide plane, and consequently displaced by $\frac{1}{2}b$ and differently oriented.

* See also p. 395

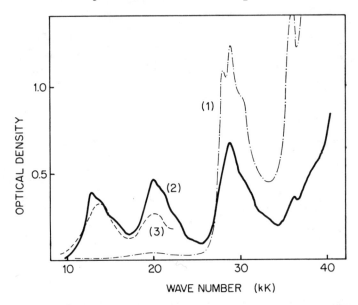

Figure 13. Pyrene:TCNE. The absorption spectra of single crystals (using polarized radiation) and solution: (1) the electric vector of the polarized incident radiation is parallel to the $b$-axis of the crystal and thus strong absorption parallel to the in-plane axes of pyrene occurs. (2) $c$-Axis spectrum, which therefore shows the two charge-transfer bands and diminished pyrene absorption bands. (3) The charge-transfer bands of the solution spectrum. [Reproduced with permission from reference 390.]

The molecular planes are again virtually parallel (angle 2°) and the interplanar distance is 3.23 Å. The overlap diagram is given in Figure 11b, and the similarity to the other two members of this group is quite striking. The polarized absorption spectrum of single crystals has been measured[390] and can be related to the crystal structure in the manner already described for pyrene:TCNE.

### 4. Copper oxinate:TCNQ
[$a = 12.00, b = 7.54, c = 7.12$ Å, $\alpha = 112.5, \beta = 88.8, \gamma = 96.8°$, $P\bar{1}, Z = 1$; group 6c; cell oriented as in reference 327].

The crystal structure of this molecular compound has been determined by Williams and Wallwork[327] and shows some unusual features. Each of the component molecules is at a centre of symmetry in the crystals of the molecular compound and can thus be considered as being composed of two equal parts. It is these "halves" that interact with one another and not the components as a whole; this is why we discuss it here rather than with the metal 8-quinolinolate or TCNQ molecular compounds. The crystal structure is shown in Figure 14 and the overlap diagram in Figure 15. The arrangement

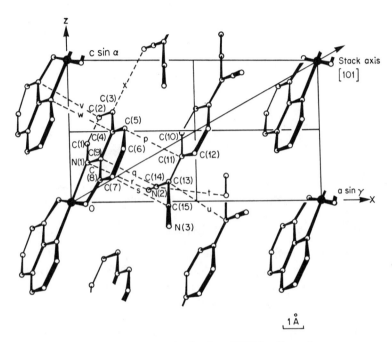

Figure 14. Copper bis-8-quinolinolate:TCNQ. Crystal structure projected along the *b*-axis, showing short intermolecular contacts. The distances (Å) indicated are: $p = 3.236$, $q = 3.267$, $r = 3.303$, $s = 3.314$, $t = 3.387$, $u = 3.451$, $v = 3.457$, $w = 3.410$, $x = 3.467$, all ± ca. 0.007 Å. [Reproduced with permission from reference 327.]

of adjacent stacks is such that not only do donor and acceptor molecules overlap within the mixed stacks but also that donor molecules in one stack overlap with donor molecules in an adjacent stack, and similarly, although to a lesser extent, for the smaller acceptor molecules. This arrangement results from the rather shallow inclination of the almost planar molecules to the stack axis [101]. The donor–acceptor interplanar distances are 3.24–3.31 Å (see caption to Figure 14) whereas the donor–donor and acceptor–acceptor distances are 3.41–3.47 Å.

The physical properties of this molecular compound have not been studied; it would be interesting to know how they are influenced by the somewhat unusual component arrangement.

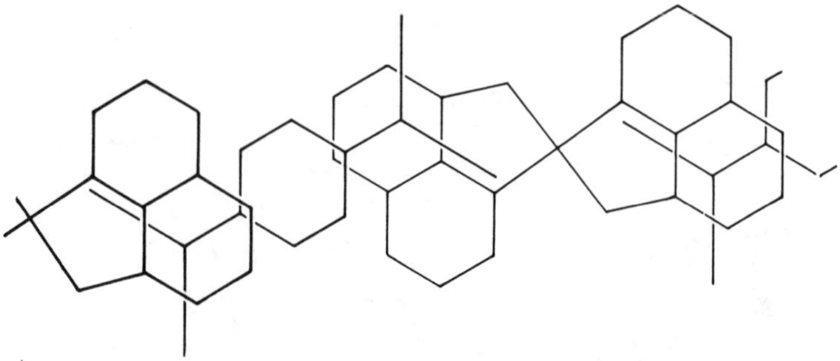

Figure 15. Copper bis-8-quinolinolate:TCNQ. Molecular overlap as viewed in a direction perpendicular to the mean molecular planes. [Reproduced with permission from reference 327.]

### 5. HMB:TCNE

Equimolar[392] and 2:1[393] molecular compounds of HMB and TCNE have been prepared and studied by infrared spectroscopy, with both unpolarized and polarized radiation. The equimolar molecular compound was studied together with the ionic salts $K^+$ $TCNE^-$ and $Na^+$ $TCNE^-$ and, from a comparison of the average bathochromic shift of 9 cm$^{-1}$ for the C≡N and C═C modes of TCNE in the molecular compounds with the 130 cm$^{-1}$ bathochromic shift found in the salts, it was concluded that there was less than 10% electron-transfer to TCNE in the HMB:TCNE molecular compound.

The 2:1 molecular compound was shown to have a "sandwich" arrangement within the molecular stacks, similar to that found in some other 2:1 molecular compounds. No crystallographic study has been reported for either of these molecular compounds [but see p. 395].

### 6. Ferrocene:TCNE
$[a = 7.77, b = 7.85, c = 6.78$ Å, $\alpha = 113.6, \beta = 96.7, \gamma = 77.0°,$
$P\bar{1}, Z = 1;$ group 6a]

The results of the crystal-structure determination at room temperature (Adman, Rosenblum, Sullivan, and Margulis[289]) show, in agreement with $^{57}$Fe Mössbauer studies (Collins and Pettit[290]), that there is no special interaction between metal atom and TCNE, as had been proposed earlier (Rosenblum, Fish, and Bennett[288]). The usual quasi-hexagonal packing of mixed stacks is found instead (Figure 16). The planes of the TCNE molecules are parallel to the cyclopentadienyl rings and the distance between the components is 3.14 ± 0.1 Å,

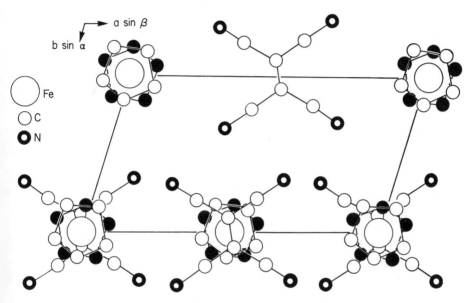

Figure 16. Ferrocene:TCNE. Projection of crystal structure on to (001) plane. Components have been omitted at certain positions in order to clarify the diagram. This diagram was kindly supplied by Dr. T. N. Margulis (see reference 289).

which seems rather low. There are strong indications of disorder in the crystals; a fairly small fraction of the accessible reflections was measured and the Debye–Waller factors of the carbon and nitrogen atoms are high (4.4–11.1 Å$^2$). Indeed, those of the ferrocene rings are higher than those of TCNE (5.0–11.1 Å$^2$, compared with 4.4–5.6 Å$^2$), and a Fourier synthesis shows that the ferrocene rings are much less well defined than the TCNE molecules. However, it is not clear whether this disorder is due to intrinsic rotation of the cyclopentadienyl rings of the ferrocene molecules or to rotation of the ferrocene molecules as a whole. The quality of the crystals was rather poor (Margulis, personal communication), so that there does not seem much hope of getting more detailed results.

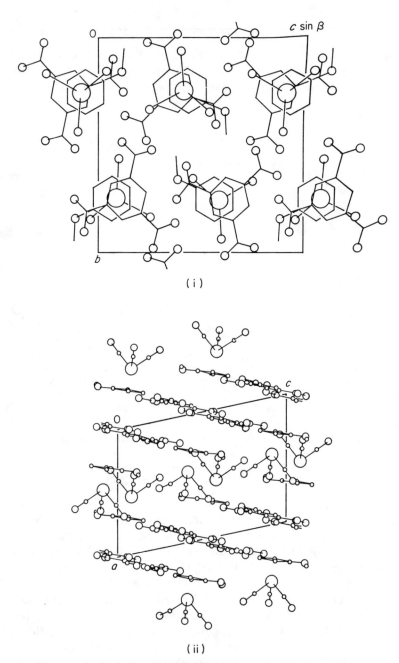

(i)

(ii)

Figure 17. Tricarbonylchromiumanisole:TNB. The crystal structure viewed in projection along (i) the $a$-axis (ii) the $b$-axis. [Reproduced with permission from reference 292.]

274

## D. Molecular Compounds between Donors containing "Benzene" Regions and Acceptors that are Substituted Benzenes

### 1. Tricarbonylchromiumanisole: TNB
$[a = 10.10, b = 13.42, c = 13.87 \text{ Å}, \beta = 101.8°, Z = 4, P2_1/c]$

Tricarbonylchromiumanisole:TNB (Carter, McPhail, and Sim[292]) has the usual mixed stacks, arranged in quasi-hexagonal close packing (Figure 17). There is no indication of disorder in the structure, the atoms in the electron-density syntheses (Figure 18) all being well

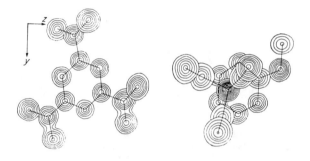

Figure 18. Tricarbonylchromiumanisole:TNB: Final three-dimensional electron-density distribution shown by means of superimposed contour sections drawn parallel to (100). Contour interval 1 eÅ$^{-3}$ except around the chromium atom where the interval is 3 eÅ$^{-3}$. [Reproduced with permission from reference 292.]

defined. The mean spacing between the planes of anisole and TNB moieties is 3.41 Å, and the angle between these planes is 3°. The perpendicular distances from the carbonyl-oxygen atoms to the plane of the TNB molecule average 2.97 Å. Carter *et al.*[292] conclude that the "participation of tricarbonylchromiumanisole as electron donor in solid state charge-transfer complexes is a function both of the aromatic ring and of the tricarbonylchromium fragment." The overlap diagram of the TNB molecule and carbonyl groups is shown in Figure 19; this is rather symmetrical. However, there is a mutual displacement of centres of TNB and anisole in the other overlap diagram (Figure 20). The tricarbonylchromiumanisole molecule is itself in an eclipsed conformation.

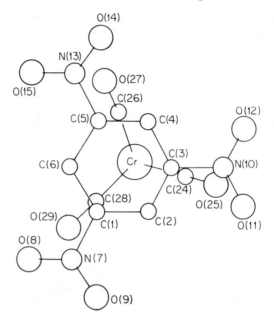

Figure 19. Tricarbonylchromiumanisole:TNB: The arrangement of TNB and tricarbonylchromium components as viewed in projection on to the plane of the TNB. [Reproduced with permission from reference 292.]

## 2. 1,3,5-Benzenetriamine: TNB

[$a = 14.06, b = 15.08, c = 6.98$ Å, $\beta = 103.5°, Z = 4, P2_1/a$; group 3b]

These crystals are purple-black needles, stable in air. The crystal structure has been determined by Iwasaki and Saito[394] and is of the familiar mixed stack type (stack axis [001]) with quasi-hexagonal close packing of stacks. The average interplanar distances between the amine and TNB molecules alternate slightly along the stack; the values found (3.23 and 3.29 Å) are significantly shorter than the usual interplanar distances for aromatic molecules. The amine molecule and the benzene ring of TNB are both planar, and the angle between these planes is 1°. The overlap diagram (Figure 21) shows an offset of successive molecules which is rather different from the exact superposition of successive molecules found in 2,4,6-tris(dimethylamino)-*s*-triazine:TNB (Section VI-D4).

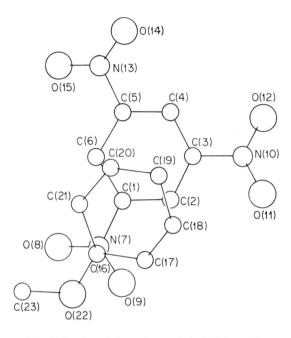

Figure 20. Tricarbonylchromiumanisole:TNB. The mutual orientation of the TNB and anisole components as viewed in projection on to the plane of the TNB. [Reproduced with permission from reference 292.]

### 3. *p*-Iodoaniline: TNB
$[a = 7.43, b = 7.39, c = 28.3 \text{ Å}, \beta = 103.4°, Z = 4, P2_1/c]$

This was the first $\pi$-molecular compound whose crystal structure was determined (Powell, Huse, and Cooke[35]) and the historical importance of the results has already been stressed. These results are not very accurate by current standards [further refinement, based on diffractometer measurement, is under way at the time of writing (Powell, personal communication, 1970)], but the usual type of quasi-hexagonal arrangement of mixed stacks is clearly shown in Figure 22. The stack axis is along [100] and the molecular planes are inclined at about 30° to this axis.

The polarized absorption spectra for the (presumably isomorphous) *p*-bromoaniline:TNB (Nakamoto[395]) show a broad charge-transfer band with a maximum at about 500 m$\mu$. The polarization ratio is close to unity, indicating some tilt of the molecular planes to the needle axis.

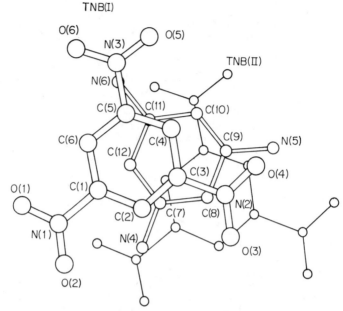

Figure 21. Overlap diagram of 1,3,5-benzenetriamine and 1,3,5-trinitrobenzene. [Reproduced with permission from reference 394.]

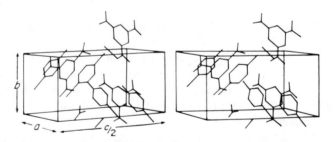

Figure 22. *p*-Iodoaniline:TNB. Stereoscopic drawing of part of the structure. The observer is looking through ($\bar{1}$00) at the right-hand half of the unit cell, with the positive direction of the *b*-axis pointing downwards. [Reproduced with permission from reference 35.]

## 4. 2,4.6-Tris(dimethylamino)-*s*-triazine : TNB
$[a = 6.72, b = 16.06, c = 9.59 \text{ Å}, \beta = 105.2°, Z = 2, P2_1/m]$

Only 25% of the reflections accessible to Cu-$K_\alpha$ radiation were recorded in the room-temperature study (Williams and Wallwork [281]), which suggests a disordered structure. Another indication of disorder was the appearance of discs of diffuse scattering in reciprocal space. Thus the results obtained can only be regarded as a first approximation. The planar molecules are in mixed stacks with stack axes along [100]. The mirror planes of the space group are normal to the molecular planes and bisect the molecules. Within a stack the two components are parallel and separated by 3.36 Å. The stacks are packed quasi-hexagonally but offset by $\frac{1}{2}a$; thus each TNB molecule is surrounded by six triazine molecules and *vice versa*. The lateral interactions between stacks are due to van der Waals forces only, and the diffuse scattering was ascribed to mistakes caused by displacements of one stack relative to another in the [100] direction. However, diffuse scattering can also be caused if the temperature at which the photographs were taken is close to a phase (or disorder–order) transformation temperature, and further examination of the source of the diffuse scattering is desirable. The overlap diagram is not reproduced here; it would show complete overlap of the two component molecules, which have very similar shapes.

## 5. Hexamethylbenzene : picryl halides

The 1:1 molecular compounds of hexamethylbenzene (HMB) with picryl chloride, bromide, and iodide were first studied crystallographically by Powell and Huse [325] and more recently by Bernstein and Herbstein.[326] Powell and Huse showed from a Patterson projection that the molecules are in layers 3.5 Å apart and normal to ⌈100⌉ (= 14 Å). Diffuse scattering shows that the true [100] periodicity is actually 42 Å, and also that the [010] axis should be trebled, from 9.0 to 27.0 Å. If the diffuse reflections are ignored, possible space groups are $A2_1am$, $Ama2$, and $Amam$. Bernstein and Herbstein showed that the diffuse reflections sharpen somewhat when the crystals are cooled to 120°K, but no phase change was found. Thus it is not clear whether the disorder is a growth effect, as seems most likely, or an equilibrium-type disorder.

The polarized absorption spectra from single crystals of HMB:picryl chloride (Nakamoto[172]) show a broad charge-transfer band with a dichroic ratio rather smaller than would be expected from the postulated crystal structure; however, an error seems to have occurred in labelling the polarizations of the spectra, so that repetition of this work seems desirable. On the other hand, the polarized infrared spectra from single crystals (Kross, Nakamoto, and Fassel[68]) agree well with the postulated crystal structure.

## E. Molecular Compounds between Donors containing "Naphthalene" Regions and Acceptors that are Substituted Benzenes

The donors discussed in this Section all contain regions with an exact or approximate naphthalene-like arrangement of atoms: naphthalene, anthracene, pyrene, perylene, acepleiadylene, azulene, skatole, and indole. Only the first five of these donors contain regions that have a strict "naphthalene" framework, but the others contain regions which are geometrically similar to naphthalene. The acceptors considered are the substituted benzenes, PMDA, TCNB, and TNB.

### 1. Naphthalene: pyromellitic dianhydride (PMDA)
$[a = 9.19, b = 13.0, c = 6.81 \text{ Å}, \beta = 104°, C2/m, Z = 2; \text{group 1}]$

A structure based on room-temperature cell dimensions and symmetry was proposed by Boeyens and Herbstein.[157] The poor photographs suggest that the crystals are disordered at room temperature and the overlap diagram (Figure 23a) probably represents only an averaged situation. [No change was detected in naphthalene:PMDA crystals on cooling to 100°K (Bernstein and Herbstein[326]), so the arrangement at 25° is presumably virtually wholly random]. No detail of interplanar angle or distance is known.

### 2. Naphthalene: TCNB
$[a = 9.39, b = 12.66, c = 6.87 \text{ Å}, \beta = 107.2°, C2/m,$
$Z = 2; \text{group 1}]$

This crystal also belongs to the quasi-isomorphous $C2/m$ group at room temperature; its crystal structure has been determined (Kumakura, Iwasaki, and Saito[356]). The stacks are translation-equivalent and are in quasi-hexagonal packing, two being displaced by $\frac{1}{2}c$ with respect to a reference stack at the origin, and four by $\frac{1}{4}c$.

The naphthalene molecules are disordered, their orientations differing by about 18°, but the TCNB molecules are completely ordered. The electron-density maps in the $(10\bar{2})$ planes are shown in Figure 24. The molecular planes are nearly parallel (angle 1°) and separated by 3.43 Å.

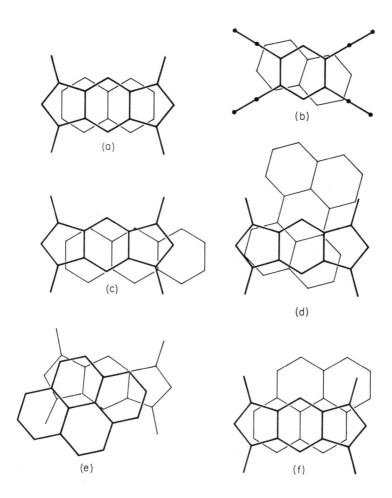

Figure 23. Overlap diagrams. (a) Naphthalene:PMDA (however, this schematic sketch is too symmetrical as these crystals are disordered at room temperature). (b) Naphthalene:TCNB. One of the two disorded orientations is shown. (c) Anthracene:PMDA. The apparent distortion of the molecules is due to incomplete refinement of the crystal structure. (d) Perylene:PMDA. (e) and (f) Pyrene:PMDA.

The overlap diagram (Figure 23b) corresponds to one of the two orientations found in the ordered structure. No cooling experiments have been done and nothing is known about the ordered structure.

Iwata, Tanaka, and Nagakura[396] estimate that the charge-transfer interaction energy of an arrangement with the long axes of the two

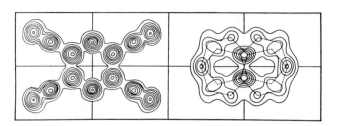

Figure 24. Naphthalene:TCNB. Electron-density syntheses in (10$\bar{2}$) planes (very nearly the molecular planes). The ordered TCNB molecule is on the left and the disordered naphthalene molecule on the right. Contours are at intervals of 1 eÅ$^{-3}$, the lowest contour being 1 eÅ$^{-3}$. [Reproduced with permission from reference 356.]

component molecules parallel and their centres one above the other is about 2 kcal/mole lower than when the centres are offset by 1.2 Å along the long axis. The lowest excited state is stabilized by a displacement of this kind. The charge-transfer interaction energy is much less sensitive to in-plane relative rotations, the energy difference between parallel and perpendicular arrangements of the long axes of the molecules being only 0.2 kcal/mole. However, Kumakura et. al.[356] point out that the parallel arrangement would have an abnormally short N$\cdots$H distance of 2.36 Å, whereas in the arrangement actually found the N$\cdots$H distance is normal (2.76 Å) (Figure 25). This intermolecular steric effect is the only explanation so far advanced why one of the components takes up two separate orientations within a stack rather than a single (averaged) orientation.

Iwata et al.[396] have also studied the polarized absorption spectra of single crystals of naphthalene:TCNB at 298°K, and the absorption and emission spectra of polycrystalline samples down to liquid-helium temperatures. No unusual feature was found.

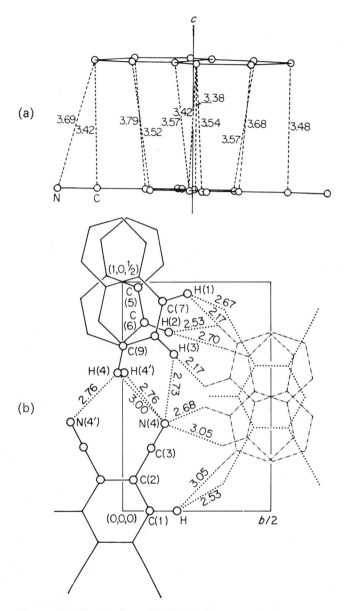

Figure 25. Naphthalene:TCNB. (a) Interatomic distances within a stack. (b) Interatomic distances between the molecules arranged on $(10\bar{2})$. The centres of molecules drawn by dotted and broken lines are placed on the positions $(\frac{1}{2}, \frac{1}{2}, 0)$ and $(\frac{1}{2}, \frac{1}{2}, \frac{1}{2})$, respectively. It is the distance N4–H4 (or N4′–H4′) which is increased to 2.76 Å in the inclined arrangement compared to the 2.36 Å for the hypothetical arrangement with the long axes of the two component molecules parallel. [Reproduced with permission from reference 356.]

283

### 3. Naphthalene : TNB
$[a = 16.19, b = 6.97, c = 14.5\text{Å}, N = 97.6°, Z = 2,$
$P2_1/a;$ group 4b]

Although only cell dimensions and space group have been reported, the structure of this compound appears to have been solved,
as an overlap diagram (Figure 26a, page 285) has been given (Wallwork[23]). The mutual disposition of the two ring systems is similar
to that in naphthalene:TCNB.

### 4. Azulene : TNB
$[a = 16.39, b = 6.66, c = 13.77\text{Å}, \beta = 96.1°, P2_1/a,$
$Z = 4;$ group 4b]

The crystal structure of this molecular compound at room temperature has been determined by Brown and Wallwork[397] and by
Hanson,[28] and the latter author has also determined the crystal
structure at $-95°$. The unit-cell dimensions (Table 20, p. 253) for
the molecular compounds of azulene, skatole, and indole with TNB
show that these crystals are isomorphous. The room-temperature
studies of azulene:TNB yield a disordered structure, with a certain
proportion of the azulene molecules in an alternative orientation
obtained by rotating the molecule by 180° in its own plane. Brown
and Wallwork[397] found that the two orientations were present in
roughly equal proportions, but Hanson found only 7% alternative
orientation in his low-temperature study. This variability suggests
that the disorder is non-equilibrium in character; additional support
is given by the difficulty of obtaining one orientation from the other
by a thermally induced equilibrium process.

There is some internal evidence in Hanson's paper for a temperature-dependent disorder–order change. For example, there is considerable low-angle diffuse scattering at room temperature; at $-140°$
this concentrates into diffuse reflections with apparent indices
$h, k, l + \frac{1}{2}$. The intensities of some reflections were also markedly
temperature-dependent. Insufficient information is available for an
explanation of these phenomena, but it seems clear that the results
given by Hanson for the structure at $-95°$ are a first approximation
only.

The stack axis is [010] and the molecular planes are parallel to
one another and almost parallel (angle 1.5°) to (010). The interplanar
separation is 3.33 Å. The overlap diagram is given in Figure 26c; the
mutual disposition of the two components is again similar to others
discussed in this Section.

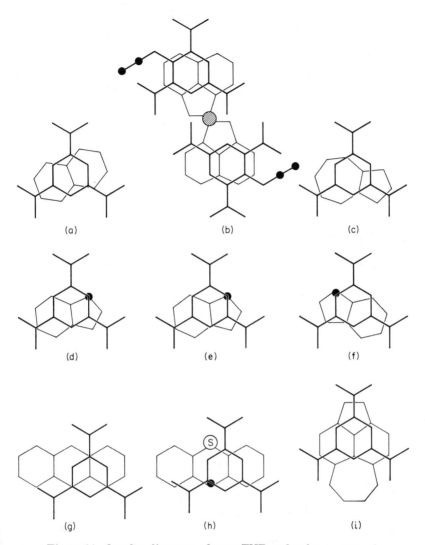

Figure 26. Overlap diagrams of some TNB molecular compounds:
(a) Naphthalene:TNB. (b) Bis-8-quinolinolatocopper:(picryl
azide)$_2$. (c) Azulene:TNB (major orientation). (d) Skatole:TNB.
(e) Indole:TNB (major orientation). (f) Indole:TNB (minor
orientation). (g) Anthracene:TNB. (h) Phenothiazine:TNB.
(i) Acepleiadylene:TNB. In some of these examples it is the
non-substituted carbon atoms of TNB that are eclipsed (approximately) by atoms of the electron-donors; however, there are other
cases, *e.g.*, (g), (h), where the substituted carbon atoms of TNB
are eclipsed.

## 5. Skatole : TNB
$[a = 16.76, b = 6.61, c = 13.45$ Å, $\beta = 95.6°$, $P2_1/a$,
$Z = 4$; group 4b]

This crystal structure was determined at $-140°$ (Hanson[365]). Despite this low temperature only about 50% of the reflections accessible to Cu-$K_\alpha$ radiation had measurable intensities. Together with the lack of definition of the trinitrobenzene molecule in the electron-density maps, this suggests that the crystals are disordered. However, this was apparently not considered during the analysis, and the published results may well represent only a first approximation to the true structure. The arrangement of the stacks is similar to that described for azulene:TNB. The skatole molecule and the benzene ring of TNB are planar and nearly perpendicular to [010]; the interplanar separation is 3.30 Å. The overlap diagram (Figure 26d) shows the same mutual disposition of the molecules as found previously.

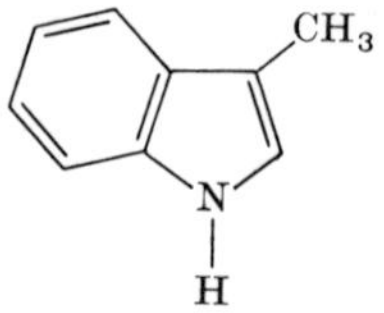

Skatole 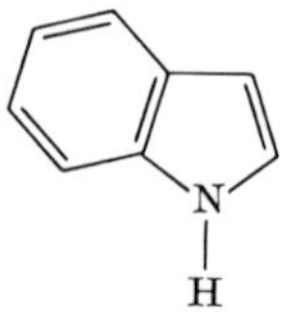        Indole

## 6. Indole : TNB
$[a = 15.87, b = 6.58, c = 13.47$ Å, $\beta = 94.8°$, $P2_1/a$,
$Z = 4$; group IVb]

This crystal structure (at $-140°$) was described together with that of skatole:TNB. Less than 50% of the accessible reflections were observed at $-140°$ and less than 25% at room temperature, a strong indication of disorder. The main difference between the two structures is that two orientations (in 2:1 ratio) were found for the indole molecules in indole:TNB (Figure 26e and f). It was claimed that the distinction made between carbon and nitrogen positions in the indole molecules was significant. If this is so, then the two orientations shown cannot be interconverted without disrupting the crystal structure, and this would suggest that the disorder is of the non-equilibrium type, probably originating during the crystallization process. Hanson[365] argued, in 1964, that the nitrogen atoms of indole and skatole and the carbon atoms at the non-substituted positions of the TNB molecules play a decisive role in the inter-molecular binding. However, no such N···C interaction can exist in azulene:TNB (analysed by Hanson[28] in 1966), which is nevertheless isomorphous with skatole:TNB and indole:TNB (Table 20, group 4b). Thus, whatever N···C interaction exists in the last two molecular compounds is not likely to be very important in determining their detailed structures.

### 7. Bis-8-quinolinolatocopper:(picryl azide)$_2$
$[a = 16.14, b = 6.90, c = 30.93 \text{ Å}, \beta = 105.6°, A2/a, Z = 4]$

The usual mixed stacks are found in this monoclinic crystal, each copper quinolinolate molecule being interleaved by *two* picryl azide molecules, one above each 8-quinolinol residue (Bailey and Prout.[332])

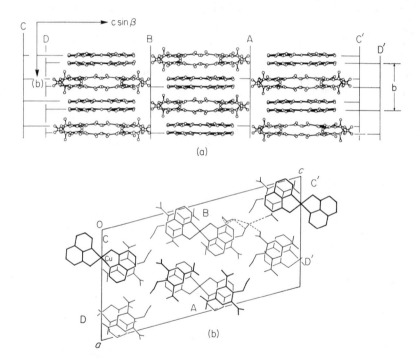

Figure 27. Bis-8-quinolinolatocopper:(picryl azide)$_2$. (a) The crystal structure projected down the $a$-axis. The atoms outlined by the thicker circles are those of molecules near the viewer. (b) The crystal structure projected down the $b$-axis, showing the relative arrangement of atoms in donor and acceptor molecules. Based on reference 332.

The quinolinolatocopper molecule has a two-fold axis of symmetry, and the two interleaved picryl azide molecules are related by this two-fold axis. The planes of the quinoline system and of the benzene ring of the picryl azide are almost parallel (angle 2°) and separated by 3.45 Å. The copper atom is not directly bonded to the picryl azide molecule, nor are there short approaches to the C–O(H) region of the 8-quinolinol unit, as in some other molecular compounds of its Cu(II) and Pd(II) coordination complexes.

The overlap diagram is shown in Figure 26b; as Bailey and Prout[332] point out, the donor–acceptor arrangement is similar to that in the TNB molecular compounds of naphthalene and other aromatic hydrocarbons. Thus it appears that the mutual disposition of the components is determined by charge-transfer interactions.

The mixed stacks are arranged in quasi-hexagonal fashion (Figure 27) with a rather complicated off-setting of the six stacks surrounding a particular reference stack A; for example, the neighbouring stacks B and B' in the [100] direction are displaced by about $b/12$. In the [001] direction the pairs of equivalent neighbouring stacks are C, C' and D, D'.

### 8. Anthracene : PMDA

$$[a = 7.6,\ b = 10.0,\ c = 7.3\ \text{Å},\ \alpha = 105,\ \beta = 115\tfrac{1}{2},$$
$$\gamma = 101°,\ P\bar{1},\ Z = 1;\ \text{group 6a}]$$

The triclinic crystal has been studied by two-dimensional Fourier methods at room temperature (Boeyens and Herbstein[301]). There is no phase change down to 100°K, and it seems unlikely that the crystal is disordered at room temperature. Because of the triclinic symmetry, all stacks are identical and some measure of interleaving occurs. The component molecules in any stack were assumed to be parallel, and their interplanar distance was measured as 3.23 Å. The overlap diagram is shown in Figure 23c; there is a strong overall resemblance to the type of overlap found between other analogous pairs of molecules.

### 9. Anthracene : TNB

$$[\text{at} -100°:\ a = 11.35,\ b = 16.27,\ c = 13.02\ \text{Å},\ \beta = 133.2°,$$
$$C2/c,\ Z = 4]$$

The anthracene:TNB phase diagram shows only an equimolar, congruently melting, molecular compound, which does not undergo any polymorphic change between 25° and the melting point of 164°.[109] The crystal structure has been determined at 25° and −100° by

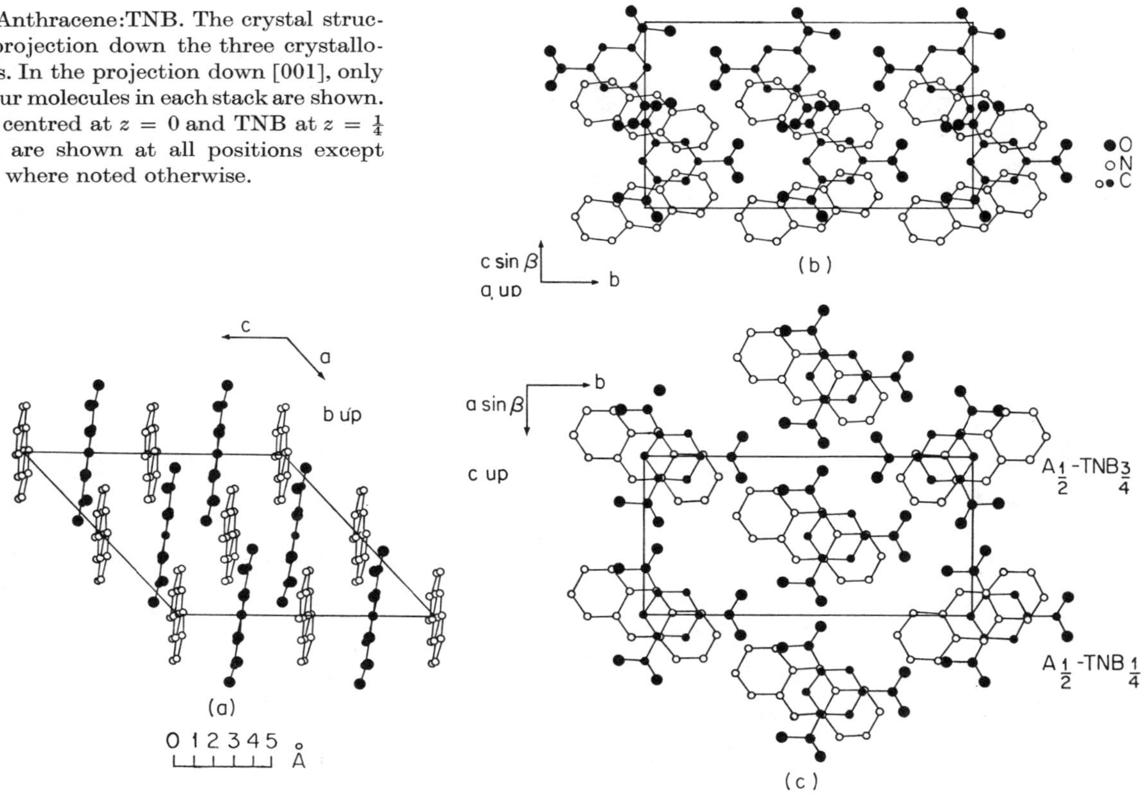

Figure 28. Anthracene:TNB. The crystal structure[314] in projection down the three crystallographic axes. In the projection down [001], only two of the four molecules in each stack are shown. Anthracene centred at $z = 0$ and TNB at $z = \frac{1}{4}$ $(A_0-TNB_{1/4})$ are shown at all positions except where noted otherwise.

Brown, Wallwork, and Wilson[314] (Figure 28). The centering of the space group and the positioning of the two component molecules at special positions make all stacks identical, a feature which facilitates interpretation of measurements of the physical properties of the crystals. Values of the Debye–Waller factors found in the room-temperature crystal-structure analysis are rather large (Figure 29a), but the corresponding values at 173°K are reasonable (Figure 29b),

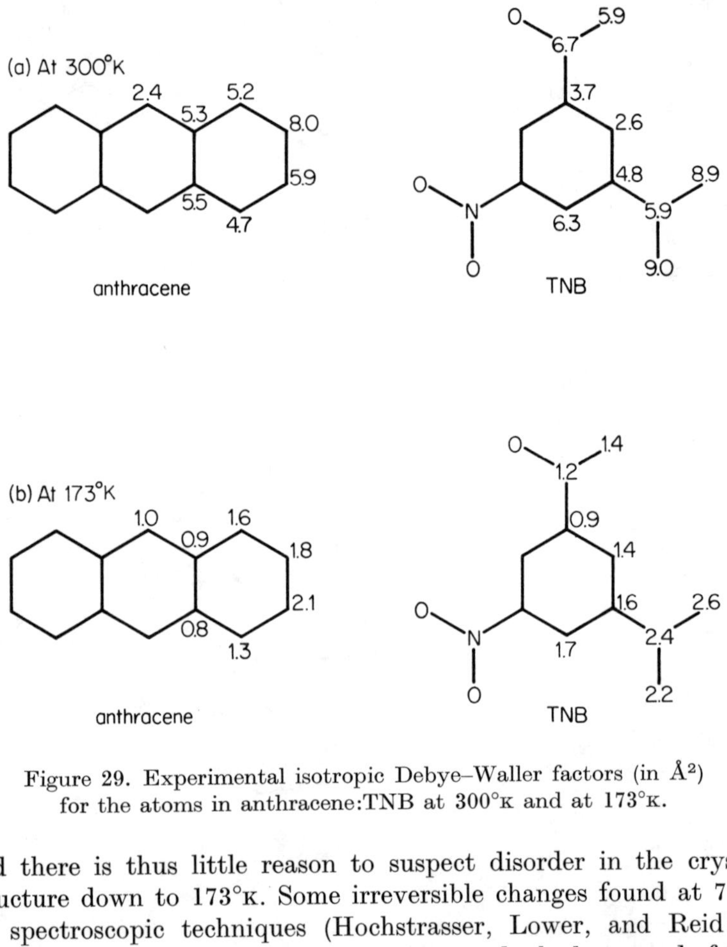

Figure 29. Experimental isotropic Debye–Waller factors (in $Å^2$) for the atoms in anthracene:TNB at 300°K and at 173°K.

and there is thus little reason to suspect disorder in the crystal structure down to 173°K. Some irreversible changes found at 77°K by spectroscopic techniques (Hochstrasser, Lower, and Reid[398]) suggest that a phase change may occur towards the lower end of this range, but further investigation is necessary.

The arrangement of the component molecules within the mixed

stacks differs from that found in other $\pi$-molecular compounds. There are two anthracene molecules (at centres of symmetry) and two TNB molecules (on two-fold axes parallel to [010]), in each 13.2 Å repeat distance along [001]; the two anthracene molecules are related by a two-fold axis and their long axes are about 60° apart, while the two TNB molecules are antiparallel and have their centroids equally and oppositely displaced from the [001] axis along [010] (Figure 28b). This alternating mutual off-set of the two components along the stack is one unusual feature; another is the angle of 8° between the mean planes of the components. However, there are no obvious centres of mutual attraction in the two component molecules and the reason for the tilt is not clear.

Qualitative examination[314] of single crystals of anthracene:TNB with the polarizing microscope shows that these are highly pleochroic, with maximum light absorption when the electric vector is parallel to the needle ([001]) axis, and that the colour changes from orange at 25° to pale yellow at −100°. The early spectroscopic studies of Nakamoto[172] in the near UV and visible range have been followed by more extensive and more quantitative work by two groups (Hochstrasser *et al.*;[398–400] Tanaka and Yoshihara[401]). Measurements of single-crystal absorption and emission spectra have been made at a number of temperatures (Figures 30 and 31); the two sets agree qualitatively but there are quantitative discrepancies. The assignments of the absorption bands are shown in Figure 30. The first charge-transfer band shows a pronounced vibrational structure, including a 250 cm$^{-1}$ interval, and a Davydov splitting of about 200 cm$^{-1}$ which is in reasonable agreement with the known crystal structure. The second charge-transfer band is shifted by about 8200 cm$^{-1}$, which is encouragingly close to the energy difference between the first and the second $\pi$-ionization potentials of anthracene (Hoyland and Goodman[402]).

Comparison of the absorption and emission spectra can help in postulating a structure for the excited (charge-transfer) state of the molecular compound. The emission and absorption spectra of crystals are approximately mirror images, and comparison of crystal and solution charge-transfer emission spectra makes it fairly certain that the observed crystal emission originates from the base level of the charge-transfer state observed in absorption. According to rigid lattice theory (McRae[403]), the polarization ratios* should be equal in

---

* Polarization ratio = $\dfrac{\text{Band intensity with radiation polarized parallel to } c}{\text{Band intensity with radiation polarized normal to } c}$.

absorption and emission. This prediction is contradicted by experiment, but the two groups differ in regard to the sign of the discrepancy. In absorption the polarization ratio is about 5:1 for both sets of measurements, but Hochstrasser *et al.* find a polarization ratio of

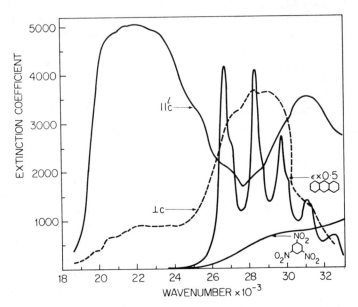

Figure 30. Absorption spectrum of single-crystal anthracene: TNB. The single-crystal spectra were taken with radiation polarized ‖ and ⊥ to the needle (c) axis; the incident radiation was normal to (110). [Reproduced with permission from reference 400.]

15:1 in emission at 300°K, which falls to 3:1 at 80°K, accompanied by a large increase in the intensity of emission. Tanaka and Yoshihara, on the other hand, find polarization ratios of 3:1 in emission at 300°K, 1.2:1 at 80°K, and 1.4:1 at 4°K. The agreement between the results is worst at 300°K but improves with reduction in temperature. Hochstrasser *et al.*[398] interpret the larger polarization ratio in emission than in absorption to mean that the emitting species in the crystal differs in orientation from the absorbing species. The dipole moment in the excited state is lined up more closely with the [001] axis as a consequence of its interaction with the surrounding permanent dipoles and is estimated to be about twenty times larger than in the ground state. This deformed region acts as a localized Davydov exciton, which is comparatively immobile, in contrast to the situation

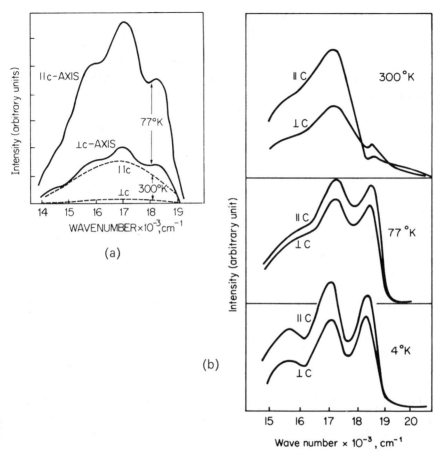

Figure 31. The emission spectra of anthracene:TNB at various temperatures. (a) Following Hochstrasser *et al.* (b) Following Tanaka and Yoshihara. [Reproduced with permission from references 398 and 401.]

in aromatic hydrocarbon crystals where the electronic energy migrates so rapidly from one molecule to the next that deformed regions are not found. It is suggested that this contrasting behaviour can perhaps explain the much weaker photoconductivity found in $\pi$-molecular compounds than in aromatic hydrocarbons.

Tanaka and Yoshihara explain the reduction in polarization ratio in emission as due to an increased contribution of the locally excited $L_a$ (short axis) state of anthracene to the excited state of the molecular compound. The rather large shift of 1700 cm$^{-1}$ in the position of the 0–0 band from absorption to emission is ascribed to a decrease of intercomponent distance (3.28 Å interplanar distance in the ground state) by 0.2 Å in the excited state. Thus two different descriptions of the excited region in crystalline anthracene:TNB have been proposed, one suggesting a change in component orientation, the other a change in component position.

The interpretation[68] of the polarized IR spectrum of single-crystal anthracene:TNB is in good agreement with the crystal structure.

The spectroscopy of anthracene:TNB has been studied in greater depth than for most other π-molecular compounds. Definitive studies of the emission spectra and their polarization ratio are still necessary. Further, generalizations should be made with some caution because the molecular arrangement in anthracene:TNB differs, as we have seen, in some respects from that found in most other π-molecular compounds.

### 10. Phenothiazine: TNB (and probably biphenyl: BTF)
[$a = 7.01$, $b = 15.14$, $c = 17.15$ Å, *Pbcn*, $Z = 4$; group 5a]

Phenothiazine is a strong electron-donor which forms molecular compounds with TNB and a number of other electron-acceptors. The equimolar TNB molecular compound (black needles, red in thin section) has been studied by Fritchie.[269] The molecular planes in different stacks are arranged in the herring-bone fashion (Figure 32) characteristic of many aromatic hydrocarbon crystals. The $R$ factor is rather large (12%), considering that the intensities were measured on a diffractometer and corrected for absorption, and this suggests that disorder may be present in the crystals. This is supported by the abnormal pattern of anisotropic Debye–Waller factors found for the atoms of the phenothiazine molecule; the Debye–Waller factors of atoms of TNB appear to be normal. The phenothiazine molecule has an approximately butterfly shape.[378] In a particular stack, the apices of the V-shaped molecules point up and down at random, in conformity with the site-group symmetry $C_2$-2 and simulating molecular planarity. The interplanar spacing of 3.37 Å between phenothiazine and TNB molecules is somewhat larger than usual. It is not clear whether the disorder is dynamic or static. The overlap diagram (Figure 26h) is remarkably similar to that of anthracene:TNB (Figure 26g), thus demonstrating that any special influence of the heteroatoms must be small. Fritchie[269] has reported a phase change at about 250°K associated with a doubling of the *a*-axis; we have repeated these experiments, using photographic methods, but do not find any phase change down to 150°K, although the crystals are damaged and split by the cooling.

The cell dimensions of biphenyl:BTF (Table 20) suggest strongly that it and phenothiazine:TNB have similar structures. The dimensions and molecular symmetry of biphenyl allow it to be situated

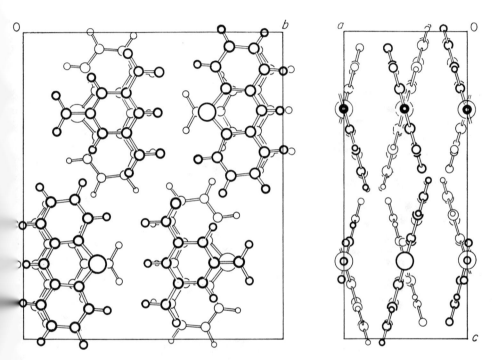

Figure 32. Phenothiazine:TNB crystal structure. Hydrogen atoms were not located and are assumed to be in the mean molecular planes. The large circle is sulphur. The molecules in any one stack are almost parallel (angle between mean planes 0.5°). [Reproduced with permission from reference 269.]

at the phenothiazine position in phenothiazine:TNB. Moreover, BTF has similar dimensions to TNB and an approximate two-fold axis passing through any pair of *para*-nitrogen atoms. Thus, it can be accommodated on the TNB sites with only minor disorder. A somewhat different structure was proposed by Boeyens and Herbstein[157] on the assumption that the biphenyl and (disordered) BTF molecules were at centres of symmetry.

## 11. Perylene: PMDA
$[a = 17.0, b = 7.16, c = 10.13$ Å, $P2_1/a$, $Z = 2$; group 4a]

This monoclinic crystal has been studied by three-dimensional methods at room temperature (Boeyens and Herbstein[301]) (Figure 33). The stack axes lie along [010] and each stack (*e.g.*, that at $(\frac{1}{2}, y, \frac{1}{2})$ is surrounded by six others in quasi-hexagonal array. Four of the neighbouring stacks (those at $(0, y, 0)$, $(1, y, 0)$, $(0, y, 1)$, $(1, y, 1)$) are derived from the reference stack by the action of the

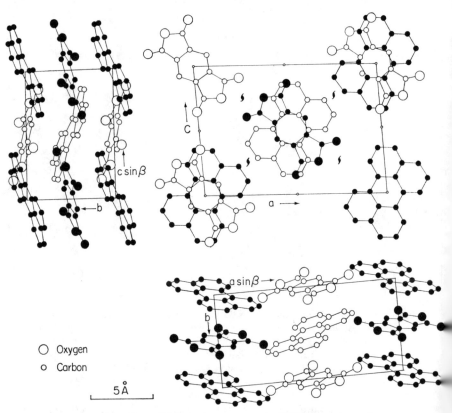

○ Oxygen

o Carbon

5 Å

Figure 33. Perylene:PMDA. Arrangement of PMDA and perylene molecules in the unit cell as seen in projection down the three crystallographic axes. [Reprinted from *Journal of Physical Chemistry* **69**, 1965, p. 2165. Copyright 1965 by the American Chemical Society. Reprinted by permission of the copyright owner.]

screw axes parallel to [010], *i.e.*, they are displaced by $\frac{1}{2}b$ along the stack axes and differ in orientation. The other two stacks (those at $(\frac{1}{2}, y, -\frac{1}{2})$ and $(\frac{1}{2}, y, 1\frac{1}{2})$) are translationally equivalent to the reference stack. The shortest C–C and C–O approaches between molecules in different stacks are 3.69 and 3.39 Å, respectively, indicating only van der Waals forces between the stacks. The perylene molecule and the benzene ring of PMDA are parallel, with an interplanar distance of 3.33 Å. The overlap diagram is shown in Figure 23d; there is a marked resemblance to the naphthalene:TCNB overlap diagram, extending even to the angle of about 15° between the long axes of the "naphthalene region" and TCNB. There is no evidence for disorder in the room-temperature structure, although a phase change occurs on cooling. However, nothing is known of the low-temperature structure as the crystals shatter to powder at about 100°K.

## 12. Pyrene:TCNB

The crystal structure of pyrene:TCNB at room temperature has been determined by Prout and his co-workers.[404] The pyrene molecule is disordered, while the TCNB molecule is ordered. Thus there are clear similarities to, say, naphthalene:TCNB, but details are not yet available.

## 13. Pyrene:PMDA

[At 300°K: $a = 13.89$, $b = 9.33$, $c = 7.34$ Å, $\beta = 93.5°$,
$P2_1/a$, $Z = 2$; group 3a]
[At 110°K: $a = 13.67$, $b = 9.13$, $c = 14.40$ Å, $\beta = 91.5°$,
$P2_1/n$, $Z = 4$]

The crystal structures of pyrene:PMDA at 110°K and 300°K have been determined (Herbstein and Snyman[303]). There is a disorder-to-order transformation at about 200°K, and this substance is of particular interest as the only example so far where both ordered and disordered structures are known.

The molecular arrangements in the ordered and the disordered structures are shown in three different projections in Figure 34a, b, c. The stack axes lie along [001] and any one stack [*e.g.*, that at $(\frac{1}{2}, \frac{1}{2}, z)$] is surrounded by six others in quasi-hexagonal array. In the ordered phase four of these stacks [those at $(0, 0, z)$, $(1, 0, z)$, $(0, 1, z)$, $(1, 1, z)$] are derived from the reference stack by the action of $n$-glide planes perpendicular to [010], *i.e.*, they are displaced by $c/2$ along the stack axis and differ in orientation. The other two stacks (those at

(i)

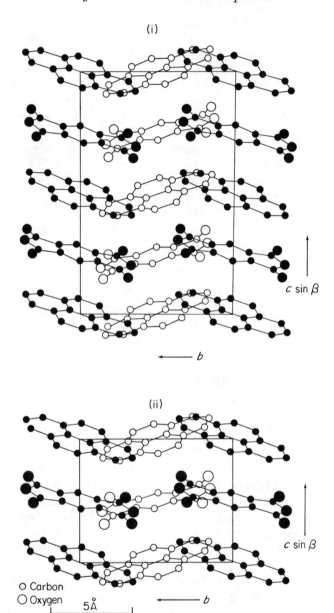

(ii)

O Carbon
O Oxygen    5Å

Figure 34a. Pyrene:PMDA. The projection down [100] in (i) the low-temperature structure and (ii) the room-temperature structure.

(i)

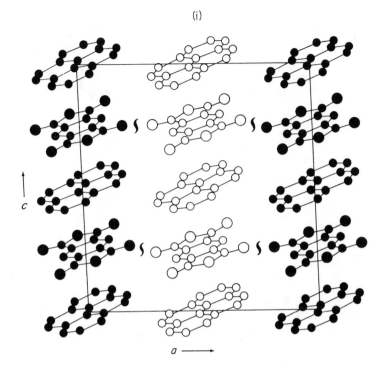

(ii)

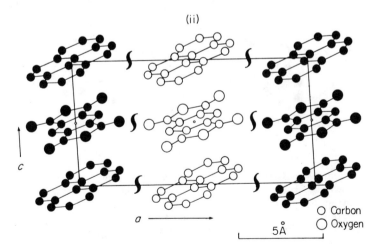

○ Carbon
○ Oxygen

5Å

Figure 34b. Pyrene:PMDA. The projection down [010] in (i) the low-temperature structure and (ii) the room-temperature structure.

(i)

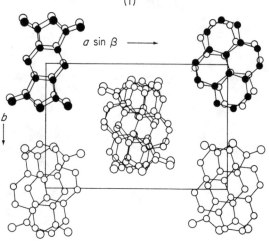

(ii)

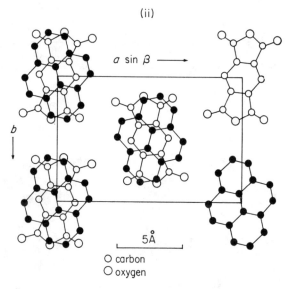

○ carbon
○ oxygen

Figure 34c. Pyrene:PMDA. The projection down [001] in (i) the low-temperature structure and (ii) the room-temperature structure. In (i) the atoms of the PMDA molecules near $\frac{1}{4}c$ are represented by shaded circles and those near $\frac{3}{4}c$ by open circles. For clarity two molecules have been left out at each corner of this projection. In (ii) one pyrene and one PMDA molecule have been left out for clarity. [Reproduced with permission from reference 303.]

$(1\frac{1}{2}, \frac{1}{2}, z)$ and $(\frac{1}{2}, 1\frac{1}{2}, z)$) are translationally equivalent to the reference stack. Thus any particular pyrene molecule (*e.g.*, that centred at $\frac{1}{2}, \frac{1}{2}, 0$) is surrounded by four others at $0,0,0$, etc. (which differ in orientation from the central molecule) and two identical pyrene molecules at $\frac{1}{2}, \frac{1}{2}, 0$ and $\frac{1}{2}, 1\frac{1}{2}, 0$. In the disordered phase there are also six stacks surrounding a particular reference stack, two being identical with the reference stack and the other four derived from it by the action of the $a$-glide planes perpendicular to [010]. These four stacks differ in orientation from the reference stack but are not shifted along [001] with respect to it. Thus perylene:PMDA and pyrene:PMDA differ in the relative displacements of close-packed stacks along the stack axes.

The molecules within a stack are parallel to within about 1°. In the ordered phase the interplanar distances between the two independent pyrene molecules and the interleaving PMDA molecules are slightly different (3.33 and 3.30 Å) but individual distances of closest approach between atoms do not differ appreciably. Thus it seems unlikely that the intermolecular forces vary between different pairs of molecules, nor can one discern any tendency to grouping into closer pairs of molecules along the stack. The overlap diagrams for the two crystallographically independent pyrene molecules in a stack are different and are shown in Figures 23e and f. In both diagrams the PMDA molecule overlaps only half of a pyrene molecule— a "naphthalene region"—and thus there is a general resemblance to the situation in perylene:PMDA. The two independent pyrene molecules are mutually rotated by about 12° while alternate PMDA molecules are shifted off the stack axis in opposite directions. The overlap diagram for the disordered phase will be the average of the two separate diagrams for the ordered phase.

## 14. Acepleiadylene: TNB
$[a = 8.79, b = 16.06, c = 6.52$ Å, $\alpha = 92.9, \beta = 102.2,$
$\gamma = 99.0°, P\bar{1}, Z = 2$; group 4b]

The crystal structure of this molecular compound at $-150°$ was determined by Hanson.[27] About 80% of the accessible reflections had measureable intensities and the thermal motion found, although anisotropic, was reasonably small. Thus disorder appears unlikely. Both molecules are practically parallel to the plane normal to [001] and thus to each other. The average interplanar spacing is $c/2$ or 3.26 Å and the overlap diagram is shown in Figure 26i. The same general disposition of TNB molecule to "naphthalene region" is found as in the other examples of this Section.

### 15. The equimolar perylene: TNB and pyrene: TNB molecular compounds

Unit-cell dimensions and space groups have been reported[262] for perylene:TNB (black needles) and pyrene:TNB (yellow needles) (see Table 20), but the crystal structures have not been determined.* Pyrene:TNB appears to be quasi-isomorphous with acepleiadylene: TNB. Although perylene:TNB has been classified with anthracene:BTF (because they have the same space group) there are appreciable differences in cell dimensions and hence, presumably, in component arrangement.

The observed single-crystal spectra[390] for perylene:TNB (Figure 35) are compatible with the unit-cell dimensions if the reasonable

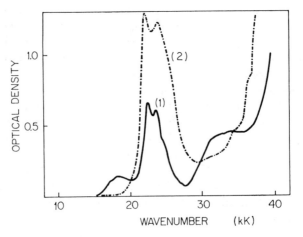

Figure 35. Absorption spectra of single-crystal perylene:TNB. The incident radiation was polarized (1) parallel to [001] and (2) perpendicular to [001]. [Reproduced with permission from reference 390].

assumption is made that mixed stacks occur along the [001] axis. The absorption band at 18.5 kK is strongly polarized along [001] and is the charge-transfer band. The similarly polarized weak band at 33 kK was tentatively assigned to the second charge-transfer band.

* [Added in proof:] The crystal structure of pyrene:TNB has been determined[221] but details are not yet available.

The strong band in the 20–25 kK region is assigned to a local excitation associated with the $p$-band of perylene.

The stack axis for pyrene:TNB is almost certainly along [001], and this is again compatible with the spectroscopic results[390] (Figure 36). The first and second charge-transfer bands, both polarized

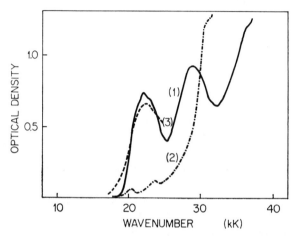

Figure 36. Absorption spectra of single-crystal pyrene:TNB. The incident radiation was polarized (1) parallel to [001] and (2) perpendicular to [001]. The charge-transfer band of the solution spectrum is shown by (3). [Reproduced with permission from reference 390.]

parallel to [001], are at 22 kK (also in solution) and at 29 kK (not observed in solution). The strong band at 31 kK, polarized perpendicular to the [001] axis, was assigned to the $p$-band of pyrene.

## F. Molecular Compounds of Aromatic Hydrocarbons and $p$-Benzoquinone

$p$-Benzoquinone has been reported to form molecular compounds with only two aromatic hydrocarbons, namely, pyrene and dibenz-[$a,e$]anthracene. Pyrene:$p$-benzoquinone crystallizes in the tetragonal system ($a = 7.7$, $c = 25.7$ Å) with non-centrosymmetric space group $P4_1$ (Herbstein and Regev;[305] Sasvari[405]). As centrosymmetric molecules very seldom or never (Kitaigorodskii[406]) crystallize in non-centrosymmetric space groups this result may indicate that some sort of pairing of the two components has occurred to form a non-centrosymmetric "unit". The actual crystal structure is unknown, so this suggestion is speculative; there is no known example of a π-molecular compound where appreciable "pairing" of the two components occurs.

### G. Molecular Compounds of Aromatic Hydrocarbons and Haloanils (also HMB:TCNB)

#### 1. Perylene:fluoranil

$$[a = 18.5, b = 7.49, c = 6.97 \text{ Å}, \beta = 112.3°, P2_1/a,$$
$$Z = 2; \text{ group 3a}]$$

This structure has the mixed stacks packed quasi-hexagonally (Figure 37) with four of the six stacks shifted by a half-translation and the other two identical with the reference stack (Hanson[360]). The overlap diagram (Figure 40a, p. 308) shows each carbonyl group overlapping a benzene ring of the perylene molecule in a manner similar to that found in quinhydrone.

Crystals of perylene:fluoranil are bluish-black; the polarized absorption spectra of single crystals have been measured by Kuroda et al.[390] (Figure 38). The bands at 14.2 and 28.7 kK are assigned, because of their polarization ratios, to charge-transfer bands. There is good agreement between the charge-transfer band in solution and the first charge-transfer band of the crystal, but the latter shows in addition vibrational structure seldom found in solution spectra. The band between 20 and 25 kK, with vibrational structure, is assigned to the $p$-band of perylene [transition moment along the long ($L$) axis of the molecule] and the band at 29.4 kK is identified as the $\alpha$-band of perylene [transition moment along the short ($M$) axis of the molecule].

The semiconduction and photoconduction of single crystals have been measured (Kokado, Hasegawa, and Schneider[407]). Conductivity parallel to [001], the stack axis, was found to be about three times as large as that in the perpendicular direction (dark conductivity along [001] is $1.5 \times 10^{-14}$ reciprocal ohm-cm). No difference was found in the semiconduction activation energy (0.73 ev) in the two directions.

#### 2. Pyrene:fluoranil.

Polarized absorption spectra for single crystals of pyrene:fluoranil have been reported,[390] but no crystallographic information is available for this compound.

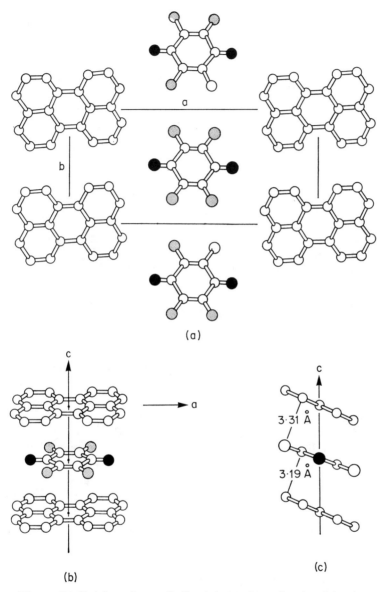

Figure 37. Perylene:fluoranil. Crystal structure showing (a) one layer, near the plane $z = 0$, (b) projection down [010] of one stack, and (c) projection down [100] of one stack (lack of parallelism of molecules exaggerated). [Reproduced with permission from reference 360.]

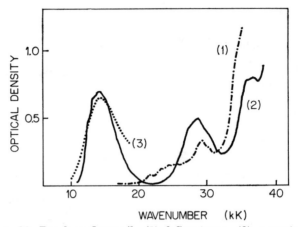

Figure 38. Perylene:fluoranil. (1) *b*-Spectrum, (2) *c*-spectrum, (3) charge-transfer band of solution spectrum. (1) and (2) are single-crystal spectra, with polarized radiation. [Reproduced with permission from reference 390.]

### 3. Pyrene:chloranil
$[a = 13.83, b = 9.04, c = 7.65$ Å$, \beta = 96°, Z = 2,$
$P2_1/a$; group 3a]

The earlier determination of the crystal structure of pyrene:chloranil by Pepinsky,[220] the results of which were only briefly reported, has now been confirmed by Tickle and Prout.[221] There seems no reason to believe that the crystals are disordered. There is good agreement between the two sets of results (Figure 39) except that the overlap diagram given by Pepinsky is wrong (presumably owing to a computing error) and is superceded by the overlap diagram shown in Figure 40b. There is overlap of a carbonyl group of chloranil with one of the rings of the pyrene molecule in a manner similar to the situation found in perylene:fluoranil (Figure 40a). The interplanar distance of 3.45 Å indicates weak charge-transfer. Polarized absorption spectra of single crystals have been reported over a narrow frequency range between 15 and 17 kK (Chakrabarti and Basu[408]). Considerable fine structure was observed in the spectra, but the difference in absorption of radiation polarized parallel and perpendicular to the needle axis of the crystals was much smaller than for, say, perylene:fluoranil. As this result appears incompatible with the crystal structure of pyrene:chloranil, further study would be desirable. Similar spectroscopic measurements were made on anthracene:chloranil, anthracene:bromanil, anthracene:iodanil, and pyrene:bromanil; crystallographic information is not available for any of these molecular compounds.

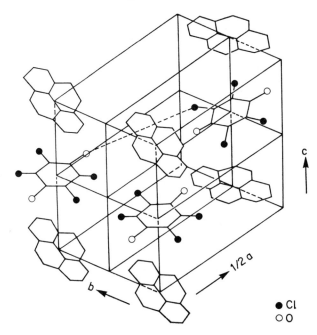

Figure 39. Pyrene:chloranil. Perspective view of the structure showing alternate stacking of the molecules. The shortest O–Cl distance (3.21 Å) is illustrated by a broken line. [Reproduced with permission from reference 220.]

### 4. HMB:chloranil
[$a = 15.26, b = 8.44, c = 7.30$ Å, $\beta = 103°$, $P2_1/a, Z = 2$; group 3a]

The structure of HMB:chloranil at room temperature has been determined. Further refinement (Jones and Marsh [409]) has shown that the distortions of the molecules claimed in the original determination (Harding and Wallwork [361]) are not significant (Wallwork; [23] Wallwork and Harding [410]). The crystals have the stack axis along [001]. The intensities of only 185 reflections served as basis for the structure analysis and rather high Debye–Waller factors were obtained for the peripheral atoms of the molecules (5–6 Å²). This might indicate high thermal motion, or disorder, or both, or perhaps only the poor quality of the crystals used for the measurements. On the basis of

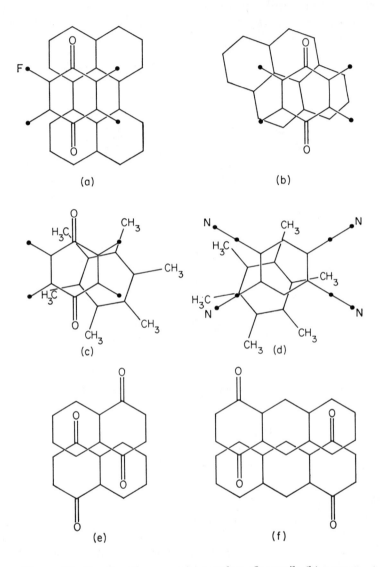

Figure 40. Overlap diagrams: (a) perylene:fluoranil; (b) pyrene: chloranil; (c) HMB:chloranil; (d) HMB:TCNB; (e) naphtho-quinone; (f) 1,4-anthraquinone.

NQR measurements[411] on HMB:chloranil at 77°K (no signals were obtained at room temperature) it has been suggested that less than 10% of an electronic charge was transferred from donor to acceptor; the chloranil molecules are reoriented more slowly than the measured line width of 1.9 kc and may indeed be stationary. On the other hand, NMR measurements[412] over the temperature range 150–400°K show that the HMB molecules are rapidly reoriented about their hexad axes, with simultaneous rotation of the methyl groups. Thus HMB:chloranil is an example of a molecular compound where the orientation of one component changes very much faster than that of the other. An abrupt change in the spin-lattice relaxation time at 45°c has been interpreted as a phase change, the molecular motion in both phases appearing to be very similar. No parallel diffraction study has yet been reported.

According to the diffraction studies the planes of the two components are almost parallel and separated by the relatively large distance of 3.51 Å. The overlap diagram (Figure 40c) shows considerable displacement of superposed molecules. The polarized absorption spectrum (Nakamoto[395]) shows a broad charge-transfer band polarized in the direction of the stack axis.

Prout and Wallwork[413] have pointed out that parallel orientation of aromatic rings and $C{=}O$ groups is found in many one-component molecular crystals and in molecular compounds such as perylene: fluoranil and quinhydrone and its analogues. The important factor appears to be the presence of polar (*e.g.*, carbonyl) and polarizable (*e.g.*, aromatic rings) groups in the same crystal, irrespective of whether these groups are in the same or different molecules, and Gaultier, Hauw, and Breton-Lacombe[414] have drawn attention to this type of overlap in one-component crystals; two examples (Figure 40e, f) show that the mutual arrangement is very similar to that in perylene:fluoranil.

The parallel orientation of aromatic rings and $C{=}O$ groups is fairly well-defined in pyrene:chloranil but does not appear at all in HMB:chloranil. Indeed the overlap diagram of the latter suggests that mutual orientation of the two components is a result of optimal mutual avoidance of bulky substituents.

## 5. HMB:TCNB
[$a = 14.90$, $b = 8.92$, $c = 7.41$ Å, $\beta = 104°$, $Z = 2$, $P2_1/a$; group 3a]

This crystal structure (at room temperature) has been determined by Niimura, Ohashi, and Saito[362] (Figure 41). The rather large values

of interplanar spacing found in HMB:TCNB and HMB:chloranil
(3.51 Å) may be ascribed to the thickness of the hexamethylbenzene
molecule (in hexamethylbenzene itself[415] the interplanar distance is
3.66 Å).

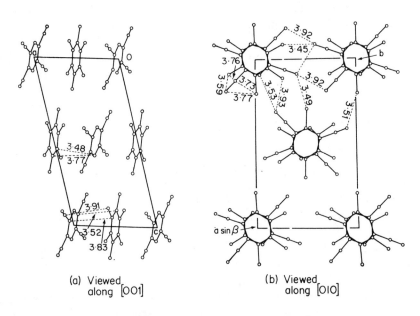

(a) Viewed along [001]

(b) Viewed along [010]

Figure 41. HMB:TCNB. Molecular arrangement. The molecular
planes are nearly parallel (angle 2°) and 3.54 Å apart. [Repro-
duced with permission from reference 362.]

The overlap diagram (Figure 40d) is again reminiscent of those
found for HMB:chloranil and pyrene:chloranil. The TCNB molecule
is displaced so that its (short) central axis is roughly over one of
the C–C bonds in HMB. It is not known whether there is an appreciable static contribution to the libration of the HMB molecule about
its six-fold axis indicated by the Debye–Waller factors.

Reference to Tables 7 and 2 shows that the ionization energies of
hexamethylbenzene and pyrene are respectively 8.0 and 7.6 ev, while
the electron affinities of chloranil and TCNB are respectively 1.4 and
0.4 ev. Thus the charge-transfer interaction between pairs of these
molecules would be expected to decrease in order pyrene:chloranil >
HMB:chloranil > pyrene:TCNB > HMB:TCNB.

## H. Crystal Structures of Quinhydrone and Related Molecular Compounds

**1. Comparison of similar structures.**—The available crystallographic information is given in Table 22 and some of the principal results in Table 23. We discuss the resemblances between the 1:1 and 2:1 molecular compounds before turning to the differences between them. Resorcinol:$p$-benzoquinone (stack axis $\sim 14$ Å) differs in some respects from the 1:1 molecular compounds with $\sim 7$ Å periodicity along the stack axis; it is discussed separately below and some of the following remarks do not apply to it.

The four 1:1 molecular compounds with $\sim 7$ Å stack axes have the same type of mixed stacks as are found in the analogous $\pi$-molecular compounds discussed above. The 2:1 molecular compounds have triads of two donor molecules sandwiching a central acceptor molecule.

Differences between the various molecular compounds arise from the different methods of arrangement of the mixed stacks or triads and, in particular, from the ways in which neighbouring stacks or triads can be hydrogen-bonded to one another. The close resemblance between the arrangements of the component molecules within stacks or triads is clearly shown by comparison of the overlap diagrams (Figure 42). Within a stack all molecules of a particular type (*e.g.*, $p$-benzoquinone) are necessarily parallel, but the two components are not required to be parallel to one another and small interplanar angles of 2–3° are found in the four equimolar compounds examined (Table 23). In the two quinhydrone polymorphs the average distances between the components in a stack must be equal; however, in ($p$-chlorophenol):$p$-benzoquinone the interplanar distances alternate, the values being 3.27 and 3.35 Å. This difference is too small to be ascribed to any appreciable difference in intercomponent bonding and is probably due to a difference in the relative dispositions of $p$-benzoquinone molecules and halogen atoms on the two sides of a $p$-chlorophenol molecule, as suggested by Shipley and Wallwork.[419] In the three isomorphous 2:1 compounds the two phenol molecules astride a $p$-benzoquinone molecule in a triad are related by a centre of symmetry and are thus necessarily mutually parallel and equidistant from the central benzoquinone molecule. There are similar angles between component planes in phenoquinone and in $(p\text{-ClC}_6\text{H}_4\text{OH})_2:\text{C}_6\text{H}_4\text{O}_2$, the larger value in the latter perhaps being due to the bulky halogen atom.

Table 22. Crystallographic results for quinhydrone and related molecular compounds.[a] ($V_{mc}$ = volume in molecular compound; $V_{add}$ = sum of component volumes.)

| Molecular compound | $a$ (Å) / $\alpha$ | $b$ (Å) / $\beta$ | $c$ (Å) / $\gamma$ | $Z$ | Space group | Volume of formula unit | | Ref. |
|---|---|---|---|---|---|---|---|---|
| | | | | | | $V_{mc}$ (Å³) | $V_{add}$ (Å³) | |
| *Group A* | | | | | | | | |
| β-[p-C₆H₄(OH)₂:C₆H₄O₂] (β-quinhydrone) | 7.65 / 107½° | 5.96 / 122° | 6.77 / 90° | 1 | $P\bar{1}$ | 244 | 268 | 416 |
| α-[p-C₆H₄(OH)₂:C₆H₄O₂] (α-quinhydrone) | 7.65 / — | 6.00 / 110° | 11.59 / — | 2 | $P2_1/c$ | 250 | 268 | 337,417 |
| m-C₆H₄(OH)₂:C₆H₄O₂ | 14.63 / — | 5.976 / — | 11.529 / — | 4 | $Pnca$ | 252 | 273 | 328 |
| p-C₆H₄(OH)₂:C₁₀H₆O₂[b,c] | 7.70 / — | 6.16 / — | 27.4 / — | 4 | $P2_12_12_1$ | 325 | — | 418 |
| *Group B* | | | | | | | | |
| p-ClC₆H₄OH:C₆H₄O₂ | 7.90 / 107° | 12.25 / 58° | 6.80 / 107° | 2 | $P\bar{1}$ | 263 | 285 | 419 |
| p-BrC₆H₄OH:C₆H₄O₂ | 7.86 / 104° | 12.49 / 59 | 6.84 / 106° | 2 | $P\bar{1}$ | 274 | — | 419 |
| *Group C* | | | | | | | | |
| (C₆H₅OH)₂:C₆H₄O₂ (phenoquinone) | 11.152 / — | 5.970 / 100.0° | 11.499 / — | 2 | $P2_1/c$ | 377 | 400.5 | 336,337 |
| (p-CH₃C₆H₄OH)₂:C₆H₄O₂[d] | 11.83 / — | 6.2 / 97° | 24.6 / — | 4 | $C_{2h}^5$ | 449 | — | 418 |

| | | | | | | | | |
|---|---|---|---|---|---|---|---|---|
| $(p\text{-ClC}_6\text{H}_4\text{OH})_2\text{:}\,\text{C}_6\text{H}_4\text{O}_2$ | 11.70 | 6.10 | 11.83 | 2 | $P2_1/c$ | 419 | 434 | 420 |
| | — | 97° | — | | | | | |
| $(p\text{-BrC}_6\text{H}_4\text{OH})_2\text{:}\,\text{C}_6\text{H}_4\text{O}_2$ | 11.91 | 6.16 | 12.03 | 2 | $P2_1/c$ | 439 | — | 420 |
| | — | 96° | — | | | | | |
| *Group* D | | | | | | | | |
| $[1,3,5\text{-C}_6\text{H}_3(\text{OH})_3]_2\text{:}\,\text{C}_6\text{H}_4\text{O}_2{}^{b}$ | 9.72 | 10.59 | 8.43 | 2 | $P\bar{1}$ | 427 | — | 421 |
| | 88.7° | 101.7° | 69.8° | | | | | |

[a] Where necessary unit cells given by the original authors have been re-oriented so that the stack axis of all unit cells are along [100].

[b] Crystal structure has not been reported.

[c] $\text{C}_{10}\text{H}_6\text{O}_2 \equiv$ 1,4-naphthoquinone; $\text{C}_6\text{H}_4\text{O}_2 \equiv$ *p*-benzoquinone.

[d] The crystal structure has not been determined but the compound is presumably isomorphous with two following molecular compounds. The *c*-axis should probably be half the value given.

Table 23. Information about the molecular packing and molecular planes in quinhydrone and related molecular compounds.

| Crystal[a] | $\dfrac{\Delta V}{A_{add}}$ (%) | Angle between molecular planes (deg.) | Approx. perpendicular distance between molecular planes (Å) | Shortest interatomic distances C..C (Å) | C..O (Å) | Distance between ring centres (Å) |
|---|---|---|---|---|---|---|
| β-Quinhydrone | −9.0 | 3.5 | 3.22 | 3.17 | 3.25 | 3.82 |
| α-Quinhydrone | −6.7 | 2.0 | 3.19 | 3.18 | 3.26 | 3.83 |
| $m\text{-}C_6H_4(OH)_2:C_6H_4O_2$ | −7.7 | 7.3 | 3.1 | 3.21 | 3.07 | 3.66 |
| $p\text{-}ClC_6H_4OH:C_6H_4O_2$ | −7.7 | 3 | 3.27 3.35 | 3.31 | 3.24 | 3.99 |
| Phenoquinone | −5.9 | 4.5 | 3.13 | 3.16 | 3.25 | — |
| $(p\text{-}ClC_6H_4OH)_2:C_6H_4O_2$ | −3.6 | 6 | 3.20 | — | — | — |
| Thymine:$p$-benzoquinone | −3.1 | — | 3.16 | 3.49 | 3.17 | — |

$^a$ $C_6H_4O_2 \equiv p$-benzoquinone.

We now consider the packing of the molecular stacks in the crystals. The most obvious qualitative feature is the reduction in volume per formula unit in the molecular compound compared with the sum of the volumes ($V_{add}$) of the components.* The values of $\Delta V/V_{add}$

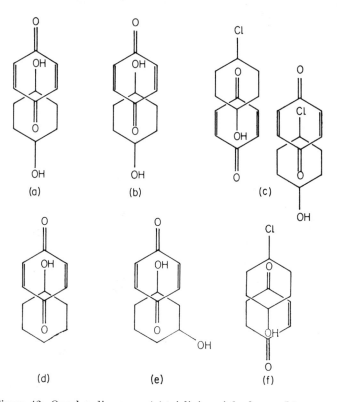

Figure 42. Overlap diagrams: (a) triclinic quinhydrone; (b) monoclinic quinhydrone; (c) ($p$-ClC$_6$H$_4$OH):(C$_6$H$_4$O$_2$), (i) $p$-chlorophenol on to the $p$-benzoquinone molecule below (smaller $y$-coordinate) and (ii) $p$-benzoquinone on to the $p$-chlorophenol below; (d) phenoquinone; (e) resorcinol:$p$-benzoquinone; (f) ($p$-ClC$_6$H$_4$OH)$_2$: (C$_6$H$_4$O$_2$). [Reproduced with permission from references 337, 416, 419, and 420.]

* Some discretion must be exercised in making this comparison as $p$-benzoquinone, hydroquinone, and $p$-bromophenol appear to have "normal" densities whereas those of phenol and $p$-chlorophenol are lower than values expected on the basis of comparison with similar compounds. Similar discretion will probably be needed in the discussion of free energies, etc., of formation of the crystalline molecular compounds from their crystalline components; at present such thermodynamic values are available only for quinhydrone.

(Table 23) are greater than in most of the compounds considered earlier (Table 20, p. 253). The most densely packed of the present group are the two quinhydrone polymorphs (Group A of Table 22). In both polymorphs the needle axis is along [100], the direction of the charge-transfer interaction. In the triclinic polymorph the mixed stacks are hydrogen-bonded by OH $\cdots$ O hydrogen bonds in the [120]

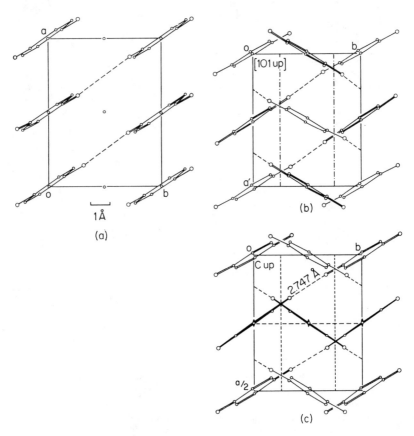

Figure 43. (a) Triclinic quinhydrone in projection down [001]. (b) Monoclinic quinhydrone in projection down [101]. (c) Resorcinol:*p*-benzoquinone in projection down [001]. (d) See p. 317.

direction. The combination of charge-transfer interaction along [100] and hydrogen-bonding along [120] gives a molecular sheet, one molecule thick, parallel to (001). In the triclinic polymorph all the molecular sheets are identical, whereas in the monoclinic polymorph the direction of the hydrogen bonds in successive sheets changes from [120] to [1$\bar{2}$0] as a result of the operation of the $c$ glide plane (Figure 43). van der Waals forces bond the sheets together. Comparison of the experimental densities shows that the triclinic form is the more

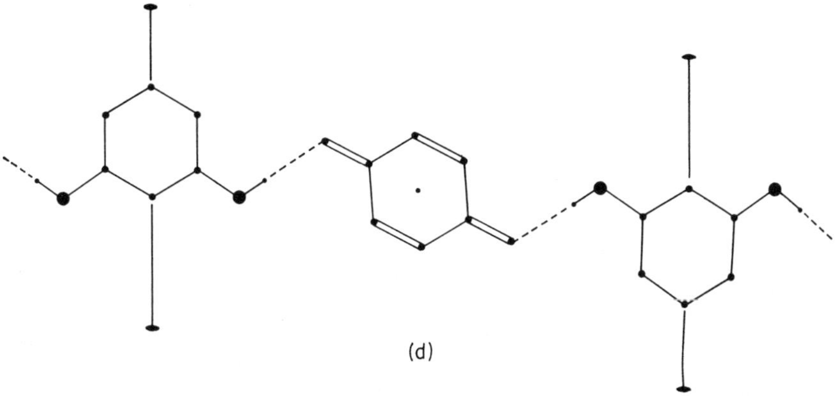

(d)

Figure 43. (d) Molecular chain in resorcinol:$p$-benzoquinone viewed down the normal to the molecular plane (cf. Figure 45c). [Reproduced with permission from references 328, 337, and 416.]

stable under ordinary conditions of temperature and pressure. A similar type of polymorphism has been found in hexadecanamide (Sakurai and Yabe[422]) and in a number of other crystals.

In $p$-chlorophenol:$p$-benzoquinone (group B of Table 23) pairs of mixed stacks are linked by hydrogen bonds into bands one unit cell wide (Figure 44). The 2:1 molecular compounds (group C of Table 23) have bands of molecules indefinitely extended along [010], the needle axis (Figure 45). Within a band a triad of D– – –A– – – D molecules is linked by charge-transfer forces operating at about 60° to the molecular plane, while a different triad of molecules is linked by hydrogen bonds in the molecular plane.

The stacks in resorcinol:$p$-benzoquinone and anthracene:TNB (Section VI-E) are formally very similar. Both have periodicities of about 14 Å, both have one component at crystallographic symmetry centres and the other on two-fold axes, and the resemblance extends to the relative orientations and tilts of the two components. The arrangement of the stacks in the two crystals differs; in anthracene:TNB all stacks are identical, whereas in resorcinol:$p$-benzo-quinone their orientations alternate along [011]. Although the structures of the stacks in resorcinol:$p$-benzoquinone and monoclinic

quinhydrone differ, there is a striking resemblance in the arrangement of the stacks in criss-crossed sheets, as shown in Figures 43b and
c. Polymorphism of resorcinol:$p$-benzoquinone similar to that in
quinhydrone could be expected but has not been reported. The
geometries of the hydrogen-bond systems in $\alpha$- and $\beta$-quinhydrone

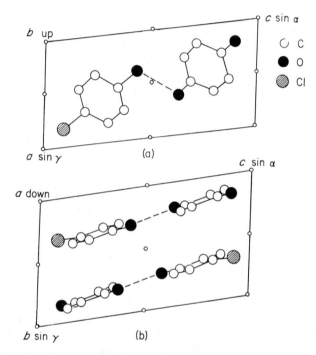

Figure 44. $p$-Chlorophenol:$p$-benzoquinone. Projections of the
crystal structure (a) down [010], (b) down [100]. In (a) a superimposed benzoquinone molecule has been omitted on the left-hand
side of the diagram and a superimposed $p$-chlorophenol molecule
on the right-hand side of the diagram. Hydrogen bonds between
the components are shown by a dashed line.

in resorcinol:$p$-benzoquinone, and in phenoquinone are very similar.
The structural similarities described above provide an interesting
example of the thrift with which the available options are used in
the construction of these crystals.

Sakurai[337,416] has attempted to estimate, from measured values
of certain structure factors, the amount of charge transferred from
hydroquinone to $p$-benzoquinone in $\alpha$- and $\beta$-quinhydrone. It was

found that 0.71 e and 0.21 e were transferred in the $\alpha$- and the $\beta$-polymorph, respectively. These values are considerably higher than estimates from charge-transfer spectra, and there seems to be no obvious reason why the two polymorphs, so similar in other respects, should differ so much in this respect. Thus, interesting as the

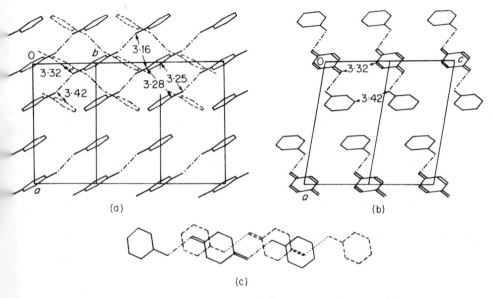

(a)  (b)

(c)

Figure 45. Phenoquinone: (a) Projection along [001]. Note the resemblance between [001] projection of phenoquinone and [101] projection of monoclinic quinhydrone. (b) Projection along [010]. (c) Arrangement of the molecular chain in phenoquinone, viewed down the normal to the molecular plane. [Reproduced by permission from reference 337.]

potentialities of the method are, the two estimated values for the quinhydrone polymorphs do not seem particularly reliable.

Further refinement has been carried out[423] for $\alpha$- and $\beta$-quinhydrone, phenoquinone, and resorcinol:$p$-benzoquinone, using aspherical atomic scattering factors for oxygen atoms. The minor changes obtained in the temperature factors are of dubious significance because of the inherent inaccuracies of the photographic intensity measurements.

The enthalpies of solution of quinone, hydroquinone, and (presumably triclinic) quinhydrone in water and acetone (Suzuki and

Seki[424]) together with some derived results, are summarized in Table 24. The three enthalpies of solution allow derivation of the enthalpy of formation of quinhydrone at 297°K, the relevant reaction being:

$$Q_{cr} + Hq_{cr} \rightleftharpoons (QHq)_{cr}$$

(Q = benzoquinone, Hq = hydroquinone, QHq = quinhydrone)

The value of $\Delta H_f$ at 0°K was estimated from the specific-heat data of Lange[371] and Schreiner,[372] and $\Delta S_f$ by assuming the validity of the third law of thermodynamics. The values at 297°K (at 0°K in paren-

Table 24. Enthalpies of sublimation, solution, and solvation for quinhydrone and its components. All values in kcal/mole at about 20°c.

| Substance | $\Delta H_{subl.}{}^a$ | $\Delta H_{sol.}$ Water[371] | $\Delta H_{sol.}$ Acetone[424] | $\Delta H_{solv.}{}^c$ Water | $\Delta H_{solv.}{}^c$ Acetone |
|---|---|---|---|---|---|
| Quinone | 16.4 ± 0.1 | 4.7 | 4.11 ± 0.02 | 11.7 | 12.3 |
| Hydroquinone | 21.5 ± 0.2 | 4.5 ± 0.1 | −0.37 ± 0.01 | 17.0 | 21.9 |
| Quinhydrone | 43.4 ± 0.4[b] | — | 9.15 ± 0.01 | — | — |

[a] Measured directly.[425]

[b] This value cannot be measured directly because of differential evaporation of the components. It was therefore obtained indirectly by summing the enthalpies of sublimation of the components and the enthalpy of formation of quinhydrone.

[c] Obtained from $\Delta H_{sol.}^{\infty} = \Delta H_{solv.}^{\infty} - \Delta H_{subl.}$, the approximation being made that the measured enthalpies of solution were equal to the heats of solution at infinite dilution.

theses) are: $\Delta G_f$ − 3.67 (− 5.26) kcal/mole, $\Delta H_f$ − 5.39 (− 5.26 kcal/ mole, and $\Delta S_f$ − 6.28 (0) e.u. Their general trend is similar to the results found for anthracene:picric acid, *i.e.*, the components are more strongly bound in the molecular compound than in their own crystals with a concomitant lowering of entropy, presumably due to a reduction in the degree of thermal vibration. The stability of quinhydrone relative to its components decreases with increasing temperature. Monoclinic quinhydrone is less stable than triclinic quinhydrone at room temperature, but it is not known whether the two polymorphs are enantiotropically or monotropically related. However, the existence of two polymorphs is in accordance with the thermodynamic properties of triclinic quinhydrone.

Three separate contributions to the bonding between the components in quinhydrone can be distinguished: (1) dispersion interactions, (2) hydrogen bonding, and (3) donor–acceptor (or dipole–induced dipole) interactions. Rough estimates of these three contributions to $\Delta H_f$ (quinhydrone) were made by Suzuki and Seki (see their paper[424] for details), but further work is needed. One possible experimental method would be to extend the measurements of heats of solution to other suitable substances among those listed in Table 22.

Specific heats for hydroquinone, benzoquinone, and quinhydrone, measured by Lange[371] and interpreted by Schreiner,[372] give no indication of polymorphic or other transformation in any of these crystalline materials in the temperature range 20–300°K; however, the measurements were rather widely spaced in temperature, and repetition by modern techniques would be essential before any conclusions could be drawn.

Polarized absorption (Nakamoto[395]) and reflection (Anex and Parkhurst,[426] Figure 46a) spectra have been measured from small single crystals of quinhydrone (the monoclinic polymorph was used by Anex and Parkhurst according to their diagram, Figure 46b). Nakamoto interpreted his results in terms of enhanced absorption, due to charge transfer, when the radiation was polarized normal to the ring planes of the two component molecules, but Anex and Parkhurst claim that the appropriate polarization direction is along the stack axis and not along the ring normals (see Figure 46b), a result verified by both techniques used by them. They point out that Mulliken's theory in fact suggests that the polarization vector for the charge-transfer band should lie along the line joining the centres of the components.

Polarized single-crystal spectra of resorcinol:$p$-benzoquinone have been studied by Amano (quoted by Ito *et al.*[328]). Intense charge-transfer bands were observed with $a$-polarized light, while the crystal was almost transparent to $b$-polarized light. This is compatible with the crystal structure provided that the charge-transfer moment is along the stack axis, as in quinhydrone.

### 2. Thymine:$p$-benzoquinone[197]
[$a$ = 12.43, $b$ = 6.86, $c$ = 12.64 Å, $Z$ = 4, *Pbmn*; see Table 18]

The crystal structure of this complex at room temperature has been determined by Sakurai.[346] The volume per formula unit is 3.1% less than the sum of the molecular volumes of the components (Table 23). The thymine molecules lie in the mirror planes of the space group at $c = \frac{1}{4}, \frac{3}{4}$, and the benzoquinone molecules are sandwiched between these thymine layers and are at centres of symmetry (Figure 47). The thymine molecules are connected by hydrogen

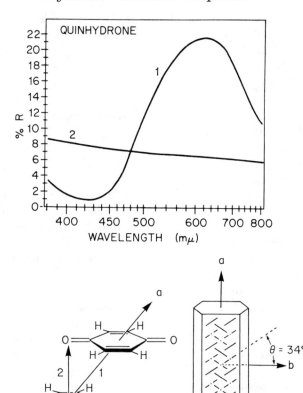

Figure 46. (a) Quinhydrone:reflection spectra obtained on the (001) face. (1) Light polarized with its electric vector parallel to *a*. (2) Light polarized with its electric vector parallel to *b*. The abscissa is linear in energy. (b) Comparison of Anex–Parkhurst (1) and Nakamoto (2) polarization assignments for the charge-transfer band in crystalline quinhydrone relative to a pair of typical molecules selected from one of the molecular stacks in the crystals. Right: Typical morphology of, and molecular orientation in, a quinhydrone crystal: –·–, benzoquine; ——, hydroquinone. The centre stack is displaced one half unit-cell length along *c* from the two outer stacks. [Reprinted from *Journal of the American Chemical Society*, **85**, 1963, p. 3302. Copyright 1963 by the American Chemical Society. Reprinted by permission of the copyright owner.]

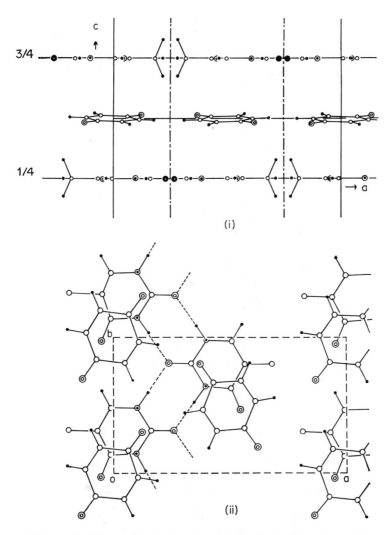

Figure 47. Thymine:*p*-benzoquinone: (i) Projection of crystal structure down [010]. (ii) Projection of crystal structure down [001] (there is similar overlap of the individual projected component molecules, but it should be remembered that the benzoquinone molecules at 0,0,0 and $\frac{1}{2},\frac{1}{2},0$ are tilted in opposite senses to the (001) plane]. The overlap diagram here represents the projection of *p*-benzoquinone on to the plane of the thymine molecule. [Reproduced with permission from reference 346.]

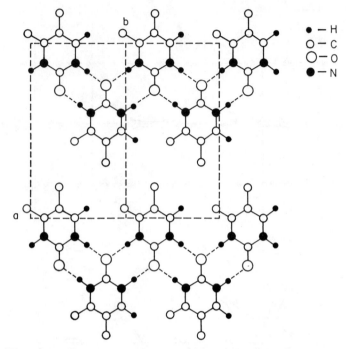

Figure 48. The arrangement of thymine molecules at $z = \frac{1}{4}$ in thymine:$p$-benzoquinone (redrawn from Figure 47b).

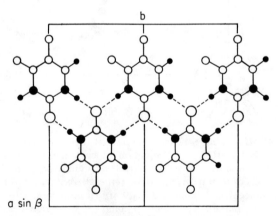

Figure 49. The arrangement of thymine molecules about $z = \frac{1}{4}$ in anhydrous thymine.[427]

bonds to form ribbons along [010] (cf. thymine [427]) (Figures 48 and 49) (cf. Section V-E). However, in thymine the ribbons are buckled, whereas they are quite flat in the molecular compound. The spacing between the two planar components is 3.16 Å, significantly less than the 3.25 Å interplanar separation found in thymine monohydrate [428] (in thymine itself, molecules in adjacent layers overlap only very slightly). The benzoquinone molecules are tilted at about 10° to the planes of the thymine molecules, with a mutual overlap of carbonyl groups and ring systems of the two components, similar to that found in quinhydrone and analogous molecular compounds (Figure 42).

## I. The Copper Oxinate: Chloranil Group

Bis-(8-quinolinolatocopper)(II) (copper oxinate), palladium(II) oxinate, and 8-quinolinol itself form molecular compounds with a number of electron-acceptors. These molecular compounds fall into three structural groups; most information is available about palladium oxinate:chloranil and similar molecular compounds (Table 20, group 6c) and some resemblance to the quinhydrones can be discerned. Only one example of the second group has been studied, namely, copper oxinate:(picryl azide)$_2$, which appears to be a typical charge-transfer compound similar to many others formed by TNB and its derivatives. It is discussed together with the other TNB molecular compounds (Section VI-E). Only one example of the third group is known, namely, copper oxinate:TCNQ, which is discussed in Section VI-C.

### 1. (8-Quinolinol)$_2$: chloranil (A) and palladium oxinate: chloranil (B)

[A: $a = 7.88$, $b = 8.03$, $c = 11.30$ Å, $\alpha = 129.9$,
$\beta = 116.7$, $\gamma = 68.2°$, $Z = 1$, $P\bar{1}$; group 6c.]
[B: $a = 8.17$, $b = 8.18$, $c = 9.69$ Å, $\alpha = 99.5$,
$\beta = 77.8$, $\gamma = 66.0°$, $Z = 1$, $P\bar{1}$; group 6c.]

These two molecular compounds have rather similar crystal structures (Kamenar, Prout, and Wright; [368] Prout and Wheeler [369]) and are conveniently discussed together. Both crystallize in the triclinic system and have a quasi-hexagonal arrangement of mixed stacks, the stack axes being along [100]. We have already noted (p. 259) how well the palladium oxinate:chloranil crystals diffract X-rays. The component molecules have their centres at crystallographic centres of symmetry; all stacks are crystallographically

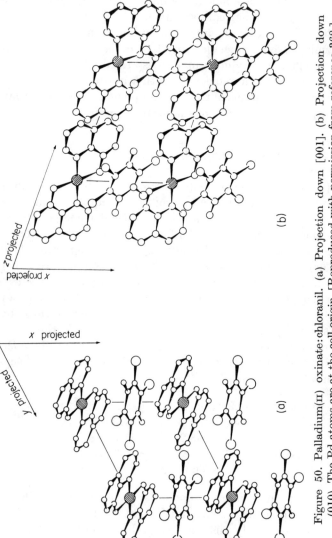

Figure 50. Palladium(II) oxinate:chloranil. (a) Projection down [001]. (b) Projection down (010). The Pd atoms are at the cell origin. [Reproduced with permission from reference 368.]

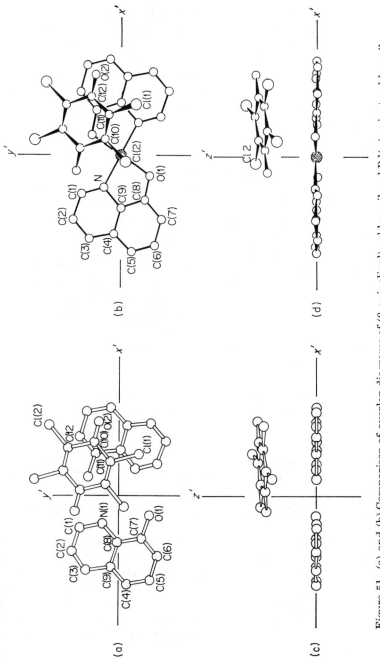

Figure 51. (a) and (b) Comparison of overlap diagrams of (8-quinolinol)$_2$:chloranil and Pd($\mathrm{II}$) oxinate:chloranil. The best plane goes through the donor molecule in both instances; the inter planar angles are $9\frac{1}{2}°$ and $15°$ respectively. Views in perpendicular directions are given in (c) and (d). [Reproduced with permission from references 368 and 369.]

identical but, because of the non-orthogonality of the triclinic cells, there is a mutual displacement of surrounding stacks with respect to the reference stack at their centre. This mutual displacement is more marked for palladium oxinate:chloranil than for (8-quinolinol)$_2$:chloranil. There is no evidence of disorder in either of these crystals. The diagrams of the palladium oxinate:chloranil structure (Figure 50) give a general idea of the molecular arrangement in the crystals of this group.

The overlap diagrams (Figures 51a, b) show a striking resemblance to the arrangement found in quinhydrone (Figure 42). The actual region of overlap is, of course, more comparable with that in phenoquinone. However, both Pd oxinate and the hydrogen-bonded dimer of 8-quinolinol provide two hydroxyl groups per chloranil molecule, and so the infinite stacks characteristic of quinhydrone rather than the triads characteristic of phenoquinone are obtained. Tilting of the chloranil molecule towards the C–O(H) region of the quinoline group is found in both molecular compounds (Figure 51); in addition, in (8-quinolinol)$_2$:chloranil the distances between one C=O group of chloranil and one C–OH region of quinoline are in the range 2.96–3.2 Å. Both these features suggest that some localized interaction occurs between the components. The metal atom may also play a role here, but it can hardly be a decisive one, otherwise there would not be a close resemblance between the two structures.

### 2. Palladium oxinate:tetracyanobenzene

$[a = 8.73, b = 7.97, c = 9.38$ Å, $\alpha = 109.6, \beta = 72.8,$
$\gamma = 107.6°, Z = 1, P\bar{1}$; group 6c.$]$

This molecular compound crystallizes in dark red triclinic needles which give diffraction patterns of excellent quality. The crystal structure at room temperature was determined by Kamenar, Prout, and Wright;[370] the usual mixed stacks are found, there is no disorder, and the general arrangement is very similar to that in the two chloranil molecular compounds discussed in the previous Section. The angle between the planes of the quinoline and the TCNB ring systems is about $4\frac{1}{2}°$, the TCNB molecules being tilted towards the C–O regions of the quinoline moieties, as in the other molecular compounds. The overlap diagram (Figure 52) does not show any particularly symmetrical mutual disposition of the two components, and the existence of at least partially localized interactions is indicated.

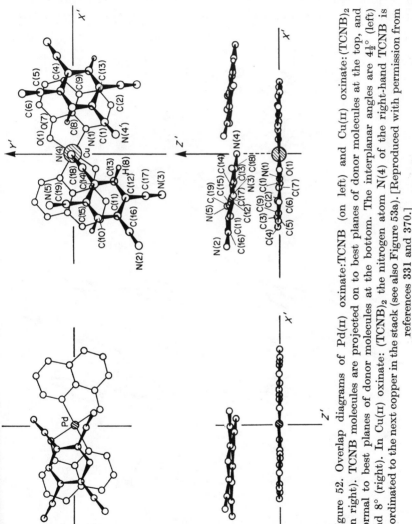

Figure 52. Overlap diagrams of Pd(II) oxinate:TCNB (on left) and Cu(II) oxinate:(TCNB)₂ (on right). TCNB molecules are projected on to best planes of donor molecules at the top, and normal to best planes of donor molecules at the bottom. The interplanar angles are $4\frac{1}{2}°$ (left) and $8°$ (right). In Cu(II) oxinate: (TCNB)₂ the nitrogen atom N(4) of the right-hand TCNB is coordinated to the next copper in the stack (see also Figure 53a). [Reproduced with permission from references 331 and 370.]

### 3. Copper oxinate (TCNB)$_2$

$$[a = 7.37, b = 7.70, c = 14.29 \text{ Å}, \alpha = 98.9, \beta = 96.3,$$
$$\gamma = 97.3°, Z = 1, P\bar{1}; \text{ group 6c}]$$

This crystal structure (Murray-Rust and Wright[331]) introduces an important new feature. The familiar quasi-hexagonal arrangement of mixed stacks is found (Figure 53), but now two molecules are

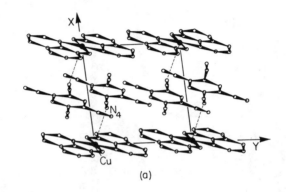

(a)

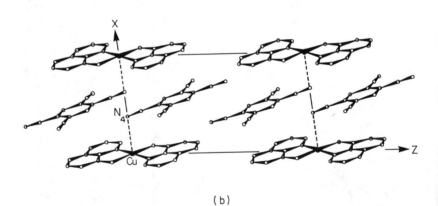

(b)

Figure 53. Copper(II) oxinate:(TCNB)$_2$. (a) Projection down $z$. (b) Projection down $y$. [Reproduced with permission from reference 331.]

interleaved between successive copper oxinate molecules. The crystals are triclinic, and the two TCNB molecules are related by different centres of symmetry from those occupied by the copper atoms; thus the two TCNB molecules take up different dispositions with respect to a given copper oxinate molecule* (Figure 53). The overlap diagram (Figure 52) shows that the arrangement of one TCNB molecule with respect to the quinoline ring system is similar to that found with other donors containing "naphthalene" regions (Section VI-E). The interaction between this TCNB and the copper oxinate molecule is presumably of the charge-transfer type. The second TCNB molecule has a nitrogen atom directly above the copper atom and only 2.95 Å away from it, indicative of a localized interaction. Thus an alternative description would be in terms of the interactions between a given TCNB molecule and its copper oxinate neighbours above and below it in the same stack; in one direction there will be a delocalized charge-transfer interaction and in the opposite direction a localized interaction with the metal atom (see the lower right-hand side of Figure 52).

### 4. Copper oxinate:benzotrifuroxan (BTF)
$[a = 9.28, b = 14.17, c = 9.11, \beta = 104.3°, Z = 2, P2_1/c]$

Copper oxinate gives monoclinic brown crystals of an equimolar molecular compound with BTF; there are mixed stacks along [100] (Prout and Powell[429]). Despite the differences in space group, the structural arrangement is essentially the same as that found in palladium oxinate:chloranil, except that the BTF molecules are disordered. As indicated in the overlap diagram (Figure 54), the positions of the carbon and nitrogen atoms of the BTF molecule could be defined reasonably well, but the oxygen positions remain uncertain. The angle between the planes of copper oxinate and BTF molecules is 10°, and the BTF molecules are tilted towards the centre of the copper oxinate molecule. As before, some localized interaction seems probable.

Palladium oxinate:BTF was studied by the same authors; at least three different types of crystal were found and all showed diffraction patterns characteristic of disordered structures.

### 5. 13,14-Dithiatricyclo[8.2.1.1⁴,⁷]tetradeca-4,6,10,12-tetraene: benzotrifuroxan.
$[a = 9.71, b = 8.01, c = 15.28$ Å, $\alpha = 102.2,$
$\beta = 96.2, \gamma = 117.9°, Z = 2, P\bar{1}]$

The triclinic crystals of the tetraene (88, p. 229) with BTF gave

---

* In the analogous copper oxinate:(picryl azide)$_2$ (see Section VI-E) the crystals are monoclinic and the two picryl azide molecules interleaved between copper oxinate molecules are related by a two-fold axis which is also the site group-symmetry element of the copper oxinate. Thus the two picryl azide molecules have the same disposition with respect to the copper oxinate molecules.

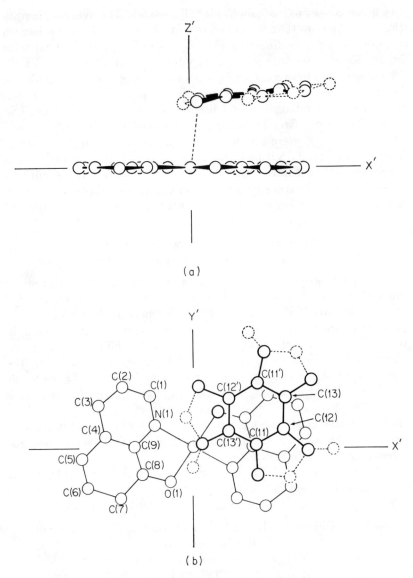

Figure 54. Copper(II) oxinate:BTF. In (a) and (b), respectively, the BTF molecule is projected on to, and perpendicular to, the plane of the copper(II) oxinate molecule. In all cases the oxygen atom sites (dotted lines) must be regarded as undetermined. [Reproduced with permission from reference 429.]

rather poor diffraction patterns.[283] The structure is similar to that found for palladium oxinate:chloranil; the BTF molecule is not disordered. The angle between thiophene rings and BTF molecules is about 10°, leading to intermolecular contacts of 3.2 Å between carbon atoms; the sulphur atoms are not involved in particularly close contacts. Presumably the main bonding within the stacks results from donor–acceptor interactions, but other factors may play a role as there is no clear orientation relation between the two components (overlap diagram in Figure 55).

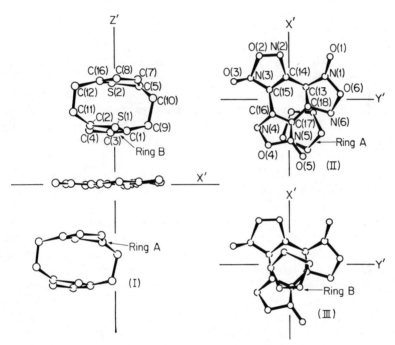

Figure 55. DDDT:BTF. Overlap diagram on the right and view normal to BTF mean plane on the left. [Reproduced with permission from reference 283.]

## J. Molecular Compounds between Aromatic Amines and Chloranil

Only *p*-phenylenediamine:chloranil (PDC) and TMPD:chloranil have been studied in detail. Full crystallographic information is lacking for PDC, but a rather detailed ESR study has been carried out (Hughes and Soos[58]); TMPD:chloranil has been studied by X-ray diffraction at room temperature as well as by ESR (over a range of temperatures) (Pott;[59] Pott and Kommandeur[60]). The explanation of their ESR results by the latter authors on the basis of a molionic lattice appears to be wrong, but some useful conclusions can be drawn by comparison of the results for PDC and TMPD: chloranil.

Some spectroscopic and ESR studies of other aromatic amine: chloranil molecular compounds are briefly summarized below.

**1. Phenylenediamine:chloranil.**—Cell dimensions have been reported for $p$-phenylenediamine:chloranil by Pepinsky[220] ($a = 15.40$, $b = 26.80$, $c = 6.92$ Å, $\beta = 108°$, $P2_1/n$). Crystal composition and density were not given but assumption of a 1:1 composition and 8 formula units per cell leads to a volume per formula unit of 340 Å³, compared with 354 Å³ for the sum of the molecular volumes of the components. However, this monoclinic material appears to be different from the crystals that were the subject of Hughes and Soos' detailed ESR study, for the latter authors state[58] that PDC has space group $P\bar{3}ml$ or $P3ml$, with the three-fold axis parallel to the needle axis of the crystal. Cell dimensions were not reported but the ESR study shows that the unit cell contains three magnetically non-equivalent stacks,* with the normals to the planes of the diamine and chloranil molecules inclined at about 6° to the needle axis.

Single-crystal optical absorption spectra (Amano *et al.*[430]) show that PDC has an ionic ground state, although band overlapping prevents as detailed an analysis of the spectra as is possible for TMPD:chloranil (see below). Infrared spectra (Matsunaga[285]) also indicate an ionic ground state. The most detailed description of PDC comes from the ESR study of Hughes and Soos; their basic assumption that donor and acceptor portions alternate face to face in mixed stacks is fully substantiated by their ESR results. Crystals were grown by vacuum-sublimation and were equimolar in composition [compositions of 2:3 and 5:3 have been reported by earlier workers for crystals grown from solution; these may, of course, be authentic compositions (see Section IV-H)].

The curve of $\chi T$ against $1/T$, shown in Figure 56, consists of two portions, corresponding to activation energies of $0.13 \pm 0.01$ ev and $0.015 \pm 0.005$ ev. The low-temperature paramagnetism is dominated by stacks with an odd number of radicals; stack lengths of a

---

* We use the term "stack," also used elsewhere in this Review, instead of "chain" used by Hughes and Soos.

few hundred radicals are estimated, with stack termination ascribed to crystal imperfections. The results for $T > 200°$K are consistent with the presence of either Wannier* or Frenkel* spin excitons; measurements at high temperatures would allow a distinction to be made but these are not practicable because irreversible changes occur in the crystals at about 350°K.

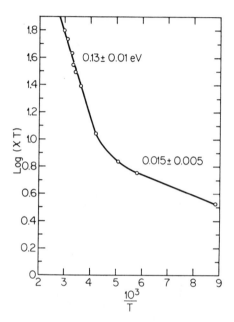

Figure 56. Plot, for PDC, of log of the product of relative intensity and temperature against reciprocal temperature. [Reproduced with permission from reference 58.]

The line-width and its dependence on temperature provide an unequivocal means of demonstrating that Wannier, and not Frenkel, spin excitons are present in PDC. Indeed, similar dependence of line width on temperature is shown by PDC, TMPD:chloranil, and TMPD:TCNQ, showing all three to be Wannier spin-exciton systems. The behaviour of Frenkel spin-exciton systems, such as TCNQ salts (Chesnut and Phillips[431]) or TMPD perchlorate (Thomas, Keller, and McConnell[432]) is quite different (see also Jones and Chesnut[433]).

The variation of ESR intensity (proportional to excitation density)

---

* In Wannier spin-exciton systems the two spins move independently, while in Frenkel spin-exciton systems the two spins are strongly coupled (see Section IV-C).

and line width with pressure (up to 8.5 kbar) were measured at a number of temperatures. The intensity decreases with pressure, as expected for Wannier spin excitons (Figure 57; see Section IV-C), and the changes in line width are also compatible with this conclusion. There is no indication of first-order phase changes in the temperature–pressure range covered in these experiments.

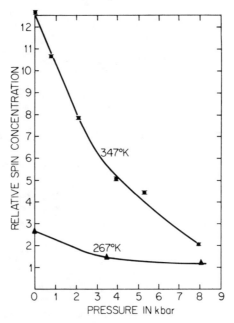

Figure 57. Plots, for PDC, of the relative spin concentration (*i.e.*, the relative excitation density $\rho$) against pressure (kbar) at 267°ᴋ and 347°ᴋ. [Reproduced with permission from reference 58.]

A single, almost axially symmetrical $g$ factor is found for the ion-radical stacks, with $g_\parallel = 2.0024 \pm 0.0002$ and $g_\perp = 2.0054 \pm 0.0002$. This is what would be expected from excitations delocalized over both types of radical. In some crystals "impurity" lines were found at low temperature ($< 200°\mathrm{K}$) with $g$ factors equal to those of (chloranil)⁻, and intensities showing a Curie-type $1/T$ dependence. These lines were ascribed to isolated chloranil anions trapped in the lattice at crystal imperfections, but with almost the same orientation as the normal chloranil anions in the mixed stacks.

A maximum of three, and a minimum of one, line appear in the ESR spectra of single crystals of PDC, depending on the orientation of the crystal with respect to the applied static field. The splittings are also temperature-dependent, and only a single line is found above 315°K, irrespective of crystal orientation. The splittings found at lower temperatures can be accounted for by postulating the presence of three magnetically non-equivalent stacks, related by rotation of $2/3\pi$ about the needle (trigonal) axis. The magnetic inequivalence results from a tilt of $\xi = 6.0 \pm 0.5°$ between the trigonal axis and the normals to the planes of the diamine and chloranil radical ions.

The splittings collapse as the temperature is raised to 315°K; two possible explanations are offered by Hughes and Soos.[58] The first is that a slow phase transition occurs over a temperature range of about 60°, it being suggested that $\xi$ falls smoothly to zero over the temperature interval 250–315°K but that the stacks remain independent. In support of this it is noted that hysteresis effects occur above 300°K, suggesting occurrence of a phase change. There is no evidence for a first-order phase change, but a change of higher order is not ruled out. However, the collapse of the splittings cannot be explained only in terms of changes in $\xi$, and Soos[434] has put forward a more detailed theory of the temperature-dependence of the $g$ tensor splittings, ascribing their behaviour to delocalization of the magnetic excitations over the (formerly) magnetically inequivalent stacks of radicals due to dipolar interaction between them, which increases with temperature. There appears to be some formal resemblance to the order–disorder transformation that occurs in pyrene:PMDA (Section VI-E), and an X-ray diffraction study would be a very useful supplement to the ESR study.

### 2. $N,N,N',N'$-Tetramethyl-$p$-phenylenediamine:chloranil
$[a = 16.32, b = 6.57, c = 8.81 \text{ Å}, \beta = 112°, Z = 2,$
$C2/m; \text{ group 2}]$

This molecular compound was first synthesized by Schlenk[195] in 1909 and then ignored for about forty years. It is of particular interest at present because the information available about its structure and properties is more detailed than for most other molecular compounds. The crystal structure of the room-temperature (monoclinic) phase is known (De Boer and Vos[435]) but not that of the low-temperature (triclinic) phase. Optical studies at room temperature show that the ground state is ionic. The ESR spectra of single crystals have been studied over a range of temperature and interpreted in terms of a molionic model of the crystal. This model is not compatible with the optical results, and reinterpretation in terms of Wannier spin excitons is necessary but has not been reported.

The central rings of the two planar, parallel, component molecules in the room-temperature phase of TMPD:chloranil completely overlap one another, with an interplanar distance of 3.28 Å. It is difficult to judge whether the crystals are disordered. On the one hand, 75% of the accessible reflections were recorded, and the electron-density maps are detailed enough to show the hydrogen atoms of the TMPD molecule; on the other hand, the Debye–Waller factors have high values, there is a phase change to triclinic below about $250°K$, and the site group symmetry of the TMPD molecule is suspiciously (but not impossibly) high (see Section V-H). There is a quasi-hexagonal close-packing of the mixed stacks, with a reference stack surrounded

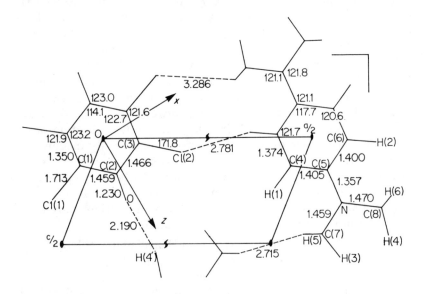

Figure 58. TMPD:chloranil. The structure at $y = 0$. The hydrogen atoms H(5) and H(6) are not in the mirror plane at $y = 0$. The orientation of the magnetic axes $x$ and $z$ is shown in the diagram. Magnetic $y$ is along [010], *i.e.*, normal to the plane of the diagram.
[Reproduced with permission from reference 435.]

by two identical stacks and four others, derived from the reference stack by action of a two-fold screw axis parallel to [010] and thus shifted by $\frac{1}{2}b$ (Figure 58). The crystal symmetry requires both molecules to be parallel to (010). The virtually exact superposition of donor and acceptor molecules is a striking feature of this structure. The only other molecular compound where this occurs is 2,4,6-tris(di-

methylamino)-*s*-triazine:TNB.[281] In both examples the two part-
ners have similar shapes, but other causes may operate as well.
Although the crystal structure of PDC is not known in detail, the
ESR results have been interpreted in terms of superposition of
N⋯N and O⋯O directions in PD and chloranil, respectively, so
that a similar superposition of components seems probable also in
PDC.

The optical absorption spectra (Figure 59) were measured by

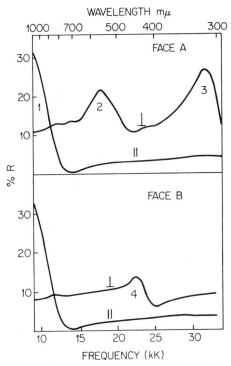

Figure 59. Polarized spectra from TMPD:chloranil single crystals.
The reflection spectra obtained for two side faces of a single
crystal.[436] The ∥ and ⊥ designations indicate that the spectra were
obtained by using polarized radiation whose electric vector
vibrated parallel or perpendicular, respectively, to the long
direction of the crystals. Very similar results were obtained for
absorption spectra.[430] The assignments of the bands was made as
follows:[430] (1) charge transfer band; (2), (3) $\pi-\pi^*$ transition
polarized along long axis of TMPD$^+$; (4) $\pi-\pi^*$ transition polarized
normal to the O⋯O axis of chloranil$^-$. [Reprinted from *Journal
of the American Chemical Society*, **88**, 1966, p. 3649. Copyright 1966
by the American Chemical Society. Reprinted by permission of the
copyright owner.]

Kainer and Überle[52] and by Pott and Kommandeur[59,60] from samples in KBr pellets, and by Amano, Kuroda, and Akamatu[430] from small single crystals, using polarized radiation; the polarized specular reflection spectra (Anex and Hill[436]) from single crystals have also been measured. In general these results agree well, with more detail in the single-crystal spectra. The broad band below 10 kK (about 1–1.5 $\mu$) is identified as a charge-transfer band by all four groups. Kainer and Überle explained their spectra as due to ionic TMPD$^+$ and chloranil$^-$ ions, and the polarized single-crystal spectra were later discussed in terms of assignments to particular transitions in these two ions (see Figure 59). The assignment of the bands at 16–18 kK and at about 31 kK to TMPD$^+$ (transition polarized along the long axis of the molecule) is in accordance with the assignments made by Albrecht and Simpson[437] (from polarized photo-oxidation studies of the TMPD cation in a rigid medium) and by Iida and Matsunaga[438] (from absorption spectra of TMPD$^+$ ClO$_4^-$ in ethanol solution). The assignment of the 22 kK (450 m$\mu$) band to chloranil$^-$ is in agreement with the results of Andre and Weill[439] and Fulton.[440] However, it should be noted that the spectra given by Andre and Weill for neutral and anionic chloranil in solution differ much more than those given by Fulton. Pott and Kommandeur[59,60] discussed their spectra in terms of contributions from TMPD and chloranil molecules and doubly-charged TMPD$^{2+}$ and chloranil$^{2-}$ ions (a molionic crystal). However, the weight of the spectroscopic evidence is against this suggestion and in favour of the occurrence of TMPD$^+$ and chloranil$^-$.

The IR spectra (Figure 60; Kainer and Otting[441]) show that neutral TMPD and chloranil molecules are not present in TMPD: chloranil (in particular the C=O band at 5.9 $\mu$ is missing). There is also no evidence for TMPD$^{2+}$ ions and the conclusion was drawn that the components were present as singly charged ions.*

The ESR spectra have been measured on single crystals over the temperature range 100–300°K and the results interpreted in terms of a molionic lattice (Pott;[59] Pott and Kommandeur[60]). The experimental results are strikingly similar to those for PDC.[58] The same dependence of line width on temperature is found; the intensity–$(1/T)$ relation shows two activation energies, 0.16 ev above about 250°K and 0.007 ev below this temperature. The $g$ factor is almost axially symmetric with $g_x = 2.0053_5$, $g_y = 2.0054_5$, and $g_z = 2.0024$ (the $z$-axis is along the stack axis). Both X-ray diffraction and ESR

* IR spectra[442] of chloranil and K$^+$ chloranil$^-$ support these conclusions.

results show that there is a phase change at about 250°K, and this is also indicated by an abrupt change of activation energy for electrical conduction in single crystals (at 268°K). An unexplained complication is that two types of crystal appear to exist, one showing the phase change at 250°K and the other at about 140°K and with certain differences of detail in their ESR spectra. Nevertheless, their X-ray diffraction patterns are identical. A detailed reinterpretation of all the ESR results is certainly desirable, but there seems little doubt that TMPD:chloranil is also a Wannier spin-exciton system.

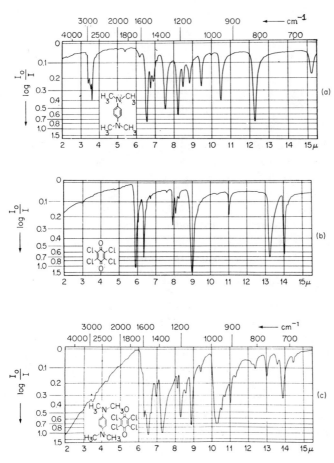

Figure 60. IR spectra (KBr disc method) of (a) TMPD, (b) chloranil, and (c) TMPD:chloranil. [Reproduced with permission from reference 441.]

**3. Comparison of dimensions of tetramethylphenylenediamine units in different environments.**—One possible method of determining the ionization state of TMPD units in different molecular compounds is by comparison of measured and theoretical bond lengths. On the basis of molecular-orbital calculations, Monkhorst and Kommandeur[443] estimate the difference between central and non-central bonds to be 0.014 Å for TMPD and 0.092 Å for TMPD$^{2+}$. The various experimental results are summarized in Table 25. Significant differences in experimental techniques make it difficult to compare the measurements directly. Nevertheless, some quinonoid character in the ring is found for all the substances studied except TMPD:TCNB; this quinonoid character correlates with C–N bond shortening. These results are compatible with cationic character for TMPD. In the remaining substance, TMPD:TCNB, the ring is aromatic and the ground state is non-ionic. In most cases (including TMPD:TCNB) the nitrogen atoms appear to be slightly but significantly pyramidal. The planarity of the nitrogens in the two exceptions may be artifacts resulting from disorder.

**4. Other molecular compounds between aromatic amines and chloranil.**—The absorption spectra of a number of aromatic amine: chloranil molecular compounds have been determined by Amano, Kuroda, and Akamatu.[187,430] Molecular compounds with non-ionic ground states could be distinguished from those with ionic ground states by using single-crystal spectra obtained with polarized radiation, and this also permitted more reliable assignment of transitions. The spectroscopic results are detailed enough to permit many useful conclusions, but the lack of crystallographic information is a disadvantage.

The UV and visible spectra, and the IR spectra, of *N,N*-dimethylaniline: chloranil, *N,N*-dimethyl-2-naphthylamine: chloranil, and 1-naphthylamine: chloranil agree in showing that the ground states of these three molecular compounds are non-ionic. The comparatively small polarization ratio for the last of the three hints at a complicated crystal structure.[430] Non-ionic ground states were also found for benzidine: chloranil and tetramethylbenzidine: chloranil.[187] The first and second charge-transfer bands are polarized in mutually perpendicular directions, suggesting that the crystal structures are different from the usual mixed-stack type.

Matsunaga,[446] using IR and diffuse reflectance spectra, has shown that 1,6-pyrenediamine: chloranil exists in three different polymorphic forms. The green form (from chloroform) and the brown

Table 25. Comparison of dimensions of tetramethyl-*p*-phenylenediamine (TMPD) units in various crystals.

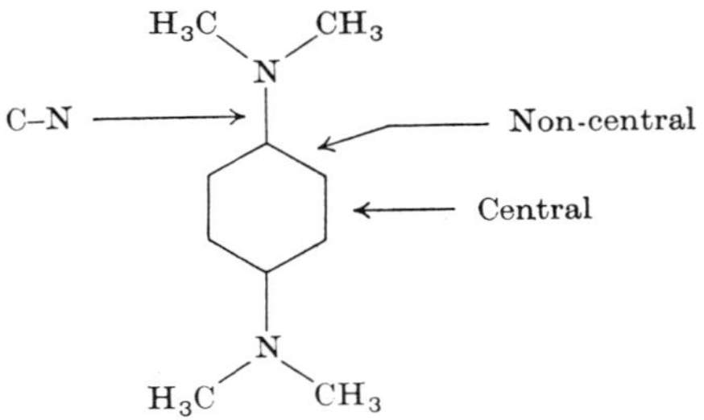

| Substance | Site-group symmetry of TMPD unit in crystal | Bond lengths[a] | | | | Configuration about nitrogen atom | Remarks | Ref. |
|---|---|---|---|---|---|---|---|---|
| | | Central (Å) | Non-central (Å) | Diff. (Å) | C–N (Å) | | | |
| TMPD⁺ Br⁻ | *mm* | 1.313 | 1.440 | 0.127 | 1.304 | Planar | $\sigma$(C–C) ≈ 0.04 Å | 444 |
| TMPD⁺ I⁻ | 2/m with mirror plane normal to ring plane | 1.361 | 1.422 | 0.061 | 1.344 | Pyramidal | Large e.s.d.'s | 445 |
| TMPD:TCNQ | | 1.374 | 1.416 | 0.042 | 1.365 | Pyramidal | Large thermal motion | 340 |
| TMPD:chloranil | 2/m with ring in mirror plane | 1.374 | 1.402 | 0.028 | 1.357 | Planar | Fairly large thermal motion | 358 |
| TMPD:(TCNQ)₂ | 1̄ | 1.367 | 1.417 | 0.050 | 1.373 | Pyramidal | Librational corrections applied | 338 |
| TMPD:TCNB | 1̄ | 1.377 | 1.380 | 0.003 | 1.430 | Pyramidal | — | 367 |

[a] In all cases $\sigma$(l) not less than 0.01–0.02 Å. Bonds in TMPD component are average values and are defined in the diagram:

form (from benzene) both have non-ionic ground states and resistivities of $10^7$ and $10^5$ ohm cm, respectively. The third form (obtained by compressing the brown form in the presence of traces of benzene) has a resistivity of only a few ohm cm. Its IR spectrum has been interpreted as indicating that it is molionic, which we have earlier indicated to be unlikely. More work is necessary.

The unambiguous assignment of transitions achieved for the TMPD:chloranil spectrum is not possible for PDC or durenediamine:chloranil because of overlap of cation and anion absorptions, especially in the 20–25 kK region. However, the ESR spectra of PDC leave no doubt that it has an ionic ground state, as has TMPD: chloranil and, probably, durenediamine:chloranil.

A number of molecular compounds between azines such as phenothiazine and acceptors such as DDQ have been studied by Matsunaga.[285] IR spectroscopy indicates that these have ionic ground states.

## K. Molecular Compounds of Aromatic Amines and 7,7,8,8-Tetra-cyanoquinodimethane (TCNQ)

**1. Comparison of some aromatic diamine:TCNQ molecular compounds with TCNQ salts.**—A number of molecular compounds of TCNQ have been studied, but unifying structural principles have so far appeared only in the comparison of the TMPD:TCNQ and TMPD:(TCNQ)$_2$ structures with those of some TCNQ salts. Other TCNQ molecular compounds are discussed elsewhere in this Review.

### TMPD:TCNQ
[$a = 9.88$, $b = 12.71$, $c = 7.72$ Å, $\beta = 97.3°$, $Z = 2$, $C2/m$; group 1]

Only 52% of the reflections accessible to Cu-$K_\alpha$ radiation could be measured at room temperature, indicative of intense thermal motion or disorder, or both. The results of the structure analysis [340] do not, however, give any clear evidence for disorder. The mixed molecular stacks are arranged quasi-hexagonally (Figure 61), the distortion being larger than usual because of the elongated shape of both components. The interplanar distance is 3.27 Å; the overlap diagram (Figure 62) shows that the TMPD moiety is displaced so that its benzene ring overlaps a C=C double bond of the TCNQ component.

The dimensions of the TMPD unit have been considered above (Section VI-J); they are compatible with other values obtained for TMPD$^+$ but cannot be very accurate, especially in the peripheral regions of the ion, because of the large thermal motion. The dimensions for the TCNQ component are compared in Table 26 (page 355) with those obtained for this component in other crystals and are compatible with its being singly ionized in TMPD:TCNQ.

The spectroscopic results are also in agreement with an ionic formulation. The visible spectra (powdered samples in Nujol mulls)

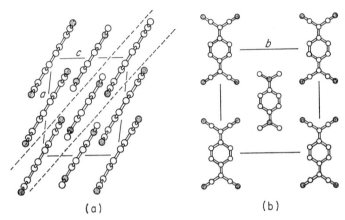

Figure 61. TMPD:TCNQ. (a) Projection down [010]. (b) Portion of structure in (a) between the broken lines seen along [001]. All the stacks are crystallographically equivalent because of the C-face centering. [Reproduced with permission from reference 340.]

show two absorption bands and this has been interpreted as evidence for an ionic ground state (Foster and Thomson[447]). The more detailed information obtained from polarized absorption spectra of single crystals supports this conclusion (Kuroda, Hiroma, and Akamutu[448]). The crystal absorption bands are broadened and shifted to higher energy compared with the solution spectra of the component ions, but identification of the transitions is still possible. The IR spectrum is the sum of the contributions of the ions and not of the molecules (Kinoshita and Akamatu[449]).

The ESR spectra have been measured on powdered samples (Kinoshita and Akamatu;[449] Ohmasa et al.[450]) and for single crystals by Hughes and Hoffman.[451,452] The value found by Hughes and Hoffman for the thermally excited singlet–triplet excitation energy $J$ of the linear Heisenberg chain was 0.068 ev, lower than values reported for other ionic charge-transfer crystals. A study of the spin-spin and spin–lattice relaxation rates was therefore possible over a

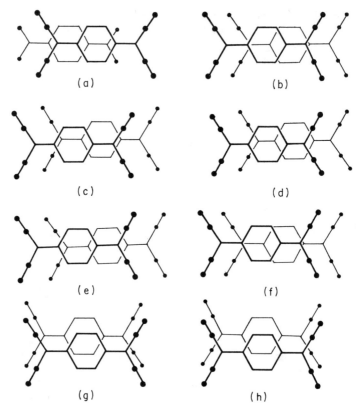

Figure 62. Overlap diagrams for TCNQ molecular compounds and ion-radical salts.
    A. Ring–external bond (R–EB) overlaps:
        (a) TMPD:TCNQ;
        (b) $(TCNQ)^-$ $(TCNQ)^-$ in $N$-methylphenazinium TCNQ;
        (c) $(TCNQ)^0$ $(TCNQ)^-$ in $Cs_2(TCNQ)_3$;
        (d) $(TCNQ)^0$ $(TCNQ)^-$ in $TPP(TCNQ)_2$;
        (e) $(TCNQ)^0$ $(TCNQ)^-$ in $TMPD(TCNQ)_2$;
        (f) $(TCNQ)^0$ $(TCNQ)^-$ in $((C_6H_5CH_3)_2Cr):(TCNQ)_2$.
    B. Ring–ring (R–R) overlaps:
        (g) $(TCNQ)^-$ $(TCNQ)^-$ in $Cs_2(TCNQ)_3$;
        (h) $(TCNQ)^-$ $(TCNQ)^-$ in $((C_6H_5CH_3)_2Cr):TCNQ$.

temperature range where the behaviour of Frenkel and Wannier exciton systems is strikingly different. It was consequently possible to characterize TMPD:TCNQ unambiguously as a Wannier exciton system. The measured $g$ tensor was shown to have molecular rather than crystal symmetry (a similar result was obtained for TMPD:

chloranil, which also has crystallographically equivalent stacks). These results provide additional support for the model of essentially non-interacting one-dimensional stacks used to represent these crystals.

### N-Methylphenazinium tetracyanoquinodimethanide
$[a = 3.868, b = 7.781, c = 15.735$ Å, $\alpha = 91.67, \beta = 92.67,$
$\gamma = 95.38°, Z = 1, P1$ or $P\bar{1}]$

This ion-radical salt has segregated stacks (Figure 63) (Fritchie[453]), as in TMPD:(TCNQ)$_2$, and thus differs strikingly from TMPD:TCNQ, which has mixed stacks. If the assumed space group $P\bar{1}$ is correct, then the cations must be disordered and the TCNQ anions equivalent. The overlap diagram is given in Figure 62b.

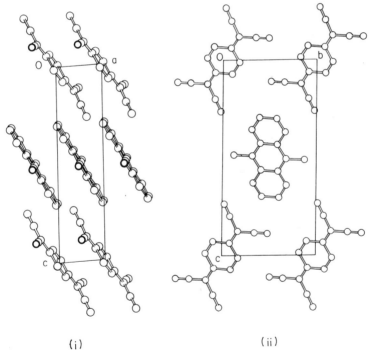

(i)                                                        (ii)

Figure 63. *N*-methylphenazinium TCNQ. (i) Projection on (010); (ii) projection on (100). The disordered *N*-methyl group is shown in both its alternate sites. [Reproduced with permission from reference 453.]

**TMPD:(TCNQ)$_2$**
$[a = 7.78, b = 15.02, c = 6.49$ Å$, \alpha = 93.5, \beta = 102.8,$
$\gamma = 83.0°, Z = 1, P\bar{1}]$

These triclinic crystals (Hanson[338]) are black, shiny, and opaque; they are easily confused with TMPD:TCNQ. In TMPD:(TCNQ)$_2$ one might expect either a formal charge of $\frac{1}{2}$ on the TCNQ units or presence of both (TCNQ)$^0$ and (TCNQ)$^-$ in the crystal. The space group is assumed to be $P\bar{1}$, which would imply the presence of crystallographically equivalent (TCNQ)$^{-0.5}$ units. The excellent agreement between observed and calculated structure factors ($R \approx 5\%$) supports the choice of space group, but the possibility that the crystals belong to $P1$ is not completely disproved. In contrast to TMPD:TCNQ, the crystals of TMPD:(TCNQ)$_2$ diffract well—80% of the reflections accessible to Cu-$K_\alpha$ were recorded and there was no indication of disorder.

The component arrangement in these crystals is quite different from that found in 2:1 or 1:2 charge-transfer molecular compounds [*e.g.*, phenoquinone (Section VI-H) or 3,4-benzopyrene:(TMU)$_2$ (Section VI-M)] but has many features in common with various TCNQ salts. There are segregated stacks of TMPD and TCNQ, the latter being essentially perpendicular to [001] while the former are inclined at an angle of about 70° to this axis (Figure 64). This allows the spatial requirements of the 1:2 stoichiometry to be met even though the components are segregated into different stacks. The TMPD units hardly overlap at all, but the TCNQ molecules overlap directly with enough mutual displacement to bring a ring of one molecule over the C=C double bond of the next (Figure 62e); mean spacing between TCNQ units is 3.24 Å. TCNQ stacks have a zig-zag arrangement, different from the mixed stacks found in most molecular compounds (Figure 65). TMPD units have the dimensions appropriate to monopositive ions while those of TCNQ are appropriate to (TCNQ)$^{-0.5}$ (see Table 26). No measurement of the physical properties of TMPD:(TCNQ)$_2$ has yet been reported.

### Ditoluenechromium:TCNQ
$[a = 7.00, b = 15.45, c = 20.5$ Å$, \beta = 97°, Z = 4, P2_1/n;$
axial orientation here and in Figure 66 differs from that
given in Table 20, group 3b]

The crystal structure of $[(C_6H_5CH_3)_2Cr]$:TCNQ consists of segregated stacks of ditoluenechromium cations and TCNQ anions, arranged in quasi-hexagonal fashion (Figure 66).[324] The interplanar

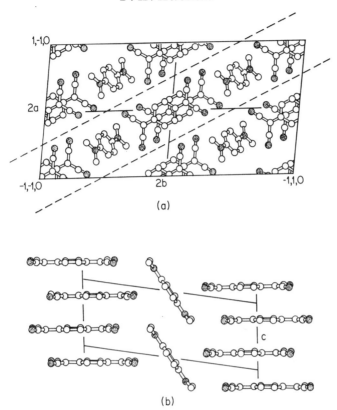

Figure 64. TMPD:(TCNQ)$_2$. (*a*) The structure viewed along the *c*-axis. (*b*) That part of the structure between the broken lines, viewed along the normal to the plane (110). [Reproduced with permission from reference 338.]

spacings between the anions in the stacks has the rather large value of 3.42 Å. The overlap of two TCNQ$^-$ anions is shown in Figure 62h and is of the ring–ring type. The bond lengths were not given in the original paper, but the double bonds are said to be longer, and the single bonds shorter, than in neutral TCNQ. There is some evidence for disorder in the cation positions.

## Ditoluenechromium:(TCNQ)$_2$
$$[a = 8.25, b = 7.76, c = 13.77 \text{ Å}, \alpha = 94.7, \beta = 92.3,$$
$$\gamma = 112.5°, Z = 1, P\bar{1}]$$

The crystals are dark violet needles elongated along [010] (Shibaeva et al.[339]). The crystal structure consists of segregated stacks of

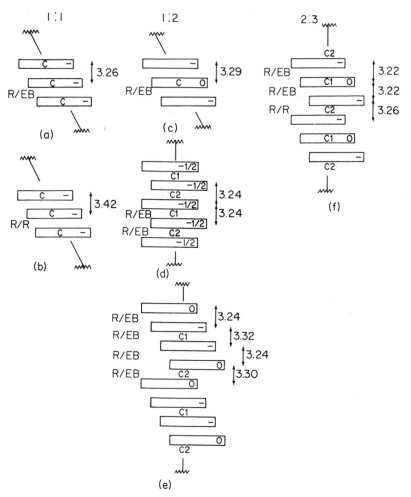

Figure 65. Summary of the types of stack packings of TCNQ moieties in various TCNQ ion-radical salts: (a) $N$-methylphenazinium TCNQ; (b) $[(C_6H_5CH_3)_2Cr]TCNQ$; (c) $[(C_6H_5CH_3)_2Cr]$ $(TCNQ)_2$; (d) TMPD $(TCNQ)_2$; (e) TEA $(TCNQ)_2$ (see Addenda, Section VIK); (f) $Cs_2(TCNQ)_3$.

R/EB and R/R refer to ring-external bond and ring/ring overlap (see Figure 62). c denotes translationally-equivalent centres of symmetry. $c_1, c_2$ denotes non-equivalent centres. $-$, $-\frac{1}{2}$, 0 denote formal charges on TCNQ units.

The stack axes are shown by heavy lines; the offset of adjacent units is shown schematically; mean interplanar distances are given in Å. The stacks shown, for example, in (a) are similar to those found in equimolar $\pi$-molecular compounds where, however, donor and acceptor molecules alternate along the stack.

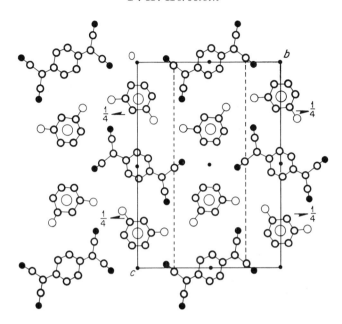

Figure 66. Ditoluenechromium:TCNQ. Projection down [100].
[Reproduced with permission from reference 324.]

ditoluenechromium cations and TCNQ units, arranged in alternate layers parallel to (001) (Figure 67). The two TCNQ units are at crystallographically independent symmetry centres and an alternating sequence of charged and neutral units along a stack:

$$--- (TCNQ)^0\ (TCNQ)^-\ (TCNQ)^0\ (TCNQ)^- \ ---$$

was inferred. The measured bond lengths were not accurate enough to provide independent support for this proposal. The interplanar spacing was $3.29 \pm 0.03$ Å, and ring–external bond overlap (Figure 62f) was found.

The cation conformations are different in $[(C_6H_5CH_3)_2Cr]$:TCNQ and $[(C_6H_5CH_3)_2Cr]:(TCNQ)_2$, indicating a small barrier to mutual rotation of the two rings of the ditoluenechromium cation. More important in the present context is the striking difference in the electrical conductivity of the two materials (Yagubskii *et al.*[454]), as shown by the following values:

1:1   specific resistance, $\rho = 2.5 \times 10^5$ ohm cm, activation energy,
      $E_a = 0.36$ ev
1:2   $\rho = 0.5$ ohm cm, $E_a = 0.06$ ev

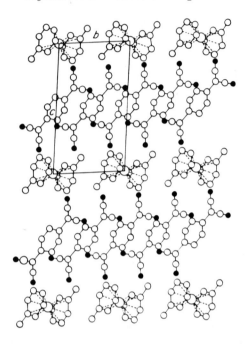

Figure 67. The [100] projection of the [(C₆H₅CH₃)₂Cr](TCNQ)₂ structure. [Reproduced with permission from reference 339.]

## Cs₂(TCNQ)₃ and tetraphenylphosphonium bis(tetracyanoquinodimethanide) TPP(TCNQ)₂

[Cs$_2$(TCNQ)$_3$: $a = 7.34$, $b = 10.40$, $c = 21.98$ Å, $\beta = 97.2°$,
$Z = 2$, $P2_1/c$]

[TPP(TCNQ)$_2$: $a = 33.005$, $b = 7.766$, $c = 15.961$ Å, $\beta = 109.31°$,
$Z = 4$, $Cc$ or $C2/c$]

The above results can be usefully compared with those obtained for the salts Cs$_2$(TCNQ)$_3$ (Arthur;[455] Fritchie and Arthur[329]) and TPP (TCNQ)$_2$ (Goldstein, Seff, and Trueblood[456]). In a formal sense both (TCNQ)$^0$ and (TCNQ)$^-$ occur in Cs$_2$(TCNQ)$_3$, and (TCNQ)$^{-0.5}$ occurs in TPP(TCNQ)$_2$.

The crystals of Cs$_2$(TCNQ)$_3$ are purple prisms.[84] The space-group determination is unambiguous, so that two crystallographically non-equivalent TCNQ molecules are present, one at centres of symmetry, the other in general positions. In fact, the molecules of both kinds have very nearly *mmm* symmetry. For convenience in description, the TCNQ at centrosymmetric sites are called centric (C) and those at general sites are called non-centric (NC). These are shown, together with the general arrangement, in Figure 68. The stacks of TCNQ moieties have the formal sequence:

$$- - - NC - C - NC - NC - C - NC - NC - C - NC - - - -$$

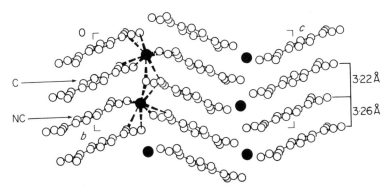

Figure 68. $Cs_2(TCNQ)_3$. [100] projection. All TCNQ molecules are centred near $x = 0$, all caesium ions (blackened) near $x = \frac{1}{2}$. The caesium coordination is roughly cubic, involving TCNQ molecules at both $x = 0$ and $x = 1$. [Reproduced with permission from reference 329.]

and the respective overlap diagrams are shown in Figure 62. Adjacent, non-centric molecules are separated by 3.26 Å and are necessarily parallel; adjacent centric and non-centric molecules are 3.22 Å apart and have a mutual inclination of 2.7°. Thus there is no tendency to "pair formation," as appeared to be the case from the earlier, less accurate analysis.[455] Comparison of the bond lengths (Table 26) in the C and NC portions with those in other crystals shows that $C \equiv (TCNQ)^0$ and $NC \equiv (TCNQ)^-$.

The crystals of $TPP(TCNQ)_2$ are black, shiny, and opaque. The space group was assumed to be $C2/c$ and the excellent results obtained in the crystal-structure analysis ($R = 0.043$) support this choice; the TCNQ units occupy a single set of eight-fold general positions —formally they are $(TCNQ)^{-0.5}$. Component arrangement is shown in Figures 69a and 69b and the overlap diagram in Figure 62d. The structure does not contain stacks of TCNQ units; instead, there are centrosymmetric TCNQ pairs, about 3.2 Å apart (as the TCNQ components are not quite planar, the interplanar distance depends

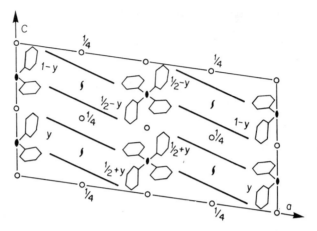

Figure 69a. TPP(TCNQ)$_2$. The structure of one unit cell viewed along $b$. TCNQ molecules, seen edgewise, are represented by straight lines. The heights along the $b$-axis are shown in terms of the $y$-positions; each position shown applies to the TCNQ and to the two phenyl groups nearest it. [Reproduced with permission from reference 456.]

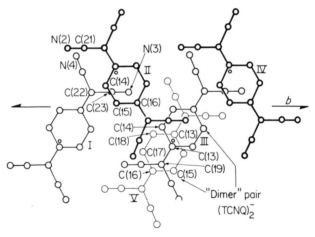

Figure 69b. TPP(TCNQ)$_2$. The TCNQ arrangement viewed along a direction which is normal to the TCNQ molecular plane and parallel to the layer of TCNQ molecules. [Reproduced by permission from reference 456.]

on the choice of reference planes). The bond distances are intermediate between those of $TCNQ^0$ and $TCNQ^-$ and this, together with their arrangement, has led to the suggestion that there are dimeric anions $(TCNQ)_2^-$ in this crystal.

Adjacent TCNQ units overlap in two different ways in the crystals considered here. The first kind of overlap has (ideally) superposition of ring and external bond (R–EB), while the second has superposition of ring and ring (R–R). As Figure 65 shows, there is no correlation between type of overlap and the formal charges on the components. The favoured arrangements of $(TCNQ)_2^0$, $(TCNQ)_2^-$, and $(TCNQ)_2^{2-}$ have been calculated by Chesnut and Mosely.[389] Binding is predicted, and there is qualitative agreement with experiment. Detailed comparison of experiment and theory would be premature: there are still too few experimental results and too few factors taken into account in the calculations.

*Comparison of dimensions of TCNQ moieties in different environments.* The bond lengths and bond angles in variously charged TCNQ moieties are summarized in Table 26, with *mmm* molecular symmetry assumed. Small, though definite, changes in bond length

Table 26. Bond lengths (Å) and bond angles in various charged TCNQ units.

| Bond or bond angle | $(TCNQ)^{0\,a}$ | $(TCNQ)^{-0.5\,b}$ | $(TCNQ)^{-\,c}$ |
|---|---|---|---|
| a | 1.344 (4) | 1.354 (2) | 1.356 (10) |
| b | 1.446 (3) | 1.434 (2) | 1.425 (7) |
| c | 1.371 (4) | 1.396 (2) | 1.401 (8) |
| d | 1.434 (4) | 1.428 (3) | 1.417 (4) |
| e $d$ | (1.14) | [(1.17)(1)] | (1.15) |
| $\alpha$ | 121.0° | 121.1° | 121.2° |
| $\beta$ | 118.0° | 117.9° | 117.4° |
| $\gamma$ | 116.3° | 115.9° | 115.2° |
| $\delta$ | 179.0° | 178.5° | 178.6° |
| Libration corrections | Partly | Yes | No |

[a] Average of values for TCNQ itself (bond lengths, but not angles, corrected for libration) and the centric unit in $Cs_2(TCNQ)_3$ (not corrected for libration).

[b] Average of results for TCNQ groups in $TMPD:(TCNQ)_2$ and $TPP(TCNQ)_2$. The non-centric unit in $Cs_2(TCNQ)_3$ gives the following values (not corrected for libration): $a$ 1.355, $b$ 1.428, $c$ 1.402, $d$ 1.418 Å (e 1.15). The averaged bond angles are also in good agreement with the more reliable values in the Table.

[c] Average of values given for $K(TCNQ)$, $N$-methylphenazinium TCNQ, and TMPD:TCNQ.

[d] Values sensitive to libration corrections.

with formal charge are detectable, and Goldstein *et al.*[456] have discussed such changes in terms of molecular-orbital calculations. The bond angles do not appear to vary with formal charge. For *mmm* symmetry the ring angles $\alpha$ and $\beta$ are related by $\alpha = 180° - 0.5\beta$. There is no such geometrical constraint on the values of $\gamma$ and $\delta$ and these angles deviate significantly from 120° and 180°, respectively.

The TCNQ units are slightly non-planar in all the crystals examined; in general, the central ring is strictly planar with the cyano groups deviating by not more than 0.02 Å from this plane. These distortions, small as they are, are ascribed to packing effects.

**2. Some other molecular compounds of TCNQ with aromatic amines.**—A number of molecular compounds have been prepared from various aromatic amines and TCNQ; their IR and ESR spectra have been studied, the latter as a function of temperature (Ohmasa, Kinoshita, Sano, and Akamatu[450]), but crystal structures are known only for TMPD:TCNQ and TMPD:(TCNQ)$_2$.

These molecular compounds could be divided into a number of classes on the basis of their IR spectra and especially the temperature-dependence of the ESR absorption. The first class (A) has an ionic ground state and an ESR intensity that falls with falling temperature and then remains constant. TMPD:TCNQ is a fairly typical member of this group, and thus it seems likely that all members of this class are Wannier spin-exciton systems, similar to PDC and TMPD:chloranil. The other cationic components are: *p*-phenylenediamine, *N,N*-dimethyl-*p*-phenylenediamine, *o*-phenylenediamine (2:1), and durenediamine (DAD) (compositions 1:1 unless stated otherwise). IR spectra[457] suggest that 1,6-pyrenediamine:TCNQ also belongs to this group, although its electrical resistivity is very low (0.5 ohm cm at 20°c). The polarized absorption spectra of single crystals of *N,N*-dimethyl-*p*-phenylenediamine:TCNQ (Kuroda, Hiroma, and Akamatu[448]) are similar to those of TMPD:TCNQ and thus support the assignment of an ionic ground state. The crystal structure is not known, but the spectroscopic results are compatible with the usual mixed-stack arrangement.

The DAD:(TCNQ)$_2$ molecular compound is the only member of class B. It has the lowest resistivity (15 ohm cm) of all the substances studied, and the ESR absorption intensity falls only by about 20% between 300°K and 120°K. At 300°K there are $3 \times 10^{23}$ spins per mole of molecular compound, and it has been suggested that these are due to the conduction electrons. The composition is intriguingly similar to that of TMPD:(TCNQ)$_2$.

The third class shows Curie-type paramagnetism; it contains only one member so far, namely, (*N*,*N*-dimethyl-*p*-phenylenediamine)$_2$: (TCNQ)$_3$. The members of the fourth class have non-ionic ground states and include *N*,*N'*-diphenyl-*p*-phenylenediamine:TCNQ, tetramethylbenzidine:TCNQ, and benzidine:TCNQ. In the last two molecular compounds the polarization directions of the first and second charge-transfer bands are mutually perpendicular[187] (a similar situation is found in the corresponding chloranil molecular compounds), and the crystal structures presumably differ from the usual mixed-stack type.

Thus, despite the chemical similarities between the various donors used, there are appreciable differences in the electrical, magnetic, and optical properties of their molecular compounds. Crystal-structure analyses would seem to be the first step necessary for interpretation of the physical properties.

### L. Comparison between Structures of Ion-radical Salts and of π-Molecular Compounds with Ionic Ground States, and between Structures of π-Molecular Compounds with Non-ionic and Ionic Ground States

In ion-radical salts either the cation or the anion is a radical, but not both; however, in the molecular compounds with ionic ground states both the cation and the anion are radicals. This formal distinction may be alleviated or complicated by physical reality; nevertheless, it is worth examining the evidence to see whether the formal distinction corresponds to structural differences between the two groups. Crystal structures are known for a number of Würster (ion-radical) salts that have TMPD or related substances as cation radical, *e.g.*, (TMPD)$^+$ClO$_4$$^-$, (TMPD)$^+$I$^-$, and *N*,*N*-dimethyl-*p*-phenylenediamine hydrobromide;[444] crystal structures are also known for some TCNQ ion-radical salts that have TCNQ as anion-radical. These ion-radical salts are characterized by segregated stacks of cations and anions, respectively, in plane-to-plane packing. The segregated stacks are found both in the simple salts, *e.g.*, (TMPD)$^+$ClO$_4$$^-$, and in the complex salts, *e.g.*, TPP$^+$(TCNQ)$_2$$^-$ and Cs$_2$(TCNQ)$_3$. The differences in detailed arrangements within the stacks presumably correspond to differences in physical properties of various Würster salts (Iida and Matsunaga[458]) as well as of simple and complex TCNQ salts (Iida[459,460]).

Mixed stacks are found in three 1:1 π-molecular compounds with ionic ground states: these are TMPD:chloranil, TMPD:TCNQ, and PDC. Ditoluenechromium:TCNQ has segregated stacks but it is uncertain how it should be classified.* Thus the available evidence favours the existence of structural differences between ion-radical salts and 1:1 π-molecular compounds with ionic ground states. Two 1:2 molecular compounds with ionic ground states have been studied. These are TMPD:$(TCNQ)_2$ and ditoluenechromium:$(TCNQ)_2$, which have segregated stacks formally similar to those in the ion-radical salts. However, one cannot generalize on the basis of two examples.

The nature of the ground state does not appear to influence the structure of 1:1 π-molecular compounds: mixed stacks are found in both cases. On the other hand, the segregated stacks of TMPD: $(TCNQ)_2$ and ditoluenechromium:$(TCNQ)_2$ are very different from the mixed-stack sandwich structures of phenoquinone or coronene:$(TMU)_2$.† However, here too generalization would be premature.

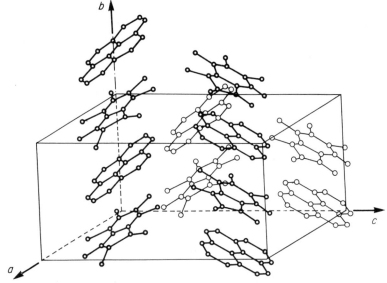

Figure 70. Pyrene:TMU. A clinographic view of the crystal structure. Interplanar distance is 3.48 Å and there is an angle of 4° between the molecular planes. [Reproduced with permission from reference 463.]

---

* The $[C_6H_6)_2Cr]$ cation-radical is paramagnetic,[461] and thus ditoluenechromium: TCNQ is formally a molecular compound with an ionic ground state. However, it is not certain whether the unpaired electrons on the cation-radicals interact with one another as do those on the anion-radicals.

† The organic, charge-transfer salt diquinolinium 2-(dicyanomethyl)-1,1,3,3-tetracyanopropenediide $[(C_9H_8N)_2^+ (C_{10}N_6)^{2-}]$ has a mixed-stack sandwich structure,[462] not very different from those of phenoquinone or coronene:$(TMU)_2$.

## M. Molecular Compounds with Tetramethyluric acid (TMU) as One Component

Pyrene:TMU. $a = 9.71$, $b = 8.00$, $c = 15.04$ Å, $\beta = 117°$,
$Z = 2$, $Pc$, needle axis [010].

3,4-Benzopyrene:$(TMU)_2$. $a = 9.33$, $b = 10.59$, $c = 10.84$ Å,
$\alpha = 119.6$; $\beta = 113.3$, $\gamma = 87.3°$, $Z = 1$, $P1$,
needle axis [010].

Coronene:$(TMU)_2$. $a = 9.36$, $b = 11.18$, $c = 10.51$ Å, $\alpha = 118.3$,
$\beta = 113.9$, $\gamma = 85.6°$, $Z = 1$, $P1$, needle axis [010].

There are striking resemblances between the TMU molecular compounds of known structure, despite the differences in composition. In pyrene:TMU[463] (Figure 70) the two components alternate along the quasi-hexagonally packed mixed stacks. The Debye–Waller factors (Figure 71) and electron-density maps indicate disorder, probably of both components. However, no diffuse scattering has been reported, and the crystals do not transform on cooling to 100°K

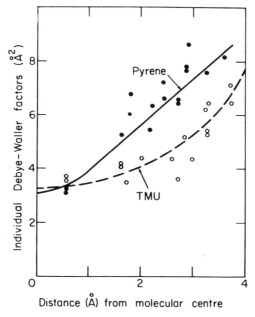

Figure 71. Pyrene:TMU at 300°K. The equivalent isotropic Debye–Waller factors of the individual atoms are shown, separate curves being drawn for each component. [Reproduced with permission from reference 303.]

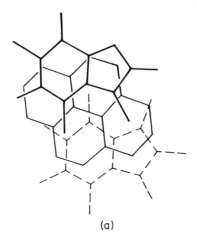

(a)

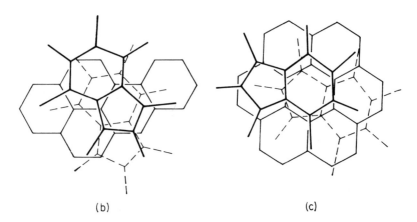

(b)                              (c)

Figure 72. Overlap diagrams for some TMU molecular compounds.
(a) Pyrene:TMU. Projection of three molecules of the same stack
on to the pyrene least-squares plane. (b) 3,4-Benzopyrene(BP):
(TMU)$_2$. Projection of three molecules in a triad on to the mean
molecular plane (TMU–BP–TMU is shown). (c) Coronene:
(TMU)$_2$. Molecular arrangement viewed in the direction perpendicular to the mean molecular plane of the coronene (CR)
molecule. (– – –) TMU, 3.45 Å under the plane of CR; (——)
3.45 Å above the plane of CR; (——)CR.

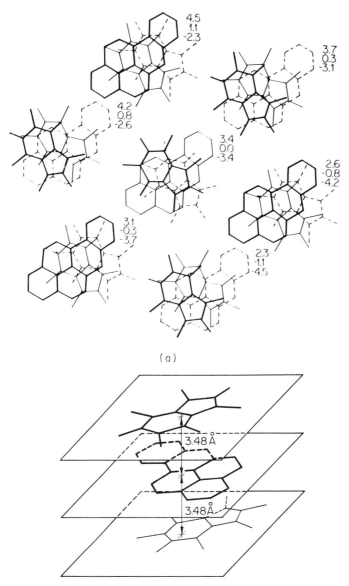

(a)

3.48 Å

3.48 Å

Figure 73. (a) 3,4-Benzopyrene (BP):(TMU)$_2$. Projection on to BP mean plane showing quasi-hexagonal arrangement of stacks. The mean coordinate (Å) of each molecule along the perpendicular to the plane of plotting is quoted. In a triad the highest molecule is shown by heavy lines, the middle one by full light lines, and the lowest by dashed lines. (b) Schematic drawing of the three molecules TMU–BP–TMU in the asymmetric unit. [Reproduced with permission from reference 334.]

361

(Bernstein and Herbstein[326]). The overlap diagram (Figure 72a) probably represents only an average situation.

The crystals of 3,4-benzopyrene:(TMU)$_2$[334] and coronene: (TMU)$_2$[335] are isomorphous (Table 17). Diffuse streaks occur in the diffraction patterns of 3,4-benzopyrene:(TMU)$_2$ but were ignored in the structure analysis; only about 20% of the reflections accessible Cu-$K_\alpha$ radiation were recorded. In the course of the analysis it was not found possible to distinguish between four orientations of the benzopyrene molecule with respect to a TMU molecule. The situation is similar to that in copper oxinate:BTF[429] where four different orientations of the BTF molecules had to be postulated. The stacks (Figure 73a) contain triads of molecules in which benzopyrene is sandwiched between two TMU molecules. The component planes are all essentially parallel and the interplanar distances are all 3.48 Å (Figure 73b). Thus trimeric sub-units are not present in the crystal.

The relatively large interplanar distances, together with spectroscopic evidence, indicate that the charge-transfer forces are weak and that the intercomponent bonding arises mainly from dipole–induced dipole forces between the polar purine molecules and the polarizable hydrocarbon molecules.

It is impossible to assess whether disorder is present in coronene: (TMU)$_2$. There is a regular interplanar spacing in the stacks of 3.45 Å, showing that there are no trimeric sub-units in the crystals. The overlap diagram (Figure 72c) shows only a general resemblance to those of the other two TMU molecular compounds, again suggesting that charge-transfer forces play only a minor role in determining the structure.

## N. Miscellaneous

### 1. Tetramethyl-*p*-phenylenediamine:tetracyanobenzene (TMPD:TCNB)

$[a = 7.654, b = 8.041, c = 7.462$ Å$, \alpha = 96.7, \beta = 85.9, \gamma = 101.3°,$
$$Z = 1, P\bar{1}]$$

Crystals of this complex are black needles, with metallic lustre, and are very stable in air. Their structure has been determined by Ohashi, Iwasaki, and Saito,[367] using three-dimensional methods. There is no sign of disorder, all atoms being well-defined in the electron-density maps (see, for example, Figure 1 of ref. 367). The usual quasi-hexagonal arrangement of mixed stacks is found. All stacks are identical and the molecules lie approximately in the

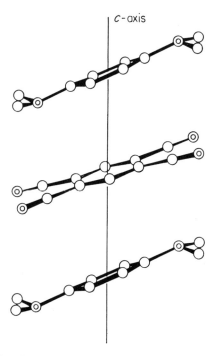

Figure 74. TMPD:TCNB. The structure, viewed along [1$\bar{1}$0]. [Reproduced with permission from reference 367.]

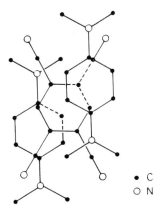

Figure 75. TMPD:TCNB. Overlapping molecules in a stack, viewed approximately normal to their mean planes. [Reproduced with permission from reference 367.]

(10$\bar{2}$) planes. The TMPD and TCNB molecules are inclined to one another at about 7°, and the average interplanar spacing is 3.40 Å (Figure 74). However, much shorter distances (3.16 Å) are found between one pair of atoms (N of TMPD and C of cyano group of TCNB) and this, together with the unusual overlap diagram (Figure 75), suggests that a localized, rather than a delocalized, interaction is involved. There is some theoretical support for this suggestion, as molecular-orbital calculations (Iwata, Tanaka, and Nagakawa[396]) for TCNB indicate that the carbon atoms of the cyano groups of TCNB have a deficiency of electron density. A detailed spectroscopic study would be useful in defining more closely any charge-transfer that may occur.

### 2. Anthracene: tetracyanoquinodimethane
[$a = 11.48$, $b = 12.95$, $c = 7.00$ Å, $\beta = 100.4°$, $Z = 2$; C2/m; group 1]

This crystal structure was determined at room temperature by Williams and Wallwork;[357] the original paper contains clear evidence for disorder in the positions of the anthracene molecules, although

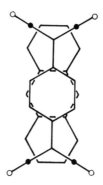

Figure 76. Anthracene:TCNQ. Overlap diagram (averaged orientation as crystals are disordered) (interplanar spacing 3.50 Å, molecules parallel). [Adapted from reference 357.]

this was apparently not recognized. The fraction of accessible reflections recorded is relatively low; electron-density and difference syntheses showed the hydrogen atoms of the TCNQ molecule but left atoms 2 and 3 (chemical numbering) of the anthracene molecules unresolved! There is no doubt that the structure is correct in general but the overlap diagram (Figure 76) gives the averaged, and not the actual, positions of the anthracene molecules with respect to the TCNQ molecules. The usual quasi-hexagonal arrangement of mixed stacks is found, with all stacks identical because of the centred space group. The optical properties have not been studied quantitatively, but the deep-green colour of the crystals and their pleochroic nature indicate that charge-transfer occurs. On the other hand, the comparatively large interplanar spacing of 3.50 Å and the lack of a clear orientation relationship between the components suggest that the charge-transfer contribution to the intercomponent binding is small.

### 3. Benzidine : 1,3,5-trinitrobenzene

The overlap diagram for this molecular compound has been reported (Wallwork;[23] Figure 77) but without crystallographic details. The polarized absorption spectra of single crystals[187] are compatible

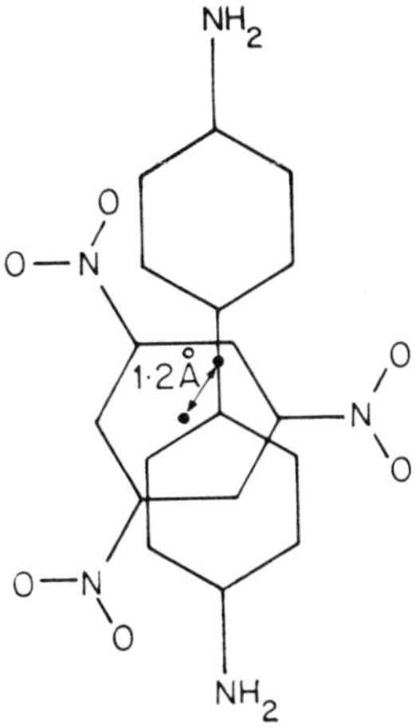

Figure 77. Benzidine:TNB. Overlap diagram. [Reproduced with permission from reference 23.]

with a mixed-stack structure. A similar conclusion has been drawn[187] for tetramethylbenzidine:(TNB)$_2$, for which no crystallographic details are available. The infrared spectra (KBr matrix)[441] show that both these molecular compounds have non-ionic ground states.

### 4. The self-complexing molecule 1-phenyl-2-(methylthio) vinyl 2,4,6-trinitrobenzenesulfonate

$$[a = 10.154, b = 10.716, c = 9.709 \text{ Å}, \alpha = 118.33, \beta = 93.83, \gamma = 99.05°, Z = 2, P\bar{1}]$$

This compound forms deep red crystals whose structure has been determined by Meyers and Trueblood.[464] The molecule has the unusual feature that it is composed of a possible electron-donating portion [the phenyl(methylthio)vinyl residue] and an electron-accepting portion (the TNB residue). The results of the structure

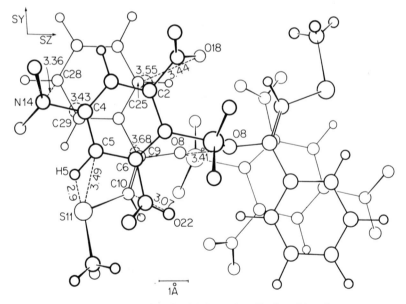

Figure 78. The two molecules in the unit cell related by the centre of symmetry shown as a small circle. The molecules are viewed in a direction approximately normal to the planes of the six-membered rings. The shorter non-bonded intermolecular distances are shown. [Reproduced with permission from reference 464.]

analysis indicate that the molecule forms a self-complexed "dimer" in the crystalline state. Two molecules related by a centre of symmetry are shown in Figure 78; the electron-donating and electron-accepting portions are superimposed with mutual orientation resembling the arrangements found in TNB molecular compounds (see Figure 26, page 285). The colour of the crystals suggests that charge-transfer occurs, but detailed spectroscopic studies are lacking. It should, however, be noted that the donor–acceptor propinquity is restricted to a pair of molecules, and the crystals do not contain stacks resembling the alternating mixed stacks found in the π-molecular compounds; at least part of the reason for this is the need to accommodate the bulky (methylthio)vinyl substituent.

## 5. The self-complexing molecule 2-methyl-3-(*N*-methylanilino-methyl)-1,4-naphthoquinone

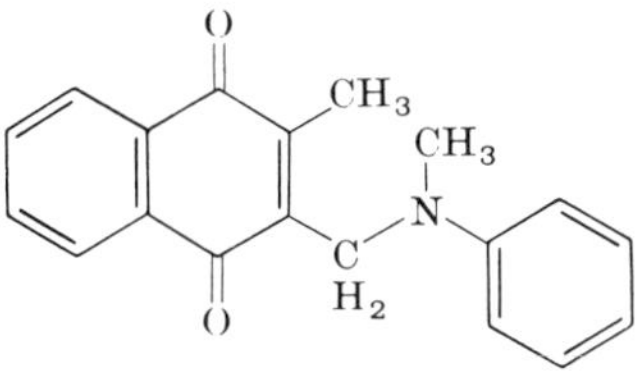

$[a = 14.22, b = 14.04, c = 7.46\text{Å}, \gamma = 104.2°, Z = 4, P2_1/n$ with unique axis $c]$

This compound has donor and acceptor groups in the same molecule. The crystal structure, determined by Prout and Castellano,[465] shows that the molecules are stacked with their planes roughly normal to the unique axis of the crystal. Despite the elongated shape of the molecule, there is a quasi-hexagonal arrangement of stacks, with each stack arranged about a screw axis in such a way that the naphthoquinone part of one molecule overlaps the aniline part of its neighbours above and below. The geometry of the overlap is only slightly different for the neighbours above and below, and the interplanar distances in both directions are 3.39 Å. The angle between overlapping naphthoquinone and aniline rings is 4.5°. The type of overlap (Figure 79) is strikingly similar to that found in HMB:chloranil (Figure 40c, page 308) and there is no direct aromatic ring–carbonyl group overlap. In contrast to the last preceding example of self-complexing, here we have infinite stacks of superimposed molecules with interactions between their donor and acceptor portions apparently large enough to determine the molecular packing. The reflectance spectra of the solid show a charge-transfer band which Prout and Castellano[465] ascribe to both intermolecular ($\pi-\pi^*$) and intramolecular ($n-\pi^*$) interactions; only the latter are invoked to explain the solution spectra.[472]

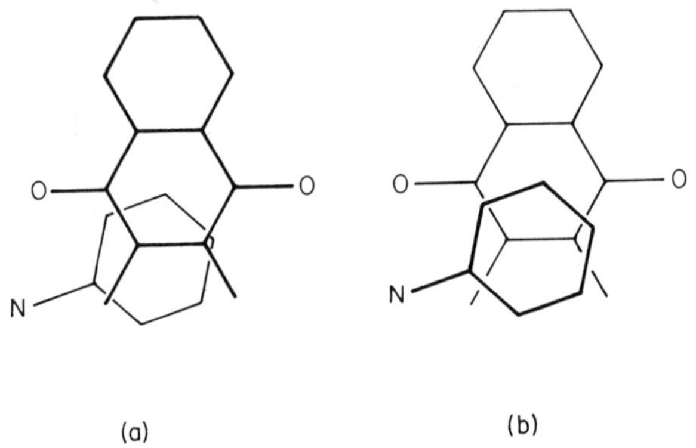

(a)                                (b)

Figure 79. Overlap diagram for 2-methyl-(3-$N$-methylanilino-methyl)-1,4-naphthoquinone. The naphthoquinone part of one molecule is sandwiched between the aniline parts of molecules above and below, and conversely. [Reproduced with permission from reference 465.]

## VII. PRESENT CONCLUSIONS AND FUTURE STUDIES

One major aim of a summary of our knowledge of a particular area should be to assess how well theoretical predictions and experimental findings match; a second should be to point out directions for further work. The current theory is that molecular compounds between electron-donors and electron-acceptors are stabilized by partial or complete electron-transfer between the excited states of the components, giving molecular compounds with non-ionic ground states (the Mulliken molecular compounds), or between the ground states of the components, giving molecular compounds with ionic ground states (the Weiss molecular compounds). Optical excitation between ground and excited states gives rise to an additional band in the spectrum in the near-UV, visible, or near-IR regions, leading often to the intense colours that distinguish these molecular compounds. McConnell and his co-workers have suggested that there is a sharp distinction between the Mulliken and the Weiss-type molecular compounds and that intermediate types will not be found; in addition, the spins of the unpaired electrons of the ionized components may be either independent (Wannier spin excitons) or coupled (Frenkel spin excitons), leading to different behaviour of their ESR spectra as functions of temperature and pressure.

The broad qualitative prediction of the formation of coloured molecular compounds between electron-donors and electron-acceptors is well supported by the chemical results summarized in the first part of this Review. One would wish to convert this qualitative prediction into a quantitative tool by comparing experimental stabilities of the various molecular compounds with the calculated strengths of the intercomponent interaction. There are problems from both theoretical and experimental points of view. Although separation of the overall interaction between the components into contributions from various sources involves some gross approximations, it has nevertheless been held that the convenience of such separation outweighs its shortcomings. The correct magnitudes of the various contributions remain an open question. Following Mulliken's explanation of the spectra of $\pi$-molecular compounds on the basis of charge transfer,[38,39] it has often been assumed (perhaps tacitly rather than actively) that the charge-transfer interaction also makes the dominant contribution to the intercomponent bonding. However, Dewar and Thompson[2] and others[384,466,467] have argued that the more important contributions come from dispersion and polarization forces. Indeed, it seems that current views on the intercomponent bonding (but not on the source of the charge-transfer spectra) are now returning to those put forward by Briegleb[10] in his first book on this field.

On the experimental side, one has to find a satisfactory way of defining the stability of the molecular compounds. Three levels of sophistication have been used, beginning with the simple test of whether a particular molecular compound has been prepared. Even at this level useful questions can be asked; for example, tetracene and pentacene are strong electron-donors ($I_D = 7.0$ and $6.6$ ev, respectively), yet their $\pi$-molecular compounds are (as yet) unknown. Why? Better criteria of molecular-compound stability can be based on chemical experience in their preparation; here the sequences of donor and acceptor strength resulting from the work of Sinomiya and of Baril and his co-workers are useful, if essentially qualitative in nature. The best criteria of molecular-compound stability are the standard free energies of formation, particularly if supplemented by values of the standard enthalpies and entropies of formation. Unfortunately the information available is severely limited, but even the little that there is contains some interesting indications. In some molecular compounds (*e.g.*, naphthalene:picric acid) $\Delta H_t^\circ$ and $\Delta S_f^\circ$ are, respectively, negative and positive; thus both enthalpy and entropy factors stabilize the molecular compound. In other examples (*e.g.*, anthracene:picric acid, quinhydrone) $\Delta H_f^\circ$ and $\Delta S_f^\circ$ are both negative, and here the enthalpy change stabilizes but the entropy change destabilizes the molecular compound. If one chooses a particular temperature, say 300°K, then it is convenient to consider these two situations as defining two groups of molecular compound. However, it is possible that a molecular compound with positive $\Delta S_f^\circ$ at one temperature will have negative $\Delta S_f^\circ$ at another, with intervention of phase or disorder–order transformations. The positive $\Delta S_f^\circ$ results partly from orientational (or perhaps positional) disorder in the molecular compound at higher temperatures; this has been demonstrated by structural studies and is discussed below. One presumes that $\Delta H_f^\circ$ does not vary greatly either with temperature or with details of crystal structure, but this remains to be proved; it is the measured enthalpy of formation that is to be compared with calculated interaction energies for different types of interaction. However, the stability of the molecular compound depends on its free energy of formation, and the entropy contribution to the free energy has so far not been accessible to theoretical study.

Thermodynamic measurements give some information about the influence of substituents in the donor or the acceptor on the properties of the molecular compound. For example, naphthalene:TNB and naphthalene:picric acid have the same melting point and $\Delta G_f^\circ$; they are also isomorphous. Presumably the OH group of the picric acid has negligible influence (at room temperature) on the properties of the respective molecular compounds. However, this is not a general rule, as is shown by comparison of anthracene:TNB and anthracene:picric acid, where melting point, $\Delta G_f^\circ$, and crystal structures all differ. Similarly, naphthalene:TNB and naphthalene:TNT differ.

The planar shapes of the component molecules and the delocalized interaction between them in π-molecular compounds lead one to expect crystal structures based on mixed stacks, with alternation of parallel donor and acceptor molecules. This prediction holds remarkably well in practice for the equimolar Mulliken-type molecular compounds. Some minor deviations from the general pattern are found; for example, the components deviate appreciably from parallelism in some of the copper and palladium oxinate molecular compounds and in anthracene:TNB. The largest deviation found so far from the usual structural type is in copper oxinate:TCNQ, where the TCNQ molecule bridges between two different copper oxinate molecules and the integrity of the mixed stacks is not preserved.

One of the basic tenets of Mulliken's theoretical approach is that adjacent donor and acceptor molecules within a particular stack will be positioned and oriented so that there is maximum overlap of the highest filled orbital(s) of the donor and the lowest unfilled orbital(s) of the acceptor. The available (quantum-mechanical) calculations suggest that the charge-transfer interaction energy is dependent on the mutual *positioning* of the donor and acceptor molecules but varies little with their mutual orientation. This assessment is not conclusive because there are other contributions to be taken into account; some of these, such as dispersion energy, would be expected to vary with mutual position but not with mutual orientation, but others, such as dipole–dipole or dipole–induced dipole interactions, would be expected to depend on both factors. The experimental results, which are inferred from the overlap diagrams presented earlier, are summarized in Table 27. About three-quarters of the $\pi$–$\pi^*$ molecular compounds show a recognizable orientation relation between adjacent donor and acceptor molecules within a molecular stack (but the very use of the word "recognizable" emphasizes the subjective nature of this classification). Theoretical correlations of mutual orientation and spectroscopic properties have been made for naphthalene:TCNE (Section VI-B); reasonable agreement was obtained but the limitations inherent in the calculations should be remembered. Roughly the same proportion of molecular compounds with orientation relationships is found among those where $\pi$–C=O interaction could be expected (the sample is too small and unrepresentative for any significance to be attached to the actual difference between the two groups). No theoretical treatment of the aromatic ring–carbonyl group interaction has appeared.

Orientational disorder is found in about 40% of the crystalline $\pi$-molecular compounds studied and this makes a significant contribution to the positive $\Delta S_\mathrm{f}^\circ$ found for many molecular compounds and thus has an important stabilizing role, particularly at higher temperatures. The unsymmetrical orientational relations that appear consistently in, say, the molecular compounds between aromatic hydrocarbons and TCNE or TCNB have not yet been explained

Table 27. Summary of crystallographic results showing orientation relation between neighbouring components within a stack and presence or absence of orientational disorder in the crystalline molecular compound at 300°K (or other temperature specified).

| Type of D:A interaction[a] | Donor (D) | Acceptor (A) | Orientation relation[b] | Orientational disorder in crystals at 300°K[b] — | Remarks |
|---|---|---|---|---|---|
| $\pi$–$\pi$* | Ferrocene | TCNE | ? | ? | Clear decision not possible |
| $\pi$–$\pi$* | Naphthalene | TCNE | Yes | Yes | |
| $\pi$–$\pi$* | Pyrene | TCNE | Yes | No | |
| $\pi$–$\pi$* | Perylene | TCNE | Yes | No | |
| $\pi$–$\pi$* | Copper oxinate | TCNQ | Yes | No | Non-standard structure |
| $\pi$–$\pi$* | TMPD | TCNQ | Yes | No | Ionic ground state |
| $\pi$–$\pi$* | Anthracene | TCNQ | No | Yes | |
| $\pi$–$\pi$* | TAB | TNB | ? | ? | Insufficient information |
| $\pi$–$\pi$* | 2,4,6-Tris(dimethylamino)-$s$-triazine | TNB | Yes | Yes | Complete overlap of the two components |
| $\pi$–$\pi$* | $p$-Iodoaniline | TNB | ? | ? | Insufficient information |
| ? | Anisoletricarbonylchromium | TNB | ? | No | |
| $\pi$–$\pi$* | Naphthalene | TNB | Yes | Probably | |
| $\pi$–$\pi$* | Azulene | TNB | Yes | Yes (140°K) | |
| $\pi$–$\pi$* | Skatole | TNB | Yes | Yes (140°K) | |
| $\pi$–$\pi$* | Indole | TNB | Yes | Yes (140°K) | |

| | | | | Localized interaction? | |
|---|---|---|---|---|---|
| π–π* | Anthracene | TNB | Yes | No (140°K) | |
| π–π* | Phenothiazine | TNB | Yes | (Yes)c | |
| π–π* | Acepleiadylene | TNB | Yes | No (140°K) | |
| ? | Copper oxinate | (Picryl azide)₂ | Yes | No | |
| ? | Copper oxinate | BTF | No | Yes | |
| π–π* | DDDT | BTF | Yes | ? | |
| π–π* | Naphthalene | PMDA | Yes | Yes | |
| π–π* | Anthracene | PMDA | Yes | No | |
| π–π* | Pyrene | PMDA | Yes | Yes (300°K) | |
| π–π* | Perylene | PMDA | Yes | No (120°K) | |
| π–π* | HMB | TCNB | No | No | |
| π–π* | TMPD | TCNB | Yes | No | |
| π–π* | Naphthalene | TCNB | Yes | Yes | |
| π–π* | Pyrene | TCNB | ? | Yes | |
| π–π* | Palladium oxinate | TCNB | No | No | |
| ? | Copper oxinate | (TCNB)₂ | Yes | No | |
| π–C=O | Hydroquinone | p-Benzoquinone | Yes | No | |
| π–C=O | p-Chlorophenol | p-Benzoquinone | Yes | No | |
| π–C=O | Thymine | p-Benzoquinone | No | ? | |
| π–C=O | (Hydroquinone)₂ | p-Benzoquinone | Yes | No | |
| π–C=O | (p-Chlorophenol)₂ | p-Benzoquinone | Yes | No | |
| π–C=O | Perylene | Fluoranil | Yes | No | |
| ? | HMB | Chloranil | No | ? | |
| π–π* | TMPD | Chloranil | Yes | ? | Ionic ground state; complete overlap of two components |

Table 27 (contd.)

| Type of D:A interaction [a] | Donor (D) | Acceptor (A) | Orientation relation [b] | Orientational disorder in crystals at 300°K [b] | Remarks |
|---|---|---|---|---|---|
| ? | Pyrene | Chloranil | No | No | |
| π–C=O | (8-Quinolinol)$_2$ | Chloranil | Yes | No | |
| π–C=O | Palladium oxinate | Chloranil | Yes | No | |
| ? | Pyrene | TMU | No | Yes | |
| ? | 3,4-Benzopyrene | (TMU)$_2$ | No | Yes | |
| ? | Coronene | (TMU)$_2$ | No | Yes | |

[a] The type of donor–acceptor interaction is inferred both from the nature of the components and from knowledge of the structure. Unambiguous definition is not always possible.

[b] The following notation describes the orientation relation found experimentally: Yes = orientation relation found. No = no orientation relation found. ? = Decision not possible on basis of available evidence. A similar notation is used to describe whether the crystals are ordered at 300°K: if structure analysis was carried out at another temperature, this is noted; thus, Yes (140°K) means that the crystals were disordered at 140°K.

[c] Different type of disorder.

theoretically. As mentioned above, the calculated interaction energies vary little with changes in the mutual orientations of donor and acceptor. First-order phase transformations also occur in a number of molecular compounds, but the crystal structures of different polymorphs have been determined only for $\alpha$- and $\beta$-quinhydrone; in these two polymorphs the components are arranged in sheets (with both hydrogen bonds and charge-transfer interactions between components), and the two polymorphs differ only in the relative arrangement of adjoining sheets. Differing arrangements of (almost identical) mixed stacks are perhaps found in the polymorphs of other molecular compounds, but this has still to be proved.

The probable dependence of detailed crystal structures on temperature, in contrast to the probable invariability of overall structure type in this family of molecular compounds, emphasizes the need for appropriate choice of temperatures (and, hopefully, pressures) at which crystal-structure analyses are performed. A convenient way of choosing the temperature for analysis is to refer to the specific heat–temperature curve of the substance being studied; in addition to showing the occurrence of phase transformations and giving some information about their nature, this curve provides the thermodynamic parameters of the substance as a function of temperature.

Compositions other than 1:1 are found among Mulliken-type molecular compounds, but less frequently than the equimolar compositions. Not much is known about the factors governing their stability or, in particular, why a 2:1 ratio, say, should be more stable than a 1:1 ratio in certain systems. The crystals studied so far are not typical $\pi$-molecular compounds, *e.g.*, phenoquinone and similar molecular compounds where hydrogen bonding is important, and coronene:$(TMU)_2$ and 3,4-benzopyrene:$(TMU)_2$ where charge transfer is probably unimportant. The only available result is for $(HMB)_2$: TCNE, where the IR spectra are interpreted in terms of a "sandwich" structure although confirmatory crystallographic results are lacking.

The polarized absorption spectra of single crystals of Mulliken-type molecular crystals are composed of the sum of the absorption spectra of the individual component molecules, with some shifts due to crystal-field effects, plus an additional band resulting from the charge-transfer. The charge-transfer band is polarized with its electric vector along the stack axis of the mixed stacks. The frequency of the charge-transfer band in the crystal is usually somewhat shifted with respect to that in solution; however, correlations between observed and calculated charge-transfer band frequencies are usually made for solution spectra and no specific calculation has been reported for crystal spectra. The only studies of charge-transfer spectra in crystals at low temperatures have been made for anthracene:TNB, the structure of which is not entirely typical of π-donor–acceptor molecular compounds. Spectroscopic studies at low temperatures, with their greatly enhanced resolution, should appreciably improve our knowledge of the details of the charge-transfer interaction between donor and acceptor.

Spectroscopic studies of small single crystals (by either absorption or reflection techniques) have been especially useful for differentiating between Mulliken and Weiss-type molecular compounds. The spectra of Weiss-type molecular compounds are made up of the charge-transfer band plus the spectra of the donor and the acceptor ions. Unequivocal proof of the existence of an ionic ground state is given by ESR spectroscopy; use of single-crystal specimens and study of the variations of the spectra with pressure and temperature allow the more subtle distinction between Wannier and Frenkel spin–exciton systems to be made. It is among these materials that the need for correlated studies by different techniques is most obviously necessary; unfortunately, instability of many of the materials may limit progress.

The summary above has attempted to highlight both achievements and problems. Where are the main gaps in our current knowledge and what paths are likely to reward exploration? First, as to materials for study: Until now primary emphasis has been on equimolar molecular compounds, mainly of symmetrical donors and acceptors. More attention should be given to unsymmetrical donors and acceptors, where dipole–dipole interactions should add to molecular-compound stability, and to compositions other than 1:1 and 2:1 (or 1:2). The main emphasis has been on Mulliken-type molecular compounds; attention should also be given to the substances on the periphery of the family tree (Figure 4, page 235), the Weiss-type molecular compounds, and the "isomeric complexes". And there are also the apparently anomalous examples of stable molecular compounds, such as HMB:TCNB and TMPD:TCNB, where there is no obvious orientation relation between the components. The crystal-structure results need to be supplemented by spectroscopic studies of single crystals.

Secondly as to techniques: The importance of crystal structure and spectroscopy needs no stress. However, execution of these studies at low temperatures (the physicists' low temperatures) is likely to repay the effort involved. Again, studies at isolated temperatures are useful but nowhere near as useful as study over a suitably chosen range of temperatures.

Thirdly, as to what should be studied: Properties have been rather neglected in comparison with structures (geometrical and electronic). There are hints of interesting electronic and catalytic properties but nothing very definite has yet emerged (conductivities of crystalline $\pi$-molecular compounds and ion-radical salts have been discussed by Le Blanc[468]). The marked anisotropy of most of the crystalline molecular compounds should surely be exploitable in some way.

Finally, it seems worthwhile emphasizing that the application of a battery of tests to a single (well chosen) substance is likely to yield better dividends than application of a single technique to many different materials. Many of the broad surveys essential in the preliminary stage of an investigation have been completed, and the time for study of a narrower field has arrived. Indeed, it is an interesting commentary on the working methods of individual scientists (including here the present author) that no single substance has yet been studied by all available techniques. The most work has perhaps been done on anthracene:TNB (specific-heat measurements are lacking), but this compound is not entirely typical of the Mulliken-type molecular compounds. But the tempo of modern science is such that there is little doubt that arrival at the top of Alice's hill will not be long delayed, however curiously the path thither winds and twists. Or is our epilogue wiser?

### Epilogue

"Would you tell me, please, which way I ought to go from here?"

"That depends a great deal on where you want to get to," said the Cat.

"I don't care much where—," said Alice.

"Then it doesn't matter which way you go," said the Cat.

"—so long as I get *somewhere*," Alice added as an explanation.

"Oh, you're sure to do that," said the Cat, "if you only walk long enough."

Alice felt that this could not be denied...

LEWIS CARROLL: *Alice in Wonderland*

# ADDENDA

## Section IV-E

**3.1. 10-Methylisoalloxazinium    bromide : sesqui-(2,7-naphthalene-diol) monohydrate,** $C_{11}H_{10}O_2N_4{}^+Br^- : 1\frac{1}{2}C_{10}H_8O_2 : H_2O$.

$$[a = 8.42, \; b = 10.50, \; c = 15.26 \text{ Å}, \; \alpha = 98.54, \; \beta = 109.9,$$
$$\gamma = 103.3°, \; Z = 2, \; P\bar{1}]$$

The structure of the black lath-like crystals of this molecular compound has been determined by Langhoff and Fritchie.[469] One naphthalenediol molecule lies above the flavine, over the phenylene and the central ring, suggesting partial transfer of charge from the diol to the lowest unoccupied orbital of the flavine. The formula has not been written in our standard donor–acceptor form because of the disposition of the remaining one-half diol. This necessarily appears centrosymmetric in the crystals and has disordered oxygen positions. It is steeply inclined to the plane of the flavin and is hydrogen-bonded to the bromide ion and water molecule. No information has been published about the overall arrangement of the components in the crystal, and it is not known whether mixed stacks are present.

## 3.2 Hydroquinone : riboflavine dihydrobromide

$$[a = 20.55, \; b = 13.69, \; c = 10.18 \text{ Å}, \; \beta = 91°9', \; Z = 4, \; P2_1]$$

Very unstable black crystals of this compound were grown from 6N-HBr solution containing an excess of diol (cf. Section IV-E-3). A crystallographic investigation[470] showed that the asymmetric unit consists of two molecules of hydroquinone, two doubly-protonated riboflavin cations, and four bromide anions. There was no evidence for water in the structure. The various units are hydrogen-bonded in a complicated fashion; what is important in the present context is the plane-to-plane association of hydroquinone molecules and riboflavine cations in pairs. It was inferred that these pairs were bonded by charge-transfer interactions. In this respect this compound resembles that described immediately above; however, there are a number of differences of detail.

## Section VI-C

### 7. [3,3]Paracyclophane : tetracyanoethylene

$$[a = 8.533, \; b = 8.538, \; c = 7.705 \text{ Å}, \; \alpha = 103.33, \; \beta = 110.78,$$
$$\gamma = 104.01°, \; Z = 1, \; P\bar{1}]$$

The usual quasi-hexagonal arrangement of mixed stacks is found, the stack axis being the body diagonal of the unit cell (Trueblood and Bernstein[471]). Each molecule occupies a crystallographic symmetry centre. The ethylene molecule is disordered and takes up two positions 90° apart in azimuth (Figure 80), which are occupied in a 3:1 ratio. This suggests that the disorder occurs during crystal growth and would not vary with temperature. The dimensions of the cyclophane molecule are not significantly affected by complex formation.

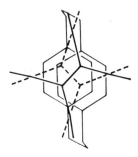

Figure 80. [3,3]Paracyclophane:TCNE overlap diagram. All three molecules are projected on to the least-squares plane of the four unsubstituted carbon atoms of the cyclophane. [Reproduced with permission from reference 471.]

## Section VI-E

### 16. 1-Bromo-2-naphthylamine:picric acid
$[a = 16.58, b = 16.99, c = 7.01$ Å, $\beta = 121.4°, Z = 4, P2_1/a$; cell re-oriented to conform to group 3b]

These two components form isomeric complexes: a red molecular compound and a yellow picrate (Hertel;[62] see Section VI-C-2). The molecular compound is stable above 117° but can be crystallized below this temperature in metastable form. However, the red crystals are transformed into yellow pseudomorphs during a period of weeks at room temperature; the yellow pseudomorphs do not give X-ray reflections and hence have no long-range order. The crystal structure of the molecular compound has been determined.[476] The familiar mixed stacks are found (stack axis [001]) and there is no hydrogen-bonding between the components. There was growth disorder in the crystals studied, one orientation of components being found to the extent of 83% and a second to the extent of 17%. Alternative orientations were found for both components. The overlap diagram for the major orientation is shown in Figures 81a and

b. The picric acid molecules have different positions with respect to naphthylamine molecules above and below in a particular stack. In one instance (Figure 81a) the same relative arrangement is found as in most TNB molecular compounds (overlap diagrams in Figure 26, page 285), while the second type has a displacement away from the usual arrangement. The angle between the ring planes is 1.7°. This material appears to be a typical molecular compound, in respect both to its physical properties and to the relative arrangement of the components.

### Section VI-I

#### 6. 8-Quinolinol : trinitrobenzene

$[a = 7.12, b = 8.06, c = 13.54 \text{ Å}, \alpha = 95.6°, \beta = 89.4°, \gamma = 92.2°;$
$Z = 2, P\bar{1}$; group 6c (cell has been reoriented to have stack
axis along [100])]

This molecular compound (Table 9, page 232) presents difficulties of classification: formally it should be in Section VI-E, but the crystallographic results of Castellano and Prout[473] show that it is very similar in structure to copper oxinate : $(TCNB)_2$ (reference 331; Section VI-I-3). The 8-quinolinol molecules are hydrogen-bonded to form centrosymmetric dimers [as in 8-quinolinol : chloranil[369] (VI-I-1)], and these dimers are interleaved by two TNB molecules, related by different centres of symmetry. The TNB molecules are disposed differently to the 8-quinolinol molecules above and below. In one instance (Figure 81c) the overlap is very similar to that between "naphthalene" groupings and TNB molecules (cf. Figure 26, page 285), while in the other (Figure 81d) there is a lateral displacement of the two molecules in the stack which largely nullifies this resemblance. A similar situation is found, for example, in 1-bromo-2-naphthylamine : picric acid[476] (Figures 81a and b). The interplanar distances between the dissimilar pairs are 3.39 and 3.42 Å.

### Section VI-K

#### 1. Triethylammonium bis-(7,7,8,8-tetracyanoquinodimethanide), $(TEA)^+ (TCNQ)_2^-$

$[a = 13.22, b = 14.44, c = 7.89 \text{ Å}, \alpha = 108.1, \beta = 103.6,$
$\gamma = 87.3°, Z = 2, P\bar{1}]$

A very brief mention of the crystal structure determined for $TEA(TCNQ)_2$ by Arthur (quoted in ref. 474) has been replaced by a full report by other workers.[475] The structure resembles that of

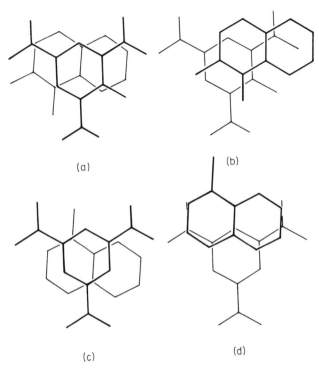

(a)                                        (b)

(c)                                        (d)

Figure 81. Overlap diagrams for 1-bromo-2-naphthylamine:picric acid (a,b) and 8-quinolinol:TNB (c,d) (cf. Figure 26 and, for c,d, also Figures 52 and 53). [Reproduced with permission from references 476 and 473.]

TMPD(TCNQ)$_2$ in that there are separate stacks (along [101] in the above unit cell) of TCNQ units separated by (somewhat disordered) TEA$^+$ ions. The schematic arrangement in the TCNQ stacks is compared with that in analogous stacks in other TCNQ ion-radical salts in Figure 65 (page 350). The two TCNQ units in TEA(TCNQ)$_2$ are crystallographically independent and each stack contains groups of four parallel units interrupted by jogs in the stacks. Jogs are found after two and three TCNQ units in TMPD(TCNQ)$_2$ and Cs$_2$(TCNQ)$_3$, respectively (Figure 65). The electrical conductivity of TEA(TCNQ)$_2$ has been reported as 4 ohm$^{-1}$ cm$^{-1}$ in a direction normal to the TCNQ molecules, 0.05 ohm$^{-1}$ cm$^{-1}$ in the TCNQ planes, and 0.001 ohm$^{-1}$ cm$^{-1}$ in a direction normal to the stacks of TCNQ and TEA$^+$.[477]

## General

Two theoretical papers[479,480] that appeared after this Review was completed have not been considered in the discussion.

## References

1. J. Fritzsche, *J. prakt. Chem.*, **73**, 282 (1858).
2. M. J. S. Dewar and C. C. Thompson, Jr., *Tetrahedron*, Suppl. 7, 97 (1966).
3. T. Matsuo and O. Higuchi, *Bull. Chem. Soc. Japan*, **41**, 518 (1968).
4. P. Pfeiffer, *Organische Molekulverbindungen*, 2nd edn., F. Enke, Stuttgart, 1928.
5. G. Briegleb, *Elektronen-Donator-Acceptor-Komplexe*, Springer, Berlin-Göttingen-Heidelberg, 1961.
6. J. Rose, *Molecular Complexes*, Pergamon Press, Oxford, 1967.
7. A. Szent-Gyorgi, *Introduction to a Submolecular Biology*, Academic Press, New York, 1960.
8. L. J. Andrews and R. M. Keefer, *Molecular Complexes in Organic Chemistry*, Holden-Day, San Francisco, 1964.
9. R. Foster, *Organic Charge Transfer Complexes*, Academic Press, London, 1969.
10. G. Briegleb, *Zwischenmolekularkräfte*, G. Brown, Karlsruhe, 1949.
11. L. J. Andrews, *Chem. Rev.*, **54**, 713 (1954).
12. L. E. Orgel, *Quart. Rev.*, **8**, 422 (1954).
13. A. N. Terenin, *Uspekhi Khim.*, **24**, 121 (1955).
14. S. P. McGlynn, *Chem. Rev.*, **58**, 1113 (1958); *Radiation Res. Suppl.*, **2**, 300 (1960).
15. D. Booth, *Sci. Progr.*, **48**, 335 (1960).
16. J. N. Murrell, *Quart. Rev.*, **15**, 287 (1961).
17. S. Nagakura, *Kagaku to Seibutsu*, **2**, 290 (1964) (in Japanese); *Chem. Abstr.*, **64**, 7385f (1966).
18. San Up Choi, *Hwahak Kwa Kongop Ui Chinbo*, **4**, 123 (1960) (in Korean); *Chem. Abstr.*, **62**, 13064h (1965).
19. S. F. Mason, *Quart. Rev.*, **15**, 287 (1961).
20. R. S. Mulliken and W. B. Person, *Ann. Rev. Phys. Chem.*, **13**, 107 (1962).
21. R. S. Mulliken, *J. Chim. Phys.*, **61**, 20 (1964).
22. H. Tsubomura and A. Kuboyama, *Kagaku to Kogyo*, **14**, 537 (1960) (in Japanese).
23. S. C. Wallwork, *J. Chem. Soc.*, **1961**, 494.
24. C. K. Prout and J. D. Wright, *Angew. Chem. Int. Ed., Engl.*, **7**, 659 (1968).
25. H. A. Bent, *Chem. Rev.*, **68**, 587 (1968).
26. R. S. Mulliken and W. B. Person, *Molecular Complexes—A Lecture and Reprint Volume*, Wiley, New York, 1969; also in *Physical Chemistry (An Advanced Treatise)*, Vol. III, edited by D. Henderson, Academic Press, New York/London, 1969.
27. A. W. Hanson, *Acta Cryst.*, **21**, 97 (1966).
28. A. W. Hanson, *Acta Cryst.*, **19**, 19 (1966).
29. A. W. Hanson, *Acta Cryst.*, **13**, 215 (1960).

30. J. M. Robertson, H. M. M. Shearer, C. A. Sim, and D. G. Watson, *Acta Cryst.*, **15**, 1 (1962).
31. G. Ferguson, I. R. Mackay, D. R. Pollard, and J. M. Robertson, *Acta Cryst.*, A, **25**, 132 (1969); I. R. Mackay, J. M. Robertson, and J. G. Sime, *Chem. Commun.*, **1969**, 1470.
32. T. M. Lowry, *Chem. & Ind. (London)*, **43**, 218 (1924).
33. G. M. Bennett and G. H. Willis, *J. Chem. Soc.*, **1929**, 256.
34. J. Martinet and L. Bornand, *Rev. gen. sci.*, **36**, 569 (1925); *Chem. Abstr.*, **20**, 861 (1926).
35. H. M. Powell, G. Huse, and P. W. Cooke, *J. Chem. Soc.*, **1943**, 153.
36. J. Weiss, *J. Chem. Soc.*, **1942**, 245.
37. W. Brackman, *Rec. Trav. Chim.*, **68**, 147 (1949).
38. R. S. Mulliken, *J. Amer. Chem. Soc.*, **72**, 600 (1952).
39. R. S. Mulliken, *J. Phys. Chem.*, **56**, 801 (1952).
40. R. S. Mulliken, *Rec. Trav. Chim.*, **75**, 845 (1956).
41. W. K. Duerksen and M. Tamres, *J. Amer. Chem. Soc.*, **90**, 1379 (1968).
42. M. Kroll, *J. Amer. Chem. Soc.*, **90**, 1097 (1968).
43. E. Hertel and H. Kleu, *Z. Phys. Chem.*, B, **11**, 59 (1930).
44. A. R. Lepley and J. P. Thelman, *Tetrahedron*, **22**, 101 (1966).
45. P. R. Hammond, *Nature*, **206**, 891 (1965).
46. P. R. Hammond, *J. Chem. Soc.*, A, **1968**, 145.
47. G. Briegleb, *Angew. Chem. Int. Ed., Engl.*, **3**, 617 (1964).
48. H. M. Rosenberg, E. Eimutis, and D. Hale, *J. Phys. Chem.*, **70**, 4096 (1966).
49. T. K. Mukherjee, *Tetrahedron*, **24**, 721 (1968).
50. J. J. Weiss, *Phil. Mag.*, **8**, 1169 (1963).
51. R. D. Kross and V. A. Fassel, *J. Amer. Chem. Soc.*, **79**, 38 (1957).
52. H. Kainer and A. Überle, *Chem. Ber.*, **88**, 1147 (1955).
53. K. M. C. Davis and M. C. R. Symons, *J. Chem. Soc.*, **1965**, 2079.
54. H. M. McConnell, B. M. Hoffman, and R. M. Metzger, *Proc. Nat. Acad. Sci. U.S.*, **53**, 46 (1965).
55. H. M. McConnell in *Molecular Biophysics*, B. Pullman and M. Weissbluth, ed., Academic Press, N.Y., 1965, p. 311.
56. P. L. Nordio, Z. G. Soos, and H. M. McConnell, *Ann. Rev. Phys. Chem.*, **17**, 237 (1966).
57. T. Sakata and S. Nagakura, *Bull. Chem. Soc. Japan*, **43**, 713 (1970).
58. R. C. Hughes and Z. G. Soos, *J. Chem. Phys.*, **48**, 1066 (1968).
59. G. T. Pott, Thesis, Groningen, Netherlands, 1966.
60. G. T. Pott and J. Kommandeur, *Mol. Phys.*, **13**, 373 (1967).
61. E. Hertel and J. van Cleef, *Ber.*, **61**, 1545 (1928).
62. E. Hertel, *Ann. Chem.*, **451**, 179 (1926).
63. R. Râscanu, *Ann. Sci. Univ. Jassy*, **26**, Pt. I, 3 (1940); *Chem. Abstr.*, **34**, 4385 (1940).
64. E. Hertel and H. Frank, *Z. Phys. Chem.*, B, **27**, 460 (1934).
65. E. Hertel and K. Schneider, *Z. Phys. Chem.*, B, **13**, 387 (1931).
66. G. Briegleb and H. Delle, *Z. Phys. Chem.*, **24**, 359 (1960).
67. G. Briegleb and H. Delle, *Z. Elektrochem.*, **64**, 347 (1960).
68. R. D. Kross, K. Nakamoto, and V. A. Fassel, *Spectrochim. Acta*, **8**, 142 (1956).

69. E. K. Andersen, *Acta Chem. Scand.*, **8**, 157 (1954).
70. K. J. Pedersen, *J. Amer. Chem. Soc.*, **56**, 2615 (1934).
71. E. Hertel and K. Schneider, *Z. Phys. Chem.*, *B*, **12**, 109 (1931).
72. S. Chatterjee, *J. Chem. Soc.*, *B*, **1967**, 1170.
73. C. C. Addison and J. C. Sheldon, *J. Chem. Soc.*, **1956**, 1941.
74. K. O. Strømme, *Acta Cryst.*, *B*, **24**, 1607 (1968).
75. L. J. Andrews and R. M. Keefer, *Adv. Inorg. Radiochem.*, **3**, 91 (1961).
76. A. Werner, *Ber. Deut. Chem. Ges.*, **42**, 4235 (1909).
77. T. Urbanski, M. Piskorz, W. Cetner, and M. Maciejewski, *Bull. Acad. Polon. Sci., Ser. Sci. Chem.*, **10**, 263 (1962).
78. W. Will, *Ber. Deut. Chem. Ges.*, **47**, 964 (1914).
79. P. Noble, Jr., W. L. Reed, C. J. Hoffman, J. A. Gallaghan, and F. G. Borgardt, *Amer. Inst. Aeronaut. Astronaut. J.*, **1**, 395 (1963).
80. T. L. Cairns, R. A. Carboni, D. D. Coffman, V. A. Englehardt, R. E. Heckert, E. L. Little, E. G. McGeer, B. C. McKusick, S. J. Middleton, R. M. Scribner, C. W. Theobald, and H. E. Winberg, *J. Amer. Chem. Soc.*, **80**, 2775 (1958).
81. O. W. Webster, *J. Amer. Chem. Soc.*, **86**, 2898 (1964).
82. B. C. McKusick and O. W. Webster, U.S. Pat. 3 214 455; *Chem. Abstr.*, **64**, 1909d (1966).
83. D. S. Acker and W. R. Hertler, *J. Amer. Chem. Soc.*, **84**, 3370 (1962).
84. L. R. Melby, R. J. Harder, W. R. Hertler, W. Mahler, R. E. Benson, and W. E. Mochel, *J. Amer. Chem. Soc.*, **84**, 3374 (1962).
85. J. Dickman, W. R. Hertler, and R. E. Benson, *J. Org. Chem.*, **28**, 2719 (1963).
86. T. Urbanski, *Chemistry and Technology of Explosives*, Vol. I, Pergamon Press, Oxford, 1964.
87. O. C. Dermer and R. B. Smith, *J. Amer. Chem. Soc.*, **61**, 748 (1939).
88. T. Asahina and T. Sinomiya, *J. Chem. Soc. Japan*, **59**, 341 (1938); *Chem. Abstr.*, **32**, 9075 (1938).
89. T. Sinomiya, *J. Chem. Soc. Japan*, **59**, 833 (1938); *Chem. Abstr.*, **32**, 9076 (1938).
90. T. Sinomiya, *J. Chem. Soc. Japan*, **59**, 922 (1938); *Chem. Abstr.*, **33**, 563 (1939).
91. T. Sinomiya, *J. Chem. Soc. Japan*, **60**, 170 (1939); *Chem. Abstr.*, **34**, 2360 (1940).
92. T. Sinomiya, *Bull. Chem. Soc. Japan*, **15**, 92 (1940).
93. T. Sinomiya, *Bull. Chem. Soc. Japan*, **15**, 137 (1940).
94. T. Sinomiya, *Bull. Chem. Soc. Japan*, **15**, 259 (1940).
95. T. Sinomiya, *Bull. Chem. Soc. Japan*, **15**, 281 (1940).
96. T. Sinomiya, *Bull. Chem. Soc. Japan*, **15**, 309 (1940).
97. S. J. Stephens, G. E. Hargis, and J. B. Entrikin, *Proc. Louisiana Acad. Sci.*, **10**, 210 (1947); *Chem. Abstr.*, **42**, 1921 (1948).
98. C. A. Buehler and A. G. Heap, *J. Amer. Chem. Soc.*, **48**, 3168 (1926).
99. R. Kremann, *Monatsh.*, **29**, 863 (1908).
100. A. Kent, *J. Chem. Soc.*, **1935**, 976.
101. J. J. Sudborough, *J. Chem. Soc.*, **109**, 1339 (1916).
102. S. T. Cadre and J. J. Sudborough, *J. Chem. Soc.*, **109**, 1349 (1916).
103. C. Weygand and T. Siebenmark, *Chem. Ber.*, **73**, 765 (1940).
104. R. Kühn and A. Winterstein, *Helv. Chim. Acta*, **11**, 144 (1928).

105. H. Shosenji and T. Matsuo, *Nippon Kagaku Zasshi*, **87**, 802 (1966); *Chem. Abstr.*, **66**, 37193d (1967).
106. T. Matsuo and H. Aiga, *Bull. Chem. Soc. Japan*, **41**, 271 (1968).
107. A. Kraak and H. Wynberg, *Tetrahedron*, **24**, 3881 (1968).
108. R. Kremann and R. Muller II, *Monatsh.*, **42**, 181 (1921).
109. A. Kofler, *Z. Elektrochem.*, **50**, 200 (1944).
110. D. Ll. Hammick, L. W. Andrew, and J. Hampson, *J. Chem. Soc.*, **1932**, 171.
111. D. Ll. Hammick and T. K. Hanson, *J. Chem. Soc.*, **1933**, 669.
112. L. A. Burkhardt, *J. Phys. Chem.*, **66**, 1196 (1962).
113. G. Gafner and F. H. Herbstein, *J. Chem. Soc.*, **1964**, 5290.
114. G. M. Bennett and R. L. Wain, *J. Chem. Soc.*, **1936**, 1108.
115. O. C. Dermer, *Proc. Oklahoma Acad. Sci.*, **23**, 160 (1941); *Chem. Abstr.*, **37**, 4376 (1942).
116. C. A. Buehler, A. Hisey, and J. H. Wood, *J. Amer. Chem. Soc.*, **52**, 1939 (1930).
117. T. Reichstein, *Helv. Chim. Acta*, **2**, 802 (1926).
118. N. P. Buu-Hoï, P. Jacquignon, and O. Roussel, *Rec. Trav. Chim.*, **82**, 370 (1963).
119. C. J. Fritchie, Jr., and B. L. Trus, *Chem. Commun.*, **1968**, 833.
120. O. L. Baril and E. S. Hauber, *J. Amer. Chem. Soc.*, **53**, 1087 (1931).
121. O. L. Baril and G. A. Megrdichian, *J. Amer. Chem. Soc.*, **58**, 1415 (1936).
122. V. H. Dermer and O. C. Dermer, *J. Org. Chem.*, **3**, 289 (1938).
123. R. Râscanu, *Ann. Sci. Univ. Jassy*, **25**, Pt. 1, 395 (1939); *Chem. Abstr.*, **34**, 394 (1940).
124. O. L. Baril, *Rep. New England Ass. Chem. Teachers*, **41**, 15 (1939); *Chem. Abstr.*, **34**, 1634 (1940).
125. R. P. Mariella, M. J. Gruber, and J. W. Elder, *J. Org. Chem.*, **26**, 3217 (1961).
126. J. W. Elder and R. P. Mariella, *J. Chem. Eng. Data*, **9**, 402 (1964).
127. N. N. Efremov, *J. Russ. Phys. Chem. Soc.*, **50**, 372, 421, 441 (1918); **51**, 353 (1919); *Chem. Abstr.*, **17**, 3327–8 (1923).
128. N. A. Pushin and P. Kozuhar, *Glasnik Khem. Drushtva Beograd*, (*Bull. Soc. Chim. Belgrade*), **12**, 101 (1947); *Chem. Abstr.*, **43**, 6066 (1949).
129. E. Ya. Mindovich, *Zhur. Fiz. Khim.*, **30**, 1082 (1956); *Chem. Abstr.*, **50**, 16327 (1956).
130. R. P. Bell and A. J. Fendley, *Trans. Faraday Soc.*, **45**, 121 (1949).
131. O. Dimroth and C. Bamberger, *Ann. Chem.*, **438**, 67 (1924).
132. S. V. Gorbachev and E. Ya. Mindovich, *Zhur. Fiz. Khim.*, **27**, 1391 (1953); *Chem. Abstr.*, **49**, 5931 (1954).
133. E. Ya. Mindovich and S. V. Gorbachev., *Zhur. Fiz. Khim.*, **27**, 1686 (1953); *Chem. Abstr.*, **48**, 4908 (1954).
134. N. N. Efremov and A. M. Tikhomirova, *Ann. Inst. Anal. Phys.-chim.* (*Leningrad*), **4**, 92 (1928); *Chem. Abstr.*, **23**, 3214 (1929).
135. N. N. Efremov and A. M. Tikhomirova, *Ann. Inst. Anal. Phys.-chim.* (*Leningrad*), **4**, 65 (1928); *Chem. Abstr.*, **23**, 2349 (1929).
136. D. Ll. Hammick and A. Hellicar, *J. Chem. Soc.*, **1938**, 761.
137. J. J. Sudborough, N. Picton, and D. D. Karve, *J. Indian Inst. Sci.*, **4**, 43 (1921); *Chem. Abstr.*, **16**, 560 (1922).
138. P. M. G. Bavin, *Can. J. Chem.*, **38**, 1017 (1960).

139. A. S. Bailey and J. R. Case, *Tetrahedron*, **3**, 113 (1958).
140. A. S. Bailey, *J. Chem. Soc.*, **1960**, 4710.
141. A. S. Bailey and J. M. Evans, *Chem. & Ind. (London)*, **1964**, 1424.
142. T. S. Cameron and C. K. Prout, *Chem. Commun.*, **1968**, 684.
143. K. Wallenfels and K. Friedrich, *Tetrahedron Letters*, **1963**, 1223.
144. A. S. Bailey, B. R. Henn, and J. M. Langdon, *Tetrahedron*, **19**, 161 (1963).
145. T. Kobayaski, S. Iwata, and S. Nagakura, *Bull. Chem. Soc. Japan*, **43**, 713 (1970).
146. A. S. Bailey, R. J. P. Williams, and J. D. Wright, *J. Chem. Soc.*, **1965**, 2579.
147. D. S. Pratt and G. A. Perkins, *J. Amer. Chem. Soc.*, **40**, 198 (1918).
148. P. Jacquignon and N. P. Buu-Hoï, *Bull. Soc. Chim. France*, **1957**, 488.
149. P. Jacquignon and N. P. Buu-Hoï, *Bull. Soc. Chim. France*, **1957**, 1272.
150. P. Jacquignon and N. P. Buu-Hoï, *Bull. Soc. Chim. France*, **1958**, 761.
151. N. P. Buu-Hoï, P. Jacquignon, and O. Roussel, *Bull. Soc. Chim. France*, **1962**, 1652.
152. N. P. Buu-Hoï and P. Jacquignon, *Experientia*, **13**, 375 (1957).
153. B. Getting, C. R. Patrick, and J. C. Tatlow, *J. Chem. Soc.*, **1961**, 1574.
154. R. Rudman, *Acta Cryst.*, *B*, **27**, 262 (1971).
155. R. Seka and H. Sedlatschek, *Monatsh.*, **47**, 516 (1926).
156. L. L. Ferstandig, W. G. Toland, and C. D. Heaton, *J. Amer. Chem. Soc.*, **83**, 1151 (1961).
157. J. C. A. Boeyens and F. H. Herbstein, *J. Phys. Chem.*, **69**, 2153 (1965).
158. Y. Nakayama, Y. Ichikawa, and T. Matsuo, *Bull. Chem. Soc. Japan*, **38**, 1674 (1965).
159. H. Meyer and H. Raudnitz, *Ber.*, **63**, 2010 (1930).
160. I. S. Mustafin, *J. Gen. Chem. (U.S.S.R.)*, **17**, 560 (1947); *Chem. Abstr.*, **42**, 890 (1948).
161. P. Jacquignon and N. P. Buu-Hoï, *Bull. Soc. Chim. France*, **1960**, 1618.
162. P. Jacquignon, N. P. Buu-Hoï, and C. Desjardin, *Bull. Soc. Chim. France*, **1962**, 312.
163. G. Gafner and F. H. Herbstein, *Nature*, **200**, 130 (1963).
164. D. Bryce-Smith and A. Gilbert, *J. Chem. Soc.*, **1965**, 918.
165. D. Bryce-Smith and M. A. Herns, *Tetrahedron*, **25**, 247 (1969).
166. P. O. Tawney, R. H. Snyder, R. P. Conger, K. A. Leibbrand, C. H. Stiteler, and A. R. Williams, *J. Org. Chem.*, **26**, 15 (1961).
167. P. Pfeiffer, F. Goebel, and O. Angern, *Ann. Chem.*, **440**, 241 (1924).
168. D. E. Williams, *J. Amer. Chem. Soc.*, **89**, 4280 (1967).
169. W. S. Rapson, D. H. Saunder, and E. T. Stewart, *J. Chem. Soc.*, **1946**, 1110.
170. K. Abe, Y. Matsunaga, and G. Saito, *Bull. Chem. Soc. Japan*, **41**, 2852 (1968).
171. D. H. Saunder, *Proc. Roy. Soc.*, *A.*, **190**, 508 (1947).
172. K. Nakamoto, *Bull. Chem. Soc. Japan*, **26**, 70 (1953).
173. D. Ll. Hammick and R. B. Williams, *J. Chem. Soc.*, **1935**, 1866.
174. D. Ll. Hammick, E. H. Reynolds, and G. Sixsmith, *J. Chem. Soc.*, **1939**, 98.
175. *Dictionary of Organic Compounds*, J. R. A. Pollock and R. Stevens, eds., 4th edn., Eyre and Spottiswoode, London, 1965, p. 1610.
176. D. Ll. Hammick and G. Sixsmith, *J. Chem. Soc.*, **1939**, 272.
177. K. H. Mertens, *Ber. Deut. Chem. Ges.*, **11**, 843 (1878).

178. S. K. Das, R. A. Shaw, and B. C. Smith, *Chem. Commun.*, **1965**, 176.
179. O. Hassel and C. Rømming, *Quart. Rev.*, **16**, 1 (1962).
180. S. K. Das, R. A. Shaw, B. C. Smith, and C. P. Thakur, *Chem. Commun.*, **1966**, 33.
181. R. Kremann, S. Sutter, F. Sitte, H. Strzelba, and A. Dolotzky, *Monatsh.*, **43**, 269 (1922).
182. D. E. Laskowski, *Anal. Chem.*, **32**, 1171 (1960).
183. D. E. Laskowski, *Anal. Chem.*, **38**, 1188 (1966).
184. B. Turcsany and F. Tudos, *Magy. Kem. Folyoirat*, **71**, 39 (1965); *Chem. Abstr.*, **62**, 16160 (1969).
185. F. Tudos and B. Turcsany, *Magy. Kem. Folyoirat*, **67**, 228 (1962); *Chem. Abstr.*, **57**, 13701 (1962).
186. R. Forster and T. J. Thomson, *Trans. Faraday Soc.*, **58**, 860 (1962).
187. T. Amano, H. Kuroda, and H. Akamatu, *Bull. Chem. Soc. Japan*, **42**, 671 (1969).
188. D. E. Laskowski, *Cancer Res.*, **27**, 903 (1967).
189. A. Pullman and B. Pullman, *Nature*, **196**, 228 (1962); B. Pullman and A. Pullman, *Nature*, **199**, 467 (1963); A. Pullman and H. Berthod, *Biochim. Biophys. Acta*, **66**, 277 (1963).
190. L. Michaelis, *Chem. Rev.*, **16**, 243 (1935).
191. Jackson and Oenslager, *Amer. Chem. J.*, **18**, 1 (1896).
192. K. H. Meyer, *Ber. Deut. Chem. Ges.*, **42**, 1149 (1909).
193. L. Michaelis and S. Granick, *J. Amer. Chem. Soc.*, **66**, 1023 (1944).
194. W. Siegmund, *Monatsh.*, **29**, 1089 (1908).
195. W. Schlenk, *Ann. Chem.*, **368**, 277 (1909).
196. E. Weitz and F. Schmidt, *J. Prakt. Chem.*, **158**, 211 (1941).
197. T. Sakurai and M. Okunuki, *Sci. Papers Inst. Phys. Chem. Res.*, **63**, No. 1 (1969).
198. P. Pfeiffer, W. Jowleff, P. Fischer, P. Monti, and H. Mulby, *Ann. Chem.*, **412**, 253 (1917).
199. J. W. Eastman, G. M. Androes, and M. Calvin, *J. Chem. Phys.*, **36**, 1197 (1962).
200. J. Thiele and F. Gunther, *Ann. Chem.*, **349**, 45 (1906).
201. A. Ottenberg, R. L. Brandon, and M. E. Browne, *Nature*, **201**, 1119 (1964).
202. K. Wallenfels, D. Hofmann, and R. Kern, *Tetrahedron*, **21**, 3231 (1965).
203. K. Wallenfels, G. Bachmann, D. Hofmann, and R. Kern, *Tetrahedron*, **21**, 2239 (1965).
204. R. L. Hansen, *J. Org. Chem.*, **33**, 3968 (1968).
205. B. T. Gorres and G. E. Gurr, Winter meeting of American Crystallographic Association, Tucson, Ariz., February 1968.
206. P. R. Hammond, *Science*, **142**, 502 (1963).
207. J. Willems, *Naturwiss.*, **32**, 324 (1944).
208. J. Willems, *Z. Naturforsch.*, **2**b, 89 (1947).
209. J. Fritzsche, *Z. Chem.*, **1869**, 114.
210. E. Schmidt, *J. Prakt. Chem.*, **9**, 241 (1874).
211. E. Bornstein, H. Schliewiensky, and G. V. Szczesny-Heyl, *Ber. Deut. Chem. Ges.*, **59**, 2812 (1926).
212. E. Bornstein, H. Schliewiensky, and G. V. Szczesny-Heyl, *Z. Angew. Chem.*, **39**, 678 (1926); *Chem. Abstr.*, **21**, 1115 (1927).

213. E. Hertel and G. H. Römer, *Z. Phys. Chem.*, *B*, **11**, 90 (1930).
214. S. Coffey and J. van Alphen in E. H. Rodd's *Chemistry of Carbon Compounds*, Vol. III$^B$, Elsevier, Amsterdam, 1961, p. 1408.
215. E. Hertel and H. Kurth, *Ber. Deut. Chem. Ges.*, **61**, 1650 (1928).
216. M. Orchin and E. O. Woolfolk, *J. Amer. Chem. Soc.*, **68**, 1727 (1946).
217. M. Orchin, L. Reggel, and E. O. Woolfolk, *J. Amer. Chem. Soc.*, **69**, 1225 (1947).
218. D. E. Laskowski, D. G. Graber, and W. C. McCrone, *Anal. Chem.*, **25**, 1400 (1953).
219. K. H. Takemura, M. D. Cameron, and M. S. Newman, *J. Amer. Chem. Soc.*, **75**, 3280 (1953).
220. R. Pepinsky, Final report under AF49(638)-1044, ASTIA no. AD-285512 (1962).
221. I. Tickle and C. K. Prout, personal communication, 1970.
222. F. E. Ray and W. Francis, *J. Org. Chem.*, **8**, 52 (1943).
223. M. S. Newman and W. B. Lutz, *J. Amer. Chem. Soc.*, **78**, 2469 (1943).
224. T. K. Mukherjee and L. A. Levasseur, *J. Org. Chem.*, **30**, 644 (1965).
225. W. Steinkopf and T. Hopner, *Ann. Chem.*, **501**, 174 (1933).
226. J. Rosenberg, *Ber. Deut. Chem. Ges.*, **18**, 1773 (1885).
227. W. A. Duncan and F. L. Swinton, *Trans. Faraday Soc.*, **62**, 1083 (1966).
228. D. V. Fenby, I. A. McLure, and R. L. Scott, *J. Phys. Chem.*, **70**, 602 (1966).
229. M. E. Daur, D. A. Horsma, C. M. Knobler, and P. Percy, *J. Phys. Chem.*, **73**, 641 (1969).
230. T. G. Beaumont and K. M. C. Davis, *J. Chem. Soc.*, *B*, **1967**, 1131.
231. R. Foster and C. A. Fyfe, *Chem. Commun.*, **1965**, 642.
232. D. F. R. Gibson and C. A. McDowell, *Can. J. Chem.*, **44**, 945 (1966).
233. E. McLaughlin and C. E. Messer, *J. Chem. Soc.*, *A*, **1966**, 1106.
234. G. M. Brooke, J. Burden, M. Stacey, and J. C. Tatlow, *J. Chem. Soc.*, **1960**, 1768.
235. P. R. Hammond, *J. Chem. Soc.*, *A*, **1968**, 145.
236. G. A. Corker and M. Calvin, *J. Chem. Phys.*, **49**, 5547 (1968).
237. D. A. Armitage and K. W. Morcom, *Trans. Faraday Soc.*, **65**, 688 (1969).
238. C. R. Patrick and G. S. Prosser, *Nature*, **187**, 1021 (1960).
239. G. K. Semin, V. I. Robas, V. D. Shteingarts, and G. G. Yakobson, *J. Struct. Chem.*, **6**, 143 (1965).
240. C. B. Knobler, personal communication (1968).
241. V. B. Smith and A. G. Massey, *Tetrahedron*, **25**, 5595 (1969).
242. R. Filler and E. W. Choe, *J. Amer. Chem. Soc.*, **91**, 1862 (1969).
243. Y. Matsunaga, *Nature*, **211**, 182 (1966).
244. H. Weil-Malherbe, *Biochem. J.*, **40**, 351 (1946).
245. J. Booth and E. Boyland, *Biochim. Biophys. Acta*, **12**, 75 (1953).
246. J. Booth, E. Boyland, and S. F. D. Orr, *J. Chem. Soc.*, **1954**, 598.
247. E. D. Bergmann in *Molecular Associations in Biology*, B. Pullman, ed., Academic Press, London and New York, 1968, p. 207.
248. B. L. van Duuren, *Nature*, **210**, 622 (1966); *J. Phys. Chem.*, **68**, 2544 (1964).
249. J. Caillet and B. Pullman in *Molecular Associations in Biology*, B. Pullman, ed., Academic Press, London and New York, 1968, p. 217.

250. J. D. Mold, T. B. Walker, and L. G. Veasey, *Anal. Chem.*, **35**, 2071 (1963).
251. H. Beinert and P. Hemmerich in *Encyclopedia of Biochemistry*, R. J. Williams and E. M. Lansford, Jr., ed., Reinhold, N.Y., 1967, p. 331.
252. G. Tollin in *Molecular Associations in Biology*, B. Pullman, ed., Academic Press, N.Y., 1968, p. 393.
253. E. M. Kosower in *Flavins and Flavoproteins*, E. C. Slater, ed., Elsevier, Amsterdam, 1966, p. 1.
254. D. E. Fleischman and G. Tollin, *Proc. Nat. Acad. Sci.*, *U.S.*, **53**, 38 (1965).
255. J. F. Pereira and G. Tollin, *Biochim. Biophys. Acta*, **143**, 79 (1967).
256. D. E. Fleischman and G. Tollin, *Biochim. Biophys. Acta*, **94**, 248 (1965).
257. R. Kuhn and R. Strobele, *Ber. Deut. Chem. Ges.*, **70**, 753 (1937).
258. D. E. Fleischman and G. Tollin, *Proc. Nat. Acad. Sci.*, *U.S.*, **53**, 237 (1965).
259. P. Kierkegaard, *Acta Cryst.*, *A*, **25**, *S*193 (1969).
260. C. J. Fritchie, Jr., B. L. Trus, and C. A. Langhoff, *Acta Cryst.*, *A*, **25**, S193 (1969).
261. D. Ll. Hammick, G. M. Hills, and J. Howard, *J. Chem. Soc.*, **1932**, 1530.
262. E. Hertel and H. W. Bergk, *Z. Phys. Chem.*, *B*, **33**, 319 (1936).
263. G. T. Morgan and J. G. Mitchell, *J. Chem. Soc.*, **1934**, 536.
264. K. Brass and E. Tengler, *Ber. Deut. Chem. Ges.*, **64**, 1650 (1931).
265. Bruni, *Chem. Ztg.*, **30**, 568 (through Beilstein, 4th edn., Vol. 5, 1922, p. 579.
266. G. Briegleb and T. Schachowdkoy, *Z. Phys. Chem.*, *B*, **19**, 255 (1932).
267. M. Orchin, *J. Org. Chem.*, **16**, 1165 (1951).
268. M. S. Newman, R. S. Darlak, and L. Tsai, *J. Amer. Chem. Soc.*, **89**, 6191 (1967).
269. C. J. Fritchie, Jr., *J. Chem. Soc.*, *A*, **1969**, 1328.
270. S. Sastry, *J. Chem. Soc.*, **109**, 270 (1916).
271. T. P. Rastogi and N. B. Singh, *J. Phys. Chem.*, **70**, 3315 (1966).
272. T. McL. Spotswood, *Austl. J. Chem.*, **15**, 278 (1962).
273. N. Wiberg, *Angew. Chem. Int. Ed.*, *Engl.*, **7**, 766 (1968).
274. H. E. Winberg, U.S. Pat. 3 239 518 to Du Pont, 8 March 1966; *Chem. Abstr.*, **64**, 15 898e (1966).
275. P. R. Hammond and R. H. Knipe, *J. Amer. Chem. Soc.*, **89**, 6063 (1967).
276. J. J. Sudborough and S. H. Beard, *J. Chem. Soc.*, **97**, 773 (1910).
277. A. Kent and D. McNeil, *J. Chem. Soc.*, **1938**, 8.
278. A. Kent, D. McNeil, and R. M. Cowper, *J. Chem. Soc.*, **1939**, 1858.
279. S. K. Das, R. A. Shaw, B. C. Smith, W. A. Last, and F. B. G. Wells, *Chem. & Ind.* (*London*), **1963**, 866.
280. S. K. Das, T. Gunduz, R. A. Shaw, and B. C. Smith, *J. Chem. Soc.*, *A*, **1969**, 1403.
281. R. M. Williams and S. C. Wallwork, *Acta Cryst.*, **21**, 406 (1966).
282. J. J. Sudborough and S. H. Beard, *J. Chem. Soc.*, **99**, 209 (1911).
283. B. Kamenar and C. K. Prout, *J. Chem. Soc.*, **1965**, 4838.
284. Y. Matsunaga, *J. Chem. Phys.*, **42**, 2248 (1965).
285. Y. Matsunaga, *J. Chem. Phys.*, **41**, 1609 (1964).
286. J. C. Goan, E. Berg, and H. E. Podall, *J. Org. Chem.*, **29**, 975 (1964).

287. R. L. Brandon, J. H. Osiecki, and A. Ottenberg, *J. Org. Chem.*, **31**, 1214 (1966).

288. M. Rosenblum, R. W. Fish, and C. Bennett, *J. Amer. Chem. Soc.*, **86**, 5166 (1964).

289. E. Adman, M. Rosenblum, S. Sullivan, and T. N. Margulis, *J. Amer. Chem. Soc.*, **89**, 4541 (1967).

290. R. L. Collins and R. Pettit, *J. Inorg. Nuclear Chem.*, **29**, 503 (1967).

291. J. W. Fitch and J. J. Lagowski, *Inorg. Chem.*, **4**, 864 (1965).

292. O. L. Carter, A. T. McPhail, and G. A. Sim, *J. Chem. Soc.*, *A*, **1966**, 822.

293. A. Treibs, *Ann. Chem.*, **476**, 1 (1929).

294. H. A. O. Hill, A. J. MacFarlane, and R. J. P. Williams, *Chem. Commun.*, **1967**, 905.

295. S. Iwata, J. Tanaka, and S. Nagakura, *J. Amer. Chem. Soc.*, **88**, 894 (1966).

296. C. K. Prout, R. J. P. Williams, and J. D. Wright, *J. Chem. Soc.*, *A*, **1966**, 747.

297. D. Ll. Hammick and H. P. Hutchison, *J. Chem. Soc.*, **1955**, 89.

298. J. N. Brønsted, *Z. Phys. Chem.*, **78**, 284 (1911).

299. F. S. Brown, *J. Chem. Soc.*, **1925**, 345.

300. K. Suzuki and S. Seki, *Bull. Chem. Soc. Japan*, **28**, 417 (1955).

301. J. C. A. Boeyens and F. H. Herbstein, *J. Phys. Chem.*, **69**, 2160 (1965).

302. I. Ilmet and L. Kopp, *J. Phys. Chem.*, **70**, 3371 (1966).

303. F. H. Herbstein and J. A. Snyman, *Phil. Trans. Roy. Soc.*, *A*, **264**, 635 (1969).

304. E. Hertel and G. H. Römer, *Z. Phys. Chem.*, *B*, **11**, 77 (1931).

305. F. H. Herbstein and H. Regev, unpublished results, 1969.

306. R. Kreman, H. Hohl, and R. Muller II, *Monatsh.*, **42**, 199 (1912).

307. K. Elbs, *J. prakt. Chem.*, (ii), **47**, 44 (1893).

308. M. Ohmasa, M. Kinoshita, and H. Akamatu, *Bull. Chem. Soc. Japan*, **42**, 2402 (1969).

309. S. S. Chu, G. A. Jeffrey, and T. Sakurai, *Acta Cryst.*, **15**, 661 (1962).

310. E. Hertel and H. W. Bergk, *Z. Phys. Chem.*, *B*, **33**, 319 (1936).

311. D. M. Donaldson, J. M. Robertson, and J. G. White, *Proc. Roy. Soc.*, *A*, **220**, 311 (1953).

312. A. Camerman and J. Trotter, *Proc. Roy. Soc.*, *A*, **279**, 129 (1964).

313. J. Tanaka, *Bull. Chem. Soc. Japan*, **36**, 1237 (1964).

314. D. Brown, S. C. Wallwork, and A. Wilson, *Acta Cryst.*, **17**, 168 (1964).

315. V. M. Kozkin and A. I. Kitaigorodskii, *Zhur. Fiz. Khim.*, **27**, 1676 (1953).

316. S. L. Chorgade, *Z. Krist.*, *A*, **101**, 376 (1939).

317. S. Skraup and M. Eisemann, *Ann. Chem.*, **449**, 1 (1926).

318. A. H. Ewald, *Trans. Faraday Soc.*, **64**, 733 (1968).

319. T. Danno, T. Kajiwara, and H. Inokuchi, *Bull. Chem. Soc. Japan*, **40**, 2793 (1967).

320. A. Camerman and J. Trotter, *Acta Cryst.*, **18**, 636 (1965).

321. J. M. Robertson and J. G. White, *J. Chem. Soc.*, **1945**, 607.

322. R. E. Long, R. A. Sparks, and K. N. Trueblood, *Acta Cryst.*, **18**, 932 (1965).

323. D. A. Beckoe and K. N. Trueblood, *Z. Krist.*, **113**, 1 (1960).

324. R. P. Shibaeva, L. O. Atovmyan, and L. P. Rozenberg, *Chem. Commun.*, **1969**, 649.
325. H. M. Powell and G. Huse, *J. Chem. Soc.*, **1943**, 435.
326. T. Bernstein and F. H. Herbstein, unpublished results, 1968.
327. R. M. Williams and S. C. Wallwork, *Acta Cryst.*, **23**, 448 (1967).
328. T. Ito, M. Minobe, and T. Sakurai, *Acta Cryst.*, *B*, **26**, 1145 (1970).
329. C. J. Fritchie, Jr. and P. Arthur, Jr., *Acta Cryst.*, **21**, 139 (1966).
330. F. H. Herbstein and M. Kaftori, unpublished results, 1970.
331. P. Murray-Rust and J. D. Wright, *J. Chem. Soc.*, *A*, **1968**, 247.
332. A. S. Bailey and C. K. Prout, *J. Chem. Soc.*, **1965**, 4867.
333. E. Hertel and K. Schneider, *Z. Phys. Chem.*, *B*, **15**, 79 (1931).
334. A. Damiani, E. Giglio, A. M. Liquori, and A. Ripamonti, *Acta Cryst.*, **23**, 675 (1967).
335. A. Damiani, E. Giglio, A. M. Liquori, R. Puliti, and A. Ripamonti, *J. Mol. Biol.*, **23**, 113 (1967).
336. T. T. Harding and S. C. Wallwork, *Acta Cryst.*, **6**, 791 (1953).
337. T. Sakurai, *Acta Cryst.*, *B*, **24**, 403 (1968).
338. A. W. Hanson, *Acta Cryst.*, *B*, **24**, 768 (1968).
339. R. P. Shibaeva, L. O. Atovmyan, and M. N. Orfanova, *Chem. Commun.*, **1969**, 1494.
340. A. W. Hanson, *Acta Cryst.*, **19**, 610 (1965).
341. H. W. W. Ehrlich, *Acta Cryst.*, **10**, 699 (1957).
342. P. Cherin and M. Burack, *J. Phys. Chem.*, **70**, 1470 (1966).
343. G. A. Golder, G. S. Zhdanov, and M. M. Umanskii, *Dokl. Akad. Nauk SSSR*, **92**, 311 (1953).
344. K. Ozeki, N. Sakabe, and J. Tanaka, *Acta Cryst.*, *B*, **25**, 1038 (1969).
345. J. Trotter, *Acta Cryst.*, **13**, 86 (1960).
346. T. Sakurai, *Acta Cryst.*, *A*, **25**, S128 (1969).
347. T. Dahl and O. Hassel, *Acta Chem. Scand.*, **20**, 2008 (1966).
348. P. J. Wheatley, *Acta Cryst.*, **10**, 182 (1957).
349. T. Dahl and O. Hassel, *Acta Chem. Scand.*, **22**, 715 (1968).
350. G. Gafner and F. H. Herbstein, *Acta Cryst.*, **13**, 706 (1960).
351. F. H. Herbstein, *Acta Cryst.*, **16**, 255 (1963).
352. G. Gafner and F. H. Herbstein, *J. Chem. Soc.*, **1964**, 5290.
353. H. Rheinboldt and P. Senise, *Bol. Fac. Filosofia, Cienca Letras, Univ. Sao Paulo* **14**, *Quimica* No. **1**, 3 (1942); *Chem. Abstr.*, **40**, 2049 (1946).
354. S. K. Lower, *Mol. Liq. Cryst.*, **5**, 363 (1969).
355. R. M. Williams and S. C. Wallwork, *Acta Cryst.*, **22**, 899 (1966).
356. S. Kumakura, F. Iwasaki, and Y. Saito, *Bull. Chem. Soc. Japan*, **40**, 1826 (1967).
357. R. M. Williams and S. C. Wallwork, *Acta Cryst.*, *B*, **24**, 168 (1968).
358. J. L. de Boer and A. Vos, *Acta Cryst.*, *B*, **24**, 720 (1968).
359. I. Ikemoto and H. Kuroda, *Acta Cryst.*, *B*, **24**, 383 (1968).
360. A. W. Hanson, *Acta Cryst.*, **16**, 1147 (1963).
361. T. T. Harding and S. C. Wallwork, *Acta Cryst.*, **8**, 787 (1955).
362. N. Niimura, Y. Okashi, and Y. Saito, *Bull. Chem. Soc. Japan*, **41**, 1815 (1968).
363. F. Iwasaki and Y. Saito, *Acta Cryst.*, *A*, **25**, S130 (1969).
364. I. Ikemoto and H. Kuroda, *Bull. Chem. Soc. Japan*, **40**, 2009 (1967).

365. A. W. Hanson, *Acta Cryst.*, **17**, 559 (1964).
366. S. C. Wallwork, quoted in *Crystal Data—Determinative Tables*, 2nd ed., ACA Monograph No. 5, 1963, p. 242.
367. Y. Ohashi, H. Iwasaki, and Y. Saito, *Bull. Chem. Soc. Japan*, **40**, 1789 (1967).
368. B. Kamenar, C. K. Prout, and J. D. Wright, *J. Chem. Soc.*, **1965**, 4851.
369. C. K. Prout and A. G. Wheeler, *J. Chem. Soc.*, *A*, **1967**, 469.
370. B. Kamenar, C. K. Prout, and J. D. Wright, *J. Chem. Soc.*, *A*, **1966**, 661.
371. F. Lange, *Z. phys. Chem.*, **110**, 343 (1924).
372. E. Schreiner, *Z. phys.*, *Chem.*, **117**, 57 (1925).
373. E. F. Westrum, Jr., and J. P. McCullough in *Physics and Chemistry of the Organic Solid State*, D. Fox, M. M. Labes, and A. Weissberger, ed., Interscience, New York, 1963, Vol. 1, p. 3.
374. H. Jagodzinski in *Advanced Methods of Crystallography*, G. N. Ramachandran, ed., Academic Press, New York, 1964, p. 181.
375. W. A. Wooster, *Diffuse X-ray Reflections from Crystals*, Clarendon Press, Oxford, 1962.
376. D. W. J. Cruickshank, *Acta Cryst.*, **9**, 915 (1956).
377. A. Damiani, P. de Santis, E. Giglio, A. M. Liquori, R. Puliti, and A. Ripamonti, *Acta Cryst.*, **19**, 340 (1965).
378. J. D. Bell, J. F. Blount, O. V. Briscoe, and H. C. Freeman, *Chem. Commun.*, **1968**, 1656.
379. G. S. Zhdanov, Z. V. Zvonkova, and L. G. Vorontsova, *Sov. Phys. Cryst.*, **1**, 44 (1956).
380. J. M. Lhoste and F. Tounard, *J. Chim. Phys.*, **63**, 678 (1966).
381. M. J. Mantione in *Molecular Associations in Biology*, B. Pullman, ed., Academic Press, New York and London, 1968, p. 411.
382. P. Claverie in *Molecular Associations in Biology*, B. Pullman, ed., Academic Press, New York and London, 1968, p. 115.
383. A. I. Kitaigorodskii, *Tetrahedron*, **14**, 230 (1961).
384. M. J. Mantione, *Theoret. Chim. Acta (Berlin)*, **15**, 141 (1969).
385. H. Kuroda, T. Amano, I. Ikemoto, and H. Akamatu, *J. Amer. Chem. Soc.*, **89**, 6056 (1967).
386. W. C. Herndon and J. Feuer, *J. Amer. Chem. Soc.*, **90**, 5914 (1968).
387. R. Hoffmann, *J. Chem. Phys.*, **39**, 1397 (1963), and subsequent papers.
388. S. Wold, *Acta Chem. Scand.*, **20**, 2377 (1966).
389. D. B. Chesnut and R. W. Moseley, *Theoret. Chim. Acta (Berlin)*, **13**, 230 (1969).
390. H. Kuroda, T. Kunii, S. Hiroma, and H. Akamatu, *J. Mol. Spectr.*, **22**, 60 (1967).
391. H. Kuroda, T. Ikemoto, and H. Akamatu, *Bull. Chem. Soc. Japan*, **39**, 547 (1966).
392. J. Stanley, D. Smith, B. Latimer, and J. P. Devlin, *J. Phys. Chem.*, **70**, 2011 (1966).
393. B. Hall and J. P. Devlin, *J. Phys. Chem.*, **71**, 465 (1967).
394. F. Iwasaki and Y. Saito, *Acta Cryst.*, *A*, **25**, S130 (1969); *B*, **26**, 251 (1970).
395. K. Nakamoto, *J. Amer. Chem. Soc.*, **74**, 1739 (1952).
396. S. Iwata, J. Tanaka, and S. Nagakura, *J. Amer. Chem. Soc.*, **88**, 2813 (1967).

397. D. S. Brown and S. C. Wallwork, *Acta Cryst.*, **19**, 149 (1965).
398. R. M. Hochstrasser, S. K. Lower, and C. Reid, *J. Chem. Phys.*, **41**, 1073 (1964).
399. S. K. Lower, R. M. Hochstrasser, and C. Reid, *Mol. Phys.*, **4**, 161 (1961).
400. R. M. Hochstrasser, S. K. Lower, and C. Reid, *J. Mol. Spectr.*, **15**, 257 (1965).
401. J. Tanaka and K. Yoshihara, *Bull. Chem. Soc. Japan*, **38**, 739 (1965).
402. J. R. Hoyland and L. Goodman, *J. Chem. Phys.*, **36**, 12 (1962).
403. E. G. McRae, *J. Chem. Phys.*, **33**, 932 (1960).
404. C. K. Prout, personal communication, 1968.
405. K. Sasvari, personal communication, 1967.
406. A. I. Kitaigorodskii, *Sov. Phys. Cryst.*, **3**, 393 (1958).
407. H. Kokado, K. Hasegawa, and W. G. Schneider, *Can. J. Chem.*, **42**, 1084 (1964).
408. B. Chakrabarti and S. Basu, *J. Chim. Phys.*, **65**, 1006 (1968).
409. N. D. Jones and R. E. Marsh, *Acta Cryst.*, **15**, 809 (1962).
410. S. C. Wallwork and T. T. Harding, *Acta Cryst.*, **15**, 810 (1962).
411. D. C. Douglass, *J. Chem. Phys.*, **32**, 1882 (1960).
412. J. E. Anderson, *J. Phys. Chem.*, **70**, 927 (1966).
413. C. K. Prout and S. C. Wallwork, *Acta Cryst.*, **21**, 449 (1966).
414. J. Gaultier, C. Hauw, and M. Breton-Lacombe, *Acta Cryst.*, B, **25**, 231 (1969).
415. L. O. Brockway and J. M. Robertson, *J. Chem. Soc.*, **1939**, 1324.
416. T. Sakurai, *Acta Cryst.*, **19**, 320 (1965).
417. H. Matsuda, K. Osaki, and I. Nitta, *Bull. Chem. Soc. Japan*, **31**, 611 (1958).
418. J. E. Anderson, *Nature*, **140**, 583 (1937).
419. G. G. Shipley and S. C. Wallwork, *Acta Cryst.*, **22**, 593 (1967).
420. G. G. Shipley and S. C. Wallwork, *Acta Cryst.*, **22**, 585 (1967).
421. T. Sakurai and H. Tagawa, *Sci. Papers Inst. Phys. Chem. Res.*, **62**, 35 (1968).
422. T. Sakurai and M. Yabe, *J. Phys. Soc. Japan*, **13**, 5 (1958).
423. T. Sakurai and T. Ito, *Acta Cryst.*, B, **25**, 1031 (1969).
424. K. Suzuki and S. Seki, *Bull. Chem. Soc. Japan*, **26**, 372 (1953).
425. I. Nitta, S. Seki, H. Chihara, and K. Suzuki, *Sci. Papers Osaka Univ.*, No. 29 (1951).
426. B. G. Anex and L. J. Parkhurst, *J. Amer. Chem. Soc.*, **85**, 3301 (1963).
427. K. Ozeki, N. Sakabe, and J. Tanaka, *Acta Cryst.*, B, **25**, 1031 (1969).
428. R. Gerdil, *Acta Cryst.*, **14**, 333 (1961).
429. C. K. Prout and H. M. Powell, *J. Chem. Soc.*, **1965**, 4882.
430. T. Amano, H. Kuroda, and H. Akamatu, *Bull. Chem. Soc. Japan*, **41**, 83 (1968).
431. D. B. Chesnut and W. D. Phillips, *J. Chem. Phys.*, **35**, 1002 (1961).
432. D. D. Thomas, H. Keller, and H. M. McConnell, *J. Chem. Phys.*, **39**, 2321 (1963).
433. M. T. Jones and D. B. Chesnut, *J. Chem. Phys.*, **38**, 1311 (1963).
434. Z. G. Soos, *J. Chem. Phys.*, **49**, 2493 (1968).
435. J. L. de Boer and A. Vos, *Acta Cryst.*, B, **24**, 720 (1968).
436. B. G. Anex and H. B. Hill, *J. Amer. Chem. Soc.*, **88**, 3648 (1966).
437. A. C. Albrecht and W. T. Simpson, *J. Amer. Chem. Soc.*, **77**, 4454 (1955).

438. Y. Iida and Y. Matsunaga, *Bull. Chem. Soc. Japan*, **41**, 2535 (1968).
439. J. J. Andre and G. Weill, *Mol. Phys.*, **15**, 97 (1968).
440. A. Fulton, *Austr. J. Chem.*, **21**, 2847 (1968).
441. H. Kainer and W. Otting, *Chem. Ber.*, **88**, 1921 (1955).
442. Y. Iida, *Bull. Chem. Soc. Japan*, **43**, 345 (1970).
443. H. J. Monkhorst and J. Kommandeur, *J. Chem. Phys.*, **47**, 391 (1967).
444. J. Tanaka and N. Sakabe, *Acta Cryst.*, *B*, **24**, 1345 (1968).
445. J. L. de Boer, A. Vos, and K. Huml, *Acta Cryst.*, *B*, **24**, 542 (1968).
446. Y. Matsunaga, *Nature*, **211**, 183 (1966).
447. R. Foster and T. J. Thomson, *Trans. Faraday Soc.*, **59**, 296 (1963).
448. H. Kuroda, S. Hiroma, and H. Akamatu, *Bull. Chem. Soc. Japan*, **41**, 2855 (1968)
449. M. Kinoshita and H. Akamatu, *Nature*, **207**, 291 (1965).
450. M. Ohmasa, M. Kinoshita, M. Sano, and H. Akamatu, *Bull. Chem. Soc. Japan*, **41**, 1998 (1968).
451. R. C. Hughes and B. M. Hoffman, *Solid State Commun.*, **7**, 895 (1969).
452. B. M. Hoffman and R. C. Hughes, *J. Chem. Phys.*, **52**, 4011 (1970).
453. C. J. Fritchie, Jr., *Acta Cryst.*, **20**, 892 (1966).
454. E. B. Yagubskii, M. L. Khidekel, G. F. Shchegolev, L. I. Buravov, B. G. Gribov, and M. K. Makova, *Izv. Akad. Nauk S.S.S.R.*, *Ser. Khim.*, **1968**, 2124.
455. P. Arthur, Jr., *Acta Cryst.*, **17**, 1176 (1964).
456. P. Goldstein, K. Seff, and K. N. Trueblood, *Acta Cryst.*, *B*, **24**, 778 (1968).
457. H. Scott, P. L. Kronick, P. Chairge, and M. M. Labes, *J. Phys. Chem.*, **69**, 1740 (1965).
458. Y. Iida and Y. Matsunaga, *Bull. Chem. Soc. Japan*, **41**, 2615 (1968).
459. Y. Iida, *Bull. Chem. Soc. Japan*, **42**, 71 (1969).
460. Y. Iida, *Bull. Chem. Soc. Japan*, **42**, 637 (1969).
461. E. O. Fischer and U. Piesbergen, *Z. Naturforsch.*, **11b**, 758 (1956).
462. S. Sakanone, N. Yasuoka, N. Kasai, M. Kakudo, S. Kusabayashi, and H. Mikawa, *Bull. Chem. Soc. Japan*, **42**, 2408 (1969).
463. A. Damiani, P. de Santis, E. Giglio, A. M. Liquori, R. Puliti, and A. Ripamonti, *Acta Cryst.*, **19**, 340 (1965).
464. M. Meyers and K. N. Trueblood, *Acta Cryst.*, *B*, **25**, 2588 (1969).
465. C. K. Prout and E. E. Castellano, *J. Chem. Soc.*, *A*, **1970**, 2775.
466. J.-P. Malrieu and P. Claverie, *J. Chim. Phys.*, **64**, 735 (1968).
467. R. J. W. Le Fèvre, D. V. Radford, and P. J. Stiles, *J. Chem. Soc.*, *B*, **1968**, 1297.
468. O. H. Le Blanc, Jr., in *Physics and Chemistry of the Organic Solid State*, ed. D. Fox, M. M. Labes, and A. Weissberger, Interscience, New York, Vol. 3, p. 133, 1967.
469. C. A. Langhoff and C. J. Fritchie, Jr., *Chem. Commun.*, **1970**, 20.
470. C. A. Bear, J. M. Waters, and T. N. Waters, *Chem. Commun.*, **1970**, 702.
471. K. N. Trueblood and J. Bernstein, *Acta Cryst.*, in the press.
472. R. Carruthers, F. M. Dean, L. E. Houghton, and A. Ledwith, *Chem. Comm.*, **1967**, 1206.
473. E. E. Castellano and C. K. Prout, *J. Chem. Soc.*, *A*, **1971**, 550.
474. E. Menefee and Y.-H. Pao, *J. Chem. Phys.*, **36**, 3472 (1962).
475. H. Kobayashi, Y. Ohashi, F. Marumo, and Y. Saito, *Acta Cryst.*, *B*, **26**, 459 (1970).

476. E. Carstensen-Oeser, S. Göttlicher, and G. Habermehl, *Chem. Ber.*, **101**, 1648 (1968).
477. R. G. Kepler, P. E. Bierstedt, and R. E. Merrifield, *Phys. Rev. Letters*, **5**, **503** (1960).
478. F. K. Larsen, R. G. Little, and P. Coppens, Winter Meeting Amer. Crystallographic Assoc., Columbia, S. Carol., February, 1971.
479. T. Sakata and S. Nagakura, *Bull. Chem. Soc. Japan*, **43**, 1346 (1970).
480. P. J. Strebel and Z. G. Soos, *J. Chem. Phys.*, **53**, 4077 (1970).
481. M. Ohmasa, M. Kinoshita, and H. Akamatu, *Bull. Chem. Soc. Japan*, **44**, 391, 395 (1971).
482. I. Ikemoto, K. Yakushi, and H. Kuroda, *Acta Cryst.*, *B*, **26**, 800 (1970).
483. W. Slough, *Trans. Faraday Soc.*, **61**, 408 (1965).
484. R. H. Colton and D. E. Henn, *J. Chem. Soc.*, *B*, **1970**, 1532.

Added in proof
(1) See p. 269. Further refinement of the structure of perylene : TCNE produced only minor changes in atomic parameters. [482]
(2) See p. 272. HMB:TCNQ has been studied by Slough[483] and its structure has been determined by Colton and Henn.[484]

# Author Index for Volume IV

Numerals in square brackets are reference numbers. The preceding numerals (not in brackets) refer to pages on which the reference number is cited, with or without the author's name. For instance, K. Abe is an author in reference 170 cited on pages 205 and 386, where page 386 gives the bibliographic detail; Y. Abe is an author in references 161 and 270, cited on pages 102, 119, 120, 157, and 160, where the last-mentioned two pages contain the bibliographic detail. In a few cases the author is cited in more than one of the articles in this volume; in such cases the numerals are divided into two sets.

Bold numerals refer to authors of articles.

# Subject Index for Volume IV